Ontogenetic Perspectives on Primate Evolutionary Biology

Ontogenetic Perspectives on Primate Evolutionary Biology

Edited by

Matthew J. Ravosa and Anne M. Gomez
Duke University Medical Center, Durham, NC, U.S.A.

ACADEMIC PRESS
Harcourt Brace Jovanovich, Publishers
London   San Diego   New York   Boston
Sydney   Tokyo   Toronto

ACADEMIC PRESS LIMITED
24–28 Oval Road
London NW1 7DX

United States edition published by
ACADEMIC PRESS INC.
San Diego, CA 92101

Copyright © 1992 by Academic Press Limited

ALL RIGHTS RESERVED

No part of this book may be reproduced in any form, by photostat, microfilm, or any other means, without written permission from the publishers.

(Reprinted from *Journal of Human Evolution,* Vol. 23, Nos 1–3, 1992)

ISBN 0-12-583470-5

Printed in Great Britain by Henry Ling Ltd, Dorchester, Dorset

Contents

Preface 1

Ontogenetic patterns of sexual dimorphism in the cranium of Bornean orang-utans (*Pongo pygmaeus*)
T. J. Masterson and W. Leutenegger 3

Patterns of variation in the ontogeny of primate body size dimorphism
S. R. Leigh 27

Primate population studies at Polonnaruwa. III. Somatometric growth in a natural population of toque macaques (*Macaca sinica*)
J. M. Cheverud, W. P. J. Dittus and P. Wilson 51

Relative growth of the postcranial skeleton in callitrichines
A. B. Falsetti and T. M. Cole III 79

Relative growth and shape of the locomotor skeleton in lesser apes
W. L. Jungers and M. S. Cole 93

Ontogeny and allometry of African ape manual rays
S. E. Inouye 107

The ontogeny of chimpanzee and pygmy chimpanzee locomotor behavior: a case study of paedomorphism and its behavioral correlates
D. M. Doran 139

The allometry of behavioural development: fitting sigmoid curves to ontogenetic data for use in interspecific allometric analyses
D. T. Rasmussen and C. L. Tan 159

The applications and limitations of ontogenetic comparisons for phylogeny reconstruction: the case of the strepsirhine internal carotid artery
A. D. Yoder 183

Allometry and heterochrony in extant and extinct Malagasy primates
M. J. Ravosa 197

Primitive and derived patterns of relative growth among species of lorisidae
A. M. Gomez 219

The ontogeny of *Pan troglodytes* craniofacial architectural relationships and implications for early hominids
T. G. Bromage 235

Postnatal heterochrony of the masticatory apparatus in *Cebus apella* and *Cebus albifrons*
  T. M. Cole III 253

Ontogenetic scaling of skeletal proportions in the talapoin monkey
  B. T. Shea 283

List of Contributors

T. G. Bromage, *Hard Tissue Research Unit, Department of Anthropology, Hunter College, CUNY, 695 Park Ave., New York, NY 10021, U.S.A.*

M. S. Cole, *Department of Anatomical Sciences, School of Medicine, State University of New York at Stony Brook, Stony Brook, NY 11794-5245, U.S.A.*

T. M. Cole III, *Doctoral Program in Anthropological Sciences, State University of New York at Stony Brook, Stony Brook, NY 11794, U.S.A.*

J. M. Cheverud, *Department of Anatomy & Neurobiology, Washington University School of Medicine, St. Louis, MS 63110, U.S.A.*

W. P. J. Dittus, *Department of Zoological Research, National Zoological Park, Smithsonian Institution, Washington DC 2008, U.S.A. and Institute of Fundamental Studies, Kandy, Sri Lanka*

D. M. Doran, *Department of Anatomical Sciences, Suny at Stony Brook, Stony Brook, NY 11794-8081, U.S.A.*

A. B. Falsetti, *Department of Ecology and Evolution, Division of Biological Sciences, State University of New York at Stony Brook, Stony Brook, NY 11794-5245, U.S.A.*

A. M. Gomez, *Department of Biological Anthropology and Anatomy, Duke University Medical Center, Box 3170, Durham, NC 27710, U.S.A.*

S. E. Inouye, *Department of Anthropology, Northwestern University, Evanston, IL 60208, U.S.A.*

W. L. Jungers, *Department of Anatomical Sciences, School of Medicine, State University of New York at Stony Brook, Stony Brook, NY 11794-5245, U.S.A.*

S. R. Leigh, *Department of Anthropology, Northwestern University, Evanston, IL 60208, U.S.A.*

W. Leutenegger, *Department of Anthropology, University of Wisconsin, Madison, WI 53706, U.S.A.*

T. J. Masterson, *Department of Anthropology, University of Wisconsin, Madison, WI 53706, U.S.A.*

D. T. Rasmussen, *Department of Anthropology, Washington University, St. Louis, MS 63130, U.S.A.*

M. J. Ravosa, *Duke University Medical Center, Department of Biological Anthropology and Anatomy, Durham, NC 27710, U.S.A.*

B. T. Shea, *Departments of Cellular, Molecular and Structural Biology and Anthropology, Northwestern University, 303 East Chicago Avenue, Chicago, IL 60611, U.S.A.*

C. L. Tan, *Department of Anthropology, University of California, Los Angeles, CA 90024, U.S.A.*

P. WILSON, *Department of Anatomy & Neurobiology, Washington University School of Medicine, St. Louis, MS 63110, U.S.A.*

A. D. YODER, *Duke University Medical Center, Department of Biological Anthropology and Anatomy, Durham, NC 27710, U.S.A.*

Preface

This volume is dedicated to the memory of Carmen D. Ravosa

These edited special issues aim to provide an overview of a re-emerging focus on the importance of ontogenetic data for addressing issues regarding skeletal form and function, phylogenetic reconstruction, sexual dimorphism, phyletic size change and behavioral development in primates. The study of ontogeny offers a dynamic perspective to the study of evolutionary biology which often complements and extends the utility of research based solely on adult data. The increasing number of investigations on heterochrony in primates, for instance, places renewed emphasis on the analysis of evolutionary change as a modification of development. Therefore, in order to gain a fuller understanding of evolutionary patterns across taxa, it is essential to address the developmental or ontogenetic processes underlying such changes. Clearly, there is much information to glean from studies of ontogeny. The contents of these edited special issues represent what we consider to be a fruitful avenue of research in years to come.

The contributions can be grouped roughly into analyses of sexual dimorphism (Cheverud *et al.*, Leigh, Masterson & Leutenegger), cranial and postcranial growth and form (Bromage, Cole, Falsetti & Cole, Inouye, Jungers & Cole, Ravosa, Shea), behavioral development (Doran, Rasmussen & Tan) and phylogenetic relationships (Gomez, Yoder). The studies of sexual dimorphism detail the developmental processes resulting in size and shape differentiation between adult males and females. Such analyses are especially significant since it is increasingly evident that differences in the rate and/or duration of growth between the sexes have important socio-ecological underpinnings. The majority of cranial and postcranial analyses center on whether differences in adult skeletal morphology between closely related taxa result from the differential extension of common patterns of relative growth, i.e., ontogenetic scaling. Using this ontogenetic criterion of subtraction offers unique insight into size-related aspects of morphological change, as well as specific functional aspects of bony morphology that are not related to differences in size. The studies of behavioral ontogeny are quite different in emphasis, however. Doran uses data on locomotor behavior in chimpanzees to assess the correspondence between patterns of morphological and behavioral change during ontogeny. Rasmussen & Tan provide an intriguing analysis of how to model ontogenetic data on suckling behavior for interspecific comparisons. Lastly, there are two studies which emphasize the importance of ontogenetic data for phylogenetic analysis. Gomez discusses the implications of coinciding versus non-coinciding ontogenetic trajectories for the evolution of relative growth patterns in the lorises. Yoder utilizes data on the development of the internal carotid artery to further consider the question of cheirogaleid monophyly.

All but two of the papers in these issues were presented in a symposium entitled "Ontogenetic Perspectives on Primate Evolutionary Biology", held at the 60th annual meeting of the American Association of Physical Anthropologists (AAPA) in Milwaukee, Wisconsin, U.S.A. The two additional papers are included due to their explicit ontogenetic focus on sexual dimorphism (Cheverud *et al.*) and phylogenetic reconstruction (Yoder). While many of the papers utilize allometric approaches, this is just one means of employing ontogenetic information to investigate evolutionary questions.

A number of people have provided invaluable assistance with our endeavors. First and foremost, we would like to thank the co-editors of the *Journal of Human Evolution*, Drs Peter Andrews and William Jungers, for considering our proposal to publish the symposium. Their continued and undaunting efforts on our behalf have helped to realize the publication of these edited special issues. Many reviewers, 40 to be exact, have provided timely and insightful comments, and have greatly improved the overall quality of the papers. Lastly, we are grateful to Dr Lorna Moore and the members of the AAPA executive committee for providing a forum for our symposium.

MATTHEW J. RAVOSA
ANNE M. GOMEZ
Durham, North Carolina
U.S.A.

Thomas J. Masterson
& Walter Leutenegger
*Department of Anthropology,
University of Wisconsin, Madison, WI
53706, U.S.A.*

Received 7 June 1991
Revision received 2 December
1991 and accepted 3 January
1992

*Keywords: Pongo pygmaeus pygmaeus,
sexual dimorphism, allometry,
heterochrony, sexual selection.*

Ontogenetic patterns of sexual dimorphism in the cranium of Bornean orang-utans (*Pongo pygmaeus pygmaeus*)

Ontogenetic patterns of cranial sexual dimorphism in a large sample of the Bornean orang-utan (*Pongo pygmaeus pygmaeus*) are investigated by means of univariate, bivariate and multivariate statistical analyses. Univariate analyses of 21 linear dimensions reveal that starting at the mid-juvenile stage there is a strong tendency for an increase in number and strength of significant sex differences, all in favor of males. Significant sex differences in the viscerocranium, reflecting stronger prognathism in males, emerge before those in the neurocranium.

Two major growth allometry patterns emerge from bivariate analyses. Ontogenetic scaling is present in 10 dimensions. It is particularly strong in the neurocranium directly associated with brain size, the orbital region and the dental arcade. The heterochronic process of time hypermorphosis is most likely responsible for this pattern. The second growth pattern reflects a departure from ontogenetic scaling with males exhibiting significantly steeper slopes than females. This occurs in 10 cranial dimensions associated with secondary sexual character development: prognathism, canine size and cheek pad area. We suggest that the heterochronic process of acceleration underlies this growth pattern.

Principal components analyses reveal two major multidimensional patterns of cranial sexual dimorphism. First, sexual differences at age groups 2 and 3 are primarily the result of differences in principal component II scores, reflecting mainly shape-related differences. Second, age groups 5, 6 and 7 show a trend of stronger size-related shape differences with increasing age in the allometry vector along with decreasing differences in principal component II scores. Age group 4 shows a combination of both patterns.

Sexual rather than ecological selection best explains the underlying selective regime of the ontogenetic patterns of cranial sexual dimorphism in orang-utans. Our morphometric analyses clearly corroborate the conclusions of behavioral ecologists in support of the sexual selection theory.

Journal of Human Evolution (1992) **23**, 3–26

Introduction

In the last decade there has been a change of focus in both the studies of allometry and sexual dimorphism, from using static adult endpoints to samples containing individuals representing all ages. This is in part due to Shea's (1986) argument that the investigation of sexual dimorphism may prosper by using ontogenetic samples because (1) the entire growth pattern is the target of selection, not just the adult endpoints, and (2) adult endpoints can be the result of different developmental processes, indicating different selective regimes. The unique inferences that can be drawn from ontogenetic analyses allow the researcher to examine new hypotheses dealing with issues such as behavior, morphology and heterochrony. The increasing number of ontogenetic analyses present in the literature (e.g., Shea, 1982, 1983*a,b*, 1985*a*, 1986, 1989; Bromage, 1985, 1989; Cochard, 1985; Leutenegger & Masterson, 1989*a,b*; Masterson & Leutenegger, 1990; Simpson *et al.*, 1990; Corner & Richtsmeier, 1991; Leigh & Cheverud, 1991; Ravosa, 1991*a,b*) underscores the importance of using samples that span the entire developmental range.

The documentation of the marked degree of sexual dimorphism in the cranium, teeth, outer body and body mass of the adult orang-utan is fairly extensive (Selenka, 1898;

Table 1 Developmental age groups and sample sizes of *Pongo pygmaeus pygmaeus*

Age group	Description	Sample size[1]
Stage 1[2]	Incompletely erupted deciduous dentition	$n = 3M, 2F$
Stage 2	All deciduous teeth fully erupted	$n = 7M, 3F$
Stage 3	Deciduous dentition plus M^1 partially or fully erupted	$n = 5M, 5F, 2?$
Stage 4	M^2 partially or fully erupted	$n = 6M, 11F, 5?$
Stage 5	M^3 erupting	$n = 15M, 13F, 1?$
Stage 6	C, M^3 erupted, full permanent dentition, basilar suture open, teeth little worn	$n = 36M, 33F$
Stage 7	Full permanent dentition with basilar suture closed, moderate to heavy wear	$n = 9M, 56F$

[1] M = male, F = female, ? = unknown sex.
[2] Stages 1 and 2 were pooled for all subsequent analyses and are referred to as stage 2.

Hrdlička, 1907; Ashton & Zuckerman, 1956; Biegert, 1957; Schultz, 1962, 1969; Eckhardt, 1975; Röhrer-Ertl, 1988; Winkler *et al.*, 1988). However, few studies exist that examine sexual dimorphism in immature orang-utans (Schultz, 1941, 1962; Fooden & Izor, 1983; Winkler, 1987; Röhrer-Ertl, 1988). Therefore, the examination of a large sample spanning all developmental ages is necessary in order to understand the patterns of sexual dimorphism that pertain to both immature and adult orang-utans.

The aim of this study is to explore ontogenetic patterns of sexual dimorphism in the cranium of Bornean orang-utans (*Pongo pygmaeus pygmaeus*) as revealed by univariate, bivariate and multivariate analyses. Sexual dimorphism will also be analysed from the perspective of heterochrony, evolutionary changes in the time of development of an organism's features (Gould, 1977). Finally, possible selective regimes that may be underlying the development of the marked degree of cranial and body size dimorphism in orang-utans are examined.

Materials and methods

The morphometric analyses reported here are all based on a wild-shot sample of 212 orang-utan (*Pongo pygmaeus pygmaeus*) crania spanning all postnatal age periods. The crania are from known localities in western Borneo (Röhrer-Ertl, 1984). The entire sample is from the SELENKA collection, housed at the Anthropologische Staatssammlung, Munich, Germany. Table 1 identifies the sample according to sex and age groups. The present analyses do not include individuals of questionable sexual attribution (eight specimens).

Overall age estimates for individual crania are necessary for ontogenetic analyses. Exact chronological ages of our sample are unknown; therefore, each specimen's age was estimated on the basis of dental eruption patterns. This allows for the determination of cross-sectional growth patterns on the basis of age grouping (Tanner, 1951; Cock, 1966; Shea, 1983a). Shea

(1982, 1983a, 1985b) also used these criteria in his studies on African apes. The dental stages are defined in Table 1. Initially, seven age categories were defined; however, because of the limited number of infants in our sample, dental stages 1 and 2 were pooled for all subsequent analyses. These individuals are referred to as dental stage 2 and represent the early and late infant dental stages.

When chronological age is unknown, tooth development can provide a measure of relative age. However, such a technique may introduce some aging biases into the study. Corner & Richtsmeier (1991) pointed out that developmental age intervals may not be of equal length, dentally precocial or altricial individuals may be misclassified, and the same dental pattern may not be present for both sexes of the same age.

Several points need to be addressed in the discussion of these biases. First, there is no assumption of equal length in the developmental age intervals in the present analyses. Second, while the misclassification of dentally precocial or altricial individuals is possible, it does not seem a major concern because such individuals are presumably represented by a very small number of specimens. Finally, a review of the literature reveals a relative paucity of data pertaining to times of dental eruption in wild orang-utans. Some information, however, is available from captive studies such as Fooden & Izor (1983). In the deciduous dentition a general trend exists in that females possess earlier eruption times than males for respective teeth. The mean age reported for the complete emergence of all deciduous teeth was 361 and 404 days for females and males, respectively. This age corresponds to our dental stage 2.

Information regarding chronological age and sexual differences in eruption times in the permanent dentition of the orang-utan is less complete. The most cited study of tooth development in the great apes has been that of Dean & Wood (1981). They provided calcification and eruption times for humans and the great apes, but exclude *Pongo* due to the limited amount of data on eruption times in the permanent dentition. Two recent studies have provided new information dealing with tooth development in orang-utans. Beynon *et al.* (1991) used histological methods of analysis to provide crown formation times and an overall chronology of dental development in both *Gorilla gorilla* and *Pongo pygmaeus*. Winkler *et al.* (1991) utilized radiographic analyses to study dental morphogenesis to examine patterns of variability in dental development prior to the eruption of the permanent dentition. Although these new studies provide valuable information, no discussion of either eruption times or sex differences has yet been undertaken. Therefore, the relationship between dental stages and chronological ages can not be stated until further studies of dental emergence times are completed. We, therefore, refrain from using chronological ages in the present study when discussing dental stages.

Twenty-one linear dimensions were measured on each skull by one investigator (W.L.) with a sliding caliper, calibrated to 0·1 mm (see appendix 1 for definitions of measurements). Measurements were taken on all specimens. However, damaged or missing elements prevented obtaining all measurements from all specimens. No attempt was made to estimate or compensate for these missing values.

Ontogenetic patterns of cranial sexual dimorphism are examined by means of univariate, bivariate and multivariate statistics.

Univariate analyses
Sexual differences in male and female age group means are assessed by Student's t-test in the univariate analyses. The two-tailed hypothesis was used to reveal levels of significance for differences in the unpaired t-values. Means, standard deviations, standard errors and

coefficients of variation of the sexed samples can be found in Leutenegger & Masterson (1989a).

Bivariate analyses
The ontogenetic nature of the data set allowed us to reveal growth allometries and to analyse sexual dimorphism in terms of heterochrony. Growth allometries were analysed using the log-transformed version of Huxley's (1932) bivariate power function:

$$\log y = \log b + a \log x$$

where x and y represent the independent and dependent variables. The y-axis intercept is represented by b and a is the slope of the regression line. Basicranial length (basion to nasion) was used as the independent or x variable since it represents a good measure of overall skull size (Shea, 1983a). The slope of bivariate regression lines of log-transformed data corresponds to Huxley's coefficient of growth allometry, i.e., "a". The slope coefficient indicates the constant relative linear growth between two structures throughout the organism's growth period provided that the fitted line is statistically significant. Slopes that differ from unity, $a = 1 \cdot 0$, indicate that shape changes are correlated with changes in size (Somers, 1986). Slopes greater than 1 indicate that positive allometry exists. This implies a differential increase of y relative to x (Gould, 1966); i.e., shape changes faster than size increases. Negative allometry exists when the slope is less than 1, revealing that y/x ratios decrease with increasing absolute magnitude of x (Gould, 1966). This indicates that shape is not changing as fast as size. Isometry exists when slopes equal unity, thereby reflecting equal proportional changes in size and shape.

Currently there is some debate over which line fitting technique is best suited for allometric studies (e.g., Ricker, 1973; Gould, 1975a; Smith 1980, 1981; Harvey & Mace, 1982; Pagel & Harvey, 1988; Martin & Barbour, 1989). The standard least squares regression model was utilized in the present analyses on the assumptions that y is dependent upon x and x is measured without error. The method of least squares regression is quite robust, meaning that minor violations of the underlying assumptions do not invalidate the conclusions drawn from the analyses in a major way (Chatterjee & Price, 1977). The examination of standardized residual plots does not reveal any trends of major biases, therefore, use of the standard least squares model is justified. It has also been argued by many workers that differences between the various regression models are minimized by strong correlation coefficients found in ontogenetic comparisons. This is true in the present sample. Statistical significance for differences in male and female slope values was tested by means of F-ratios (Chatterjee & Price, 1977) with similar results being obtained by analyses of covariance (Snedecor & Cochran, 1980).

In order to analyse sexual dimorphism in terms of heterochrony, data on sexual differences in size and shape as well as the degree of sexual bimaturism is necessary. The growth allometries reported below provide the data on sexual differences in size and shape. Information on body weight growth, duration of ontogeny, timing of sexual maturity, timing of growth spurts, etc., has been collected from published reports (see Leutenegger & Masterson, 1989b, for details) and will be substituted for actual chronological ages when determining which heterochronic process may be responsible for the growth patterns observed.

Multivariate analyses
The multivariate generalization of allometry proposed by Jolicoeur (1963a,b), principal components analysis (PCA), was used to investigate the ontogeny of cranial sexual

dimorphism in Bornean orang-utans (*Pongo pygmaeus pygmaeus*). Many agree that Jolicoeur's method provides a technique to distinguish non-allometric shape differences between members of a common sample. The technique uses log-transformed data and covariance matrices to derive principal components. Principal component I is interpreted as proportion changes in relative size, as long as all variable loadings or direction cosines are positive and the majority of the sample's variance is explained by component I (Shea, 1985c). Component II and all subsequent components summarize more shape-related differences than size. For this reason many workers have referred to the first principal component as a growth trend or size component (Jolicoeur & Mosimann, 1960; Jolicoeur, 1963b; Baker, 1980; Campbell & Atchley, 1981; Cochard, 1985). We prefer the terms "allometry vector" (Hursh, 1975) or "vector of relative growth" (Shea, 1985c) since our data are ontogenetic.

Briefly, PCA is a multivariate technique that derives linear combinations of the original variables so that the maximum variation of the data set is contained within the first few linear combinations, resulting in principal component factors or scores. The original coordinate system is geometrically rotated in such a way that PCA defines new orthogonal axes, called principal axes, which describe the maximum variation found in the original data with successive axes being chosen to maximize the remaining variation (Reyment *et al.*, 1984). The eigenvalue represents the amount of variation accounted for by each principal component, while its respective eigenvector (direction cosines) designates the loading of each of the original variables on the newly determined principal component variables (Shea, 1985c). The angles between the corresponding principal axes and the original coordinate system are given by correlations between the vectors. This makes it possible to determine angles between vectors and to test for divergence between hypothetical and observed axes (Anderson, 1963; Jolicoeur, 1963b; Morrison, 1976).

All PCAs were analysed using the statistical program SAS v.6·03. The total pooled sample size was reduced to 198 individuals ($n = 77$ for the male and $n = 121$ for the female samples) because of various missing values. Two types of analyses were performed in the multivariate investigation of orang-utan cranial sexual dimorphism. First, a PCA was run on a pooled male and female sample. Second, PCAs were run on subsets of those dimensions determined to be ontogenetically scaled or lying along a common ontogenetic trajectory (Gould, 1975b; Shea, 1981, 1985a) and those exhibiting the heterochronic process of acceleration, an increase in the rate of shape change (Gould, 1977), when regressed against basicranial length.

A debate over which type of PCA should be used in multi-group analyses is currently being discussed in the literature (see Masterson & Leutenegger, 1990). Ordinary PCA methodology was applied in the present analyses. The major criticism of this technique is that the principal components are determined by a balance of within-group as well as between-group variability (Shea, 1985c; Airoldi & Flury, 1988). Given the debate over which PCA technique should be used with samples containing multiple groups, we feel justified in using the ordinary PCA technique.

A discussion of the relevant data dealing with the behavioral ecology of the wild orang-utan is necessary to investigate the underlying selective pressures of sexual dimorphism in orang-utans. In this fashion our morphometric data will be utilized to corroborate the behavioral data in support of possible selective regimes responsible for the marked degree of sexual dimorphism present in orang-utans. No formal test of any specific type of selection, sexual or ecological, has been undertaken because such a test should involve a more broadly based comparative sample demonstrating ontogenetic growth patterns across species.

Table 2 *t*-values and significance levels for sexual differences in all 21 cranial dimensions

Dimension	Age 2	Age 3	Age 4	Age 5	Age 6	Age 7
Palate width at c	0·378	0·732	1·640	7·344***	13·881***	14·864***
Palate width p. to c.	0·398	1·386	0·997	6·337***	9·995***	8·841***
Palate length	0·160	0·935	2·571*	4·777***	8·660***	11·216***
Lower facial length	0·347	1·378	2·392*	5·198***	10·403***	14·447***
Alveolar height	0·425	0·118	3·747**	4·891***	6·541***	7·268***
Bimaxillary width	0·113	1·371	−0·458	2·919**	5·259***	7·377***
Biorbital width	−0·953	0·248	0·548	3·292**	5·882***	8·287***
Interorbital width	−2·897*	−1·203	0·135	1·626	4·663***	6·015***
Orbital width	0·133	0·368	0·804	3·717***	5·286***	3·806***
Orbital height	−0·250	2·891*	1·617	2·341*	4·607***	2·307*
Postorbital constriction	−1·061	0·206	0·256	2·644*	5·317***	0·117
Bizygomatic breadth	0·199	1·382	0·359	4·266***	7·795***	19·593***
Ant. basicranial width	0·465	1·472	1·019	3·276**	8·750***	3·970***
Basioccipital length	−0·247	0·365	1·391	3·183**	5·688***	6·089***
Biauricular width	−0·319	2·151	0·200	3·350**	9·350***	15·739***
Bimastoid width	−0·332	1·585	−0·201	2·278*	6·800***	15·891***
Vault height	−0·338	1·986	0·293	2·401*	5·306***	4·558***
Max. cranial length	0·170	1·452	1·649	5·380***	9·965***	14·093***
Min. cranial length	−0·151	1·320	0·360	2·802**	6·829***	5·509***
Basicranial length	0·073	0·980	1·081	2·540*	7·135***	8·483***
Basion-prosthion	0·337	1·047	2·229	5·369***	10·151***	15·182***

*$P<0.05$, **$P<0.01$, ***$P<0.001$.

Results

Univariate analyses

The results of the univariate analyses are shown in Table 2, which provides both *t*-scores and levels of significance for differences in male and female means for each age group. In age group 2, the only significant difference is found in interorbital width ($P<0.05$). Females of this age group are slightly larger than males in nine of the 21 cranial dimensions, although no difference reaches the 5% level of significance except interorbital width. In age group 3, males are larger than females in all dimensions except interorbital width. The only significant difference is found in orbital height ($P<0.05$).

In age group 4, a pattern of cranial sexual dimorphism starts to emerge with males being significantly larger in three midsagittal dimensions of the viscerocranium: palate length, lower facial length and alveolar height. This indicates that by dental age 4 males show a stronger initial development of prognathism than females.

Significant sexual differences in favor of males are found in age group 5 for all dimensions except interorbital width. These sexual differences range from the 0·1–5% levels of significance. In age group 6, males are larger in all dimensions at the 0·1% level of significance. Finally, in age group 7, males are significantly larger at the 0·1% level, except for orbital height ($P<0.05$) and postorbital constriction ($P>0.05$).

Bivariate analyses

The null hypothesis of ontogenetic scaling, where growth regressions for both sexes fall along the same ontogenetic continuum, was tested using Huxley's (1932) bivariate power function.

Table 3 Allometric coefficients and *F*-values for sex differences in various cranial dimensions regressed against basicranial length

Dimension	Sex	n	lob b	a	95% C.I. (a)	r	F-ratio
Palate width at c.	M	77	−0·844	1·341	(1·238, 1·444)	0·948	28·093***
	F	122	−0·079	0·933	(0·823, 1·042)	0·838	
Palate width p. to c.	M	78	−0·290	1·075	(0·984, 1·167)	0·937	2·289
	F	122	−0·085	0·969	(0·865, 1·072)	0·862	
Palate length	M	78	−1·435	1·691	(1·579, 1·804)	0·960	3·719
	F	121	−1·060	1·498	(1·335, 1·662)	0·858	
Lower facial length	M	78	−1·265	1·699	(1·595, 1·802)	0·966	6·091*
	F	122	−0·863	1·488	(1·354, 1·622)	0·895	
Alveolar height	M	78	−2·156	1·793	(1·602, 1·985)	0·906	6·577*
	F	122	−1·357	1·377	(1·119, 1·635)	0·694	
Bimaxillary width	M	77	−0·499	1·271	(1·158, 1·384)	0·933	1·367
	F	122	−0·275	1·162	(1·016, 1·309)	0·821	
Biorbital width	M	78	0·240	0·837	(0·750, 0·924)	0·910	2·172
	F	122	0·434	0·740	(0·644, 0·836)	0·814	
Interorbital width	M	78	−2·812	1·954	(1·727, 2·180)	0·892	9·825*
	F	122	−1·539	1·310	(0·969, 1·651)	0·571	
Orbital width	M	78	0·297	0·628	(0·538, 0·719)	0·845	0·122
	F	122	0·342	0·604	(0·500, 0·707)	0·726	
Orbital height	M	78	0·307	0·654	(0·534, 0·774)	0·780	0·739
	F	122	0·450	0·580	(0·462, 0·698)	0·665	
Postorbital constriction	M	78	0·763	0·503	(0·420, 0·586)	0·809	6·413*
	F	121	1·137	0·313	(0·178, 0·428)	0·403	
Bizygomatic breadth	M	78	−0·917	1·551	(1·443, 1·659)	0·957	8·424*
	F	122	−0·436	1·304	(1·175, 1·433)	0·877	
Ant. basicranial width	M	78	−0·236	1·046	(0·945, 1·147)	0·921	0·380
	F	122	−0·127	0·989	(0·833, 1·145)	0·753	
Basioccipital length	M	78	−1·580	1·532	(1·429, 1·636)	0·959	0·245
	F	122	−1·481	1·482	(1·307, 1·656)	0·837	
Biauricular width	M	78	−0·121	1·112	(1·027, 1·197)	0·948	22·667***
	F	122	0·500	0·789	(0·685, 0·892)	0·808	
Bimastoid width	M	78	−1·139	1·135	(1·051, 1·219)	0·952	23·850***
	F	122	0·487	0·811	(0·711, 0·911)	0·825	
Vault height	M	78	1·113	0·436	(0·380, 0·493)	0·871	0·089
	F	122	1·137	0·421	(0·338, 0·505)	0·673	
Max. cranial length	M	78	0·036	1·149	(1·074, 1·225)	0·961	10·750*
	F	122	0·405	0·955	(0·866, 1·044)	0·889	
Min. cranial length	M	78	1·158	0·475	(0·419, 0·532)	0·887	0·364
	F	122	1·202	0·450	(0·389, 0·511)	0·801	
Basion-prosthion	M	78	−1·143	1·682	(1·595, 1·769)	0·975	5·200*
	F	122	−0·835	1·522	(1·413, 1·630)	0·930	

Log $b = y$ intercept, a = slope, 95% C.I. (a) = 95% confidence interval for slope estimate, r = correlation coefficients.
*$P < 0.05$, **$P < 0.01$, ***$P < 0.001$.

Growth allometries for the 20 cranial dimensions each regressed against basicranial length are listed in Table 3. The analyses reveal two major growth patterns (Figures 1 and 2).

The first pattern (Figure 1) shows that ontogenetic scaling is found in 10 dimensions: palate width posterior to canine, palate length, bimaxillary width, biorbital width, orbital width, orbital height, anterior basicranial width, basioccipital length, vault height and minimum cranial length. In each case male and female slopes are not significantly different at or below the 5% level of significance. Therefore, the sexes are scaled versions of each other

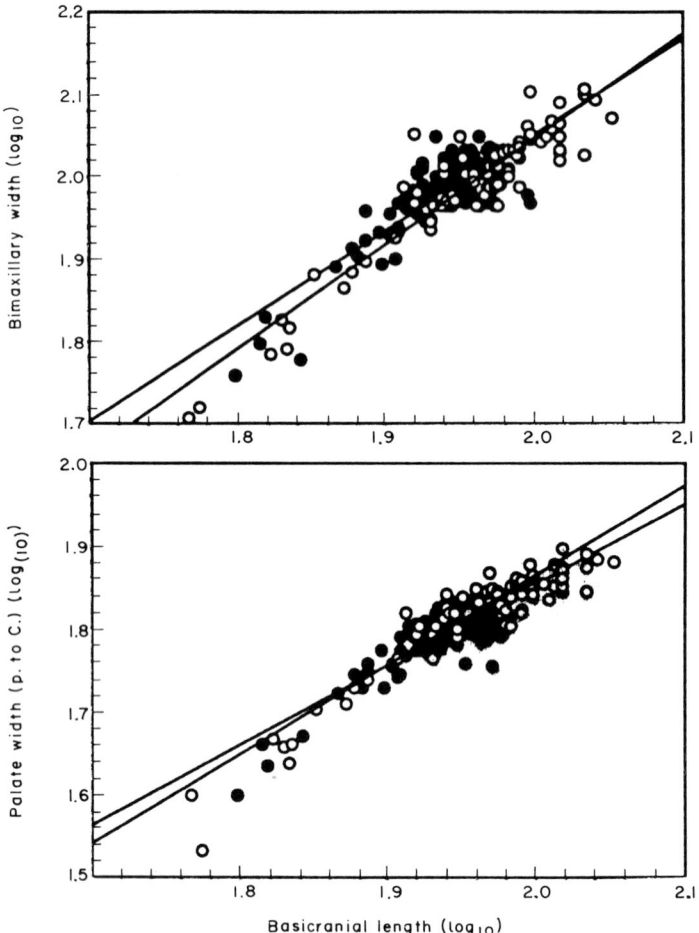

Figure 1. Two examples of the 10 cranial dimensions found to follow the overall growth trend of ontogenetic scaling. No significant sex differences in slope values are present; therefore, the sexes fall along the same ontogenetic continuum. It is suggested that time hypermorphosis may be responsible for this pattern. (●) Female; (○) male.

with no dissociation between size and shape as predicted by the null hypothesis of ontogenetic scaling.

The second growth pattern (Figure 2) shows a departure from ontogenetic scaling in the 10 remaining cranial dimensions: palate width at canine, lower facial length, alveolar height, interorbital width, postorbital constriction, bizygomatic breadth, biauricular width, bimastoid width, maximum cranial length and basion to prosthion. Male slopes are significantly steeper than corresponding female slopes, indicating that size and shape are dissociated.

Multivariate analyses

Principal component I, the allometry vector, is quite similar in magnitude when males and females are separately analysed (e.g., Baker, 1980; Handford, 1983; Reyment *et al.*, 1984; Masterson & Leutenegger, 1990). It reasons that by pooling male and female samples the

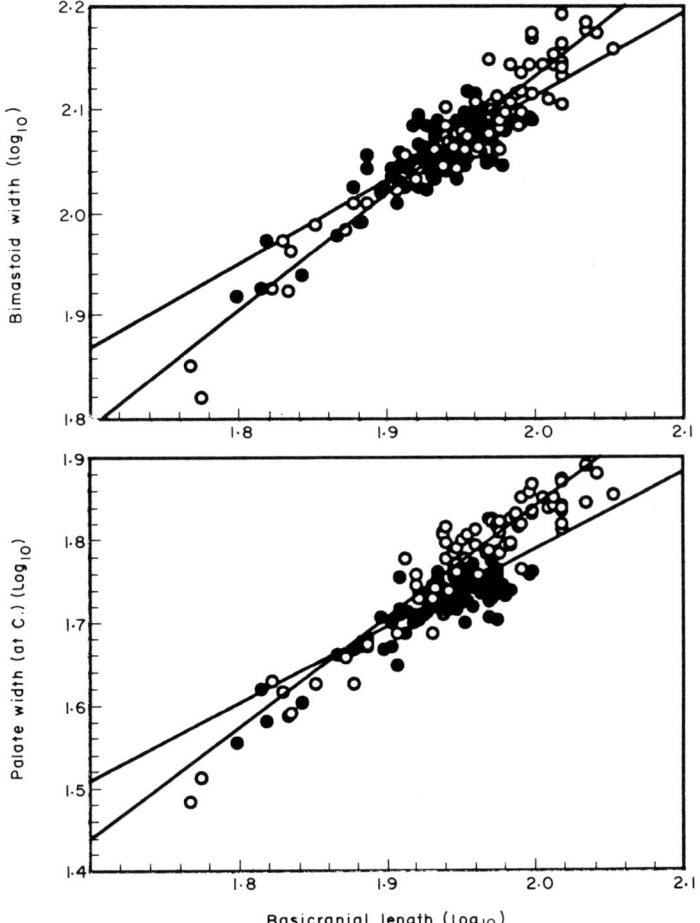

Figure 2. Two examples of the 10 cranial dimensions that follow the overall growth pattern of acceleration. Statistically significant sex differences in slope values are present, revealing that size and shape are dissociated. This pattern is associated with secondary sexual characteristics. (●) Female; (○) male.

investigator can derive a common size component, component I, with all subsequent components revealing non-allometric shape differences between the sexes (e.g. Hamilton & Johnston, 1978; Livezey & Humphrey, 1984; Somers, 1986, 1989).

The results of the pooled PCA for all 21 cranial dimensions are found in Table 4. Figure 3 shows the mean principal component score for each dental stage by sex. Principal component I accounts for 85·9% of the total sample variation. Overall size appears to be the major factor in the distribution of the skulls. In the infant age group (stage 2) both sexes possess the smallest scores, while the largest scores belong to adult male age groups 6 and 7. Age group 2 is the only case where females have larger allometry component scores. Males possess larger scores beginning in stage 3 and continuing through age group 7. Female age groups begin to cluster at stage 3, indicating that some degree of sexual dimorphism has started to appear in the early juvenile age class. This pattern continues through the mature adult stages.

The interpretation of the first component loadings (direction cosines) is as an allometry vector (Hursh, 1975) or a vector of relative growth because the variable loadings are all

Table 4 **Variable loadings on the first three principal components for an analysis on the covariance matrix of log-transformed values of the pooled sample of cranial dimensions**

Dimension	Factor I	Factor II	Factor III
Palate width at c.	0·2319	0·0821	−0·0578
Palate width p. to c.	0·1894	0·0565	0·0199
Palate length	0·3007	0·1080	−0·1366
Lower facial length	0·2987	0·0992	−0·1232
Alveolar height	0·3170	0·3536	−0·5939
Bimaxillary width	0·2243	0·0161	−0·3923
Biorbital width	0·1483	−0·0804	0·0967
Interorbital width	0·3311	−0·8938	−0·1815
Orbital width	0·1114	0·0392	0·1536
Orbital height	0·1155	0·0958	0·1630
Postorbital constriction	0·0850	−0·0251	0·0692
Bizygomatic breadth	0·2687	0·0013	0·0752
Ant. basicranial width	0·1980	0·0164	0·0573
Basioccipital length	0·2616	0·0743	0·6531
Biauricular width	0·1878	0·0329	0·0815
Bimastoid width	0·1883	0·0272	0·1249
Vault height	0·0746	0·0336	0·1190
Max. cranial length	0·2013	0·0893	−0·0242
Min. cranial length	0·0837	0·0494	0·0718
Basicranial length	0·1680	0·0204	0·1805
Basion-prosthion	0·2948	0·0839	0·0252
Total variance	0·8592	0·0463	0·0202

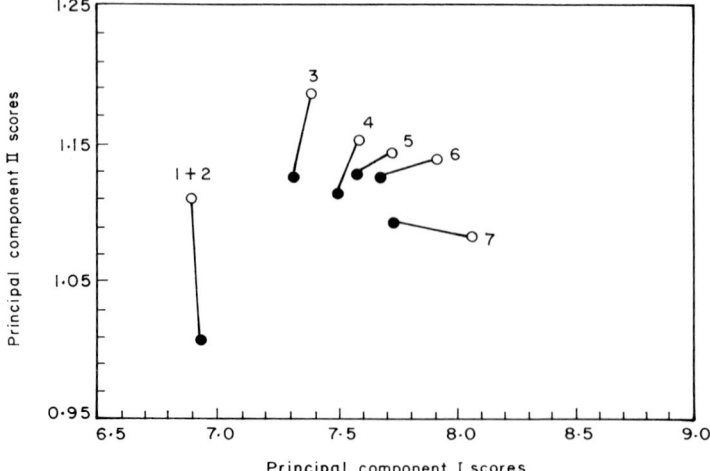

Figure 3. A plot of factor scores for a PCA of the 21 cranial dimensions for the pooled male (○) and female (●) samples showing the mean score for each dental stage by sex. An overall pattern of sexual dimorphism emerges such that younger dental stages show more shape differences, as revealed by component II, while size and shape as it is related to the allometry vector tend to differentiate the sexes at the older stages.

positive and account for a large percentage of the total sample variation (Shea, 1985c). The allometry vector reveals the direction of size increases along with shape changes that occur in each dimension. Dimensions in both the palatal and facial regions possess larger variable loadings along the allometry vector than the dimensions of the neurocranium. This agrees with the general trend of positive and negative allometric scaling in respective viscero- and neurocranial dimensions. It should be pointed out that some shape is also incorporated into the first component; therefore, the allometry vector is not just a "size" vector (Mosimann, 1970; Oxnard, 1978; Humphries et al., 1981; Shea, 1985c; Somers, 1986).

Principal component II explains 4·6% of the remaining variation. Component II substantially separates the sexes at stages 2, 3, and 4 (infant to mid-juvenile age groups). It continues to separate the sexes in the late juvenile to mature adult stages, ages 5, 6 and 7, but not on the same magnitude as for the younger age groups. Bipolar component loadings indicate that more shape variation is present in component II than the allometry vector. In PCA, bipolar loadings cause the separation of sex or age groups by shape differences due to dimensions possessing larger positive or negative loadings that either increase or decrease with respect to each other as one moves along the second component axis (Shea, 1983a, 1985c). The dimensions primarily responsible for the separation of males from females are: alveolar height, palate length, lower facial length, orbital height, maximum cranial length, basion to prosthion and palate width at the canine. All of the above dimensions possess strong positive loadings on component II. Interorbital width, biorbital width and postorbital constriction show negative loadings along this axis.

Principal component III accounts for 2·0% of the remaining variation. Component III has more bipolar loadings than component II, thus continuing to separate the sexes by shape more than size differences. Dimensions possessing strong positive loadings are: basioccipital length, basicranial length, orbital height, orbital width, bimastoid width and vault height. Alveolar height, bimaxillary width, interorbital width, palate length and lower facial length possess negative loadings.

In an effort to compare the multivariate and bivariate results, two additional analyses were run. Pooled PCAs were run on those dimensions exhibiting acceleration and ontogenetic scaling when regressed against basicranial length. This allowed for the detection of sexually divergent growth patterns in size and shape.

The results of a pooled PCA run on those 10 dimensions exhibiting acceleration in males are shown in Figure 4. The allometry vector accounts for 87·2% of the sample variation. The skulls seem to be separated by overall size according to allometric changes between both age groups and sex. The distribution of the male age groups along the allometry vector is such that increasing size with age dominates the separation of male age groups. Female age groups are also distributed by size as it relates to age; however, female age groups 4, 5, 6 and 7 cluster as in the original pooled PCA (Figure 3). The differences in allometry vector scores between these age groups are rather small compared to corresponding male ages. Therefore, males are appreciably larger per age group along the allometry vector. This reveals that sexual dimorphism along this axis has started by age group 3 and sequentially gets stronger through age group 7.

Principal component II explains 7·1% of the remaining variance. It clearly separates the sexes at age groups 2, 3 and 4. Component II also aids in the separation of age groups 5, 6 and 7 more than the original pooled component II vector. Stronger shape change exists in the accelerated dimensions as evidenced by the very large shape differences between the sexes of age groups 6 and 7. Of the dimensions exhibiting acceleration, interorbital width and

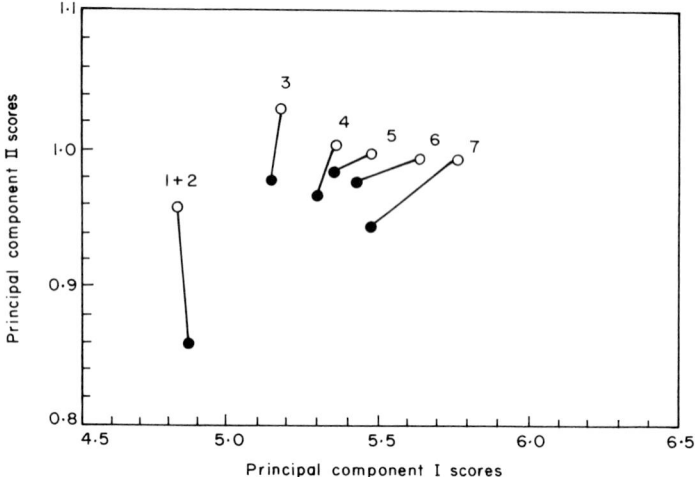

Figure 4. A plot of factor scores for a PCA of the dimensions exhibiting acceleration in bivariate comparisons against basicranial length showing the mean score for each dental stage by sex. A pattern of increased shape differences between the sexes is revealed by component II when compared to the pooled PCA. (●) Female; (○) male.

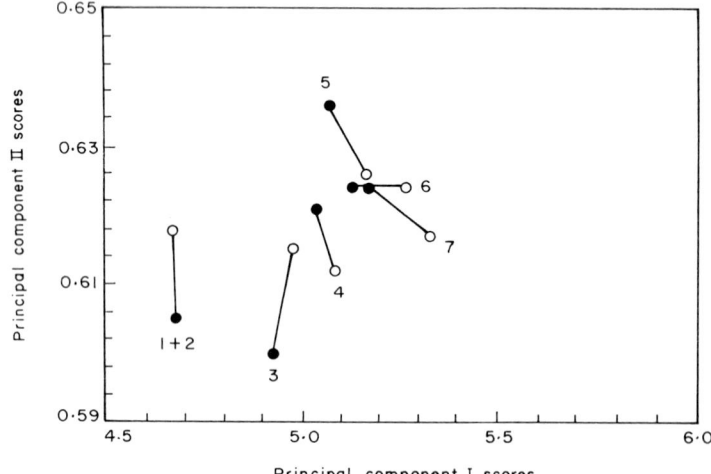

Figure 5. A plot of factor scores for a PCA of the dimensions exhibiting ontogenetic scaling in bivariate comparisons against basicranial length showing the mean score for each dental stage by sex. The sexes seem to be distributed by the allometry vector according to overall size, while shape differences between the sexes are minimal as would be expected if the sexes follow the growth pattern of ontogenetic scaling. (●) Female; (○) male.

postorbital constriction show the strongest negative loadings on component II. Alveolar height, lower facial length, basion to prosthion, palate width at the canine and maximum cranial length show the largest positive loadings. Thus, the above dimensions provide for a substantial amount of the separation between both age groups and sexes.

The results of the PCA run on those 10 dimensions in males that exhibit ontogenetic scaling are shown in Figure 5. The allometry vector explains 85·7% of the sample variance, while component II accounts for 4·1% of the remaining variance. As expected, the allometry vector distributes the age groups in an ontogenetic series from infant males and females through adult males according to overall changes in size/shape proportions. Both infant stages possess the smallest scores, while adult males possess the largest scores along this axis. Female age groups between the infant and adult male age groups tend to fall into the range of younger male age group scores. This reflects differences in size-related shape changes between the sexes.

Component II does not separate the sexes as clearly as in Figures 3 and 4 due to the presence of the smallest range of component II scores between the sexes of all three PCAs analysed. No obvious pattern is present in component II scores. The lack of a clear separation of the sexes is due to female age groups 4, 5 and 7 possessing larger component II scores than their corresponding male age groups. Thus, these data indicate that shape changes do not play a substantial role in the differentiation of the sexes within the ontogenetically scaled dimensions. Rather, component I shows the rescaling of the males as evidenced by males possessing larger component I scores for all the age groups except the infant stage. Of the dimensions exhibiting ontogenetic scaling, basioccipital length and vault height possess the strongest negative loadings on component II. Palate length, anterior basicranial width, bimaxillary width and palate width posterior to the canine possess the strongest positive loadings.

Discussion and conclusions

Univariate statistical analyses (Table 2) reveal a distinct growth pattern starting in the early juvenile stage and continuing through all subsequent age periods whereby males tend to exceed females in all 21 cranial dimensions. In conjuction, starting at the mid-juvenile stage (stage 4), there is a strong tendency for an increase in both number and strength of significant cranial sex differences in age group means. All significant differences favor males. Significant sex differences in the viscerocranium, specifically in palate length, lower facial length and alveolar height, emerge prior to those in the neurocranium. This reflects stronger initial development of prognathism in males.

In comparing our results with previous studies of cranial sexual dimorphism in orang-utans, there is full agreement that a marked degree of sexual dimorphism exists in the adult cranium (Selenka, 1898; Hrdlička, 1907; Ashton & Zuckerman, 1956; Biegert, 1957; Schultz, 1962; Röhrer-Ertl, 1988; Winkler et al. 1988). However, no such consensus exists for the extent and pattern of cranial sexual dimorphism in immature orang-utans. In a study of infant crania, Schultz (1962) found that in 14 of 16 dimensions the male mean exceeded the female mean. This would suggest the emergence of an early sexually divergent growth pattern that favors males. Such a pattern does not appear in our data until the mid-juvenile stage (dental stage 4). Schultz's results should be viewed with some caution due to both small sample sizes and the lack of significance testing.

Winkler (1987) reported a decrease in the number of significant cranial sex differences in orang-utans when proceeding from the infant to juvenile stage. Our findings show an increase in significant differences for the span of respective age groups. This discrepancy between our results and those of Winkler may have two possible explanations. First, her sample of juveniles included specimens ranging from the developmental stages of adolescence to adulthood. The permanent dentition of this category erupts in stages with periods of inactivity between successive eruption sequences (Schultz, 1941). It is therefore possible that if this category was further subdivided, such as the case with our sample, sexual differences in both timing and the degree of growth may have been revealed. Second, our sample represents a single subspecies of closely related populations, i.e., is highly homogeneous, while the sample of Winkler was based on various collections containing both subspecies. Morphological variation between subspecies may be present.

Based upon the relative aging technique of tooth emergence, our data provide no evidence for sexual differences in the timing of an adolescent growth spurt as reported for the African apes (Gavan, 1953; Watts & Gavan, 1982; Shea, 1985b). This finding is in agreement with the results of Winkler (1987). However, there is a substantial degree of sexual bimaturism in both orang-utans (Rijksen, 1978; Van der Werff Ten Bosch, 1982; Fooden & Izor, 1983) and African apes (Shea, 1985b). Sexual bimaturism in orang-utans is present in the age of attainment of sexual maturity, the age of attainment of adult weight and cessation in linear growth. Sexual maturity seems to show the least amount of bimaturism. Females reach sexual maturity at approximately 7 years of age, and males one year later (Rijksen, 1978). Adult weight and cessation in linear growth both show a marked degree of sexual bimaturism. Both are reached at approximately 8 years of age in females and 15 years of age in males (MacKinnon, 1974; Rijksen, 1978) and appear to occur at approximately the same time in captive and wild orang-utans (Fooden & Izor, 1983). However, there are some discrepancies as to the ages of maturity in wild orang-utans reported in the literature. Galdikas (1981) and Schürmann (1982) consider females as full adults when they conceive and give birth. This has been observed at approximately 12–15 years of age. Likewise, males are fully mature when they develop secondary sexual characteristics and behavior such as long calls. These characteristics appear in wild male orang-utans at ages ranging from 13–15 years (Rijksen, 1978) up to 19–20 years of age (Galdikas, 1985a). Although specific ages remain to be quantified, the marked degree of sexual bimaturism in orang-utans is clearly supported in the literature.

Bivariate analysis provides insight into patterns underlying sexual dimorphism in primates with respect to size/shape relationships and heterochrony. For example, in sexually dimorphic anthropoid primates a common pattern exists whereby males and females fall along a common intraspecific slope in certain cranial and postcranial dimensions (Wood, 1976; Shea, 1983b; Cochard, 1985; Leutenegger & Larson, 1985; Leutenegger & Masterson, 1989b). Therefore, the two sexes are ontogenetically scaled versions of each other. Any shape differences between the sexes can then be interpreted as the result of an extension of relative growth from smaller females to larger overall size in males (Shea, 1986). Thus, males are peramorphic or transcend the female form (Alberch *et al.*, 1979). The cause of such a pattern underlying adult sexual dimorphism in size and shape is sex-differentiated growth. Two different developmental processes can produce this pattern. First, females may attain sexual maturity and cease growth before males (Tanner, 1962). If the female pattern represents the ancestral growth pattern, then in heterochronic terms the males have an extended growth period in time, a process known as time hypermorphosis. The second process, rate hypermorphosis, occurs when males have a faster rate of growth than females, but possess the same time period of growth (Shea, 1983b, 1986).

As previously mentioned, ontogenetic scaling between the sexes occurs in the majority of cranial proportions of anthropoid primates. However, orang-utans show a departure from this general trend. Two major growth patterns emerge from examination of the 20 growth allometries (Figures 1 and 2).

The first pattern is strong ontogenetic scaling where growth regression of both sexes fall along a single ontogenetic continuum. Ontogenetic scaling is found in 10 cranial dimensions that relate to three cranial regions: (1) the neurocranium directly associated with brain size (anterior basicranial width, basioccipital length, vault height, minimum cranial length) because brain size itself scales ontogenetically (P. V. Tobias, pers. commun.), (2) the orbital region (bimaxillary width, biorbital width, orbital width, orbital height), and (3) the dental arcade (palate width posterior to the canine, palate length) associated with food processing functions. From a viewpoint of heterochrony, ontogenetic scaling is most likely the result of the extension of the growth period in time in males, i.e., time hypermorphosis, due to the presence of a marked degree of sexual bimaturism in orang-utans. As previously mentioned, female orang-utans reach sexual maturity, attain adult weight and cease in linear growth earlier in chronological time than males. This indicates that males grow for a longer period of time. If rate hypermorphosis is responsible, no sexual differences in the above variables should be present.

The second growth pattern indicates a departure from the null hypothesis of ontogenetic scaling in the 10 remaining cranial dimensions. The dissociation of size and shape is revealed by male slopes being significantly steeper than female slopes. Therefore, males are disproportionately larger than females. This seems to occur in cranial regions associated with secondary sexual character development. The stronger degree of prognathism found in males uncovered by univariate analyses is also revealed by departures from ontogenetic scaling in the dimensions: lower facial length, alveolar height, maximum cranial length and basion to prosthion. Non-allometric relationships in the transverse cranial dimensions (bizygomatic breadth, biauricular width, bimastoid width) may be associated with the development of the large cheek pads in mature males (Schultz, 1969; Rijksen, 1978; Rodman & Mitani, 1987). The large degree of known canine dimorphism (Swindler, 1976; Kay, 1982; Leutenegger & Shell, 1987) is reflected in the dimension palate width at the canine. In fact, statistically significant sexual differences are present in mesiodistal length ($P<0.001$), bucco-lingual width ($P<0.001$), and height ($P<0.001$) in the upper canines of the adult orang-utans of this sample (unpublished data, Leutenegger). It is most likely that these departures from ontogenetic scaling are due to the heterochronic process of acceleration, an increase in the rate of shape change in male orang-utans.

The results of the multivariate analyses coincide to a large extent with the sexually dimorphic growth patterns previously revealed in univariate and bivariate analyses. In the pooled PCA (Figure 3) principal component I, the allometry vector, reveals a pattern such that by age group 3 and continuing through age group 7 males have larger component I scores. Female age groups begin to cluster, indicating that some degree of sexual dimorphism has emerged by age group 3 and continues through age group 7. This agrees with the univariate results of males tending to be larger than females in all cranial proportions starting at age group 3. The multivariate analyses also support the tendency for increases in both number and strength of significant cranial sex differences beginning at age group 4 in the univariate analyses.

Non-allometric shape differences, as revealed by principal component II scores, cause a clear separation of the sexes within the infant to juvenile age groups (ages 2, 3 and 4) and contribute to the differentiation of the sexes in the early adolescent to mature adult stages (ages 5, 6 and 7). It is noteworthy that eight of the 11 dimensions primarily responsible for the separation of sex and age groups follow the heterochronic process of acceleration in males when regressed against basicranial length. This suggests that selection has acted to produce shape differences between the sexes that are related to the development of secondary sexual characteristics. Again, sexual differences begin in the juvenile ages only to continue through the adult ages. It is interesting that component II seems to separate the younger age groups by sex, while component I tends to separate the sexes at the older ages by size and shape variation as it is related to the allometry vector.

In the PCA of accelerated dimensions (Figure 4) males are substantially larger per age group than females along the allometry vector beginning at age group 4 and continuing through age group 7. This corresponds quite accurately with univariate results of significant differences in age group means. Principal component II scores show larger differences per age group between the sexes than was found in either the original pooled or the ontogenetically scaled samples. This most likely reflects shape differences caused by the appearance of secondary sexual characteristics.

In the PCA of those dimensions exhibiting ontogenetic scaling (Figure 5), principal component I reveals that males transcend corresponding female forms. This is due to male age groups possessing larger component I scores for all age groups except the infant stage. Sexual differences in size-related shape is inferred from the substantially smaller female component I scores. Again, this corresponds with univariate results of male means being significantly larger per age group beyond the early juvenile stage.

No clear sexually divergent growth pattern emerges from the component II scores. This is primarily because females have slightly larger component II scores in age groups 4, 5 and 7 than their male counterparts. In the clear separation of the sexes that occurs in the sample of accelerated dimensions (Figure 4), males have larger component II scores for all age groups. Again, male age groups possess larger component II scores in all ages in the original pooled PCA except age group 7. The tight range of scores seen in Figure 5 indicates that shape changes do not play a substantial role in the differentiation of the sexes within the ontogenetically scaled dimensions. This is expected if males are peramorphic due to ontogenetic scaling.

Thus, two major multidimensional patterns of cranial sexual dimorphism emerge from the data. The first pattern reveals that sexual differences at age groups 2 and 3 are primarily the result of differences in principal component II scores, reflecting mainly shape-related differences. The second pattern indicates that age groups 5, 6 and 7 show a trend of stronger size-related shape differences with increasing age in the allometry vector along with decreasing differences in principal component II scores. The latter reflects an increase in size-related shape differences between the sexes. Age group 4 shows a combination of both patterns.

As previously mentioned, Shea (1986) suggested that an investigation of the ontogenetic bases of sexual dimorphism can provide new insights and information unobtainable from studies concerned only with adult endpoints. Our morphometric analyses on a large sample spanning all developmental stages provide a good example of how valuable ontogenetic data can be in helping to determine the actions of various selective regimes.

There are two general, although not necessarily mutually exclusive, hypotheses given to explain the selective forces underlying sexual dimorphism, sexual selection and ecological selection (Rodman & Mitani, 1987) Sexual differences in reproductive strategies are the foundation of sexual selection theory. It operates either by (1) intrasexual competition for mates, which favors characters that increase an individual's reproductive success relative to members of the same sex, or (2) mate choice, which favors characters that are attractive to members of the opposite sex, i.e., intersexual or epigamic selection.

Ecological selection may also lead to the appearance of sexual differences in morphology and/or behavior in at least two ways (Rodman & Mitani, 1987). First, the reduction of feeding competition between the sexes by niche divergence can lead to the appearance of sex differences (Selander, 1972). Second, it has recently been argued that sexual dimorphism may also be based upon the constraints of energetics, gut size and dietary diversity (Demment, 1983).

Among behavioral ecologists there is a consensus that sexual selection theory, specifically intrasexual selection, best explains the marked degree of sexual dimorphism in orang-utans (Rodman, 1973, 1979; MacKinnon, 1974, 1979; Galdikas, 1978, 1979; Mitani, 1985a,b). This is primarily due to an extremely uneven distribution of parental investment. Female parental investment far exceeds male investment and operational sex ratios are heavily male-biased because female orang-utans conceive infrequently (Rodman, 1973; Mitani, 1985b). Galdikas (1978, 1981) has estimated interbirth intervals ranging from not less than 4.5 years up to 9 years. Under these conditions, sexual selection theory predicts intense competition among males and the evolution of traits that increase a male's competitive success (Rodman & Mitani, 1987).

Fieldwork provides several supportive observations favoring the male–male competition theory. First, adult males possess large home ranges that are defended with aggressive behavior. Both chasing away and active physical confrontations with other adult males are common. Galdikas (1981) observed that only three out of 12 adult males were without visible physical anomalies that centered on the hands (stiff, misshapen, or broken fingers) and head (missing eyes, torn lips, broken canines), the most vulnerable body parts during conflict. The incidence of such confrontations clearly increases if a receptive female is within the adult male's range (Galdikas, 1978, 1981, 1985a; Mitani, 1985b). In the past some workers (e.g., Galdikas, 1979; Rodman, 1984; Mitani, 1985b) proposed that large body size confers a mating advantage such that a larger male can defeat smaller males in direct competition for access to females. Mitani (1985b) provides observations in support of this hypothesis. Second, the high degree of hostility observed between adult males is not seen in encounters between adult and subadult males. Subadult males are tolerated within a resident's range because dominant males can easily displace subadult males in the presence of females (Mitani, 1985b). Finally, studies of orang-utan reproductive behavior point out that adult males are involved almost exclusively in cooperative copulations with adult females, whereas subadult males tend to be involved in uncooperative copulations (Galdikas, 1985a,b; Schürmann & van Hooff, 1986). Galdikas (1985a) characterized the adult male reproductive behavior as a "consort/combat" reproductive strategy, while a "sneak/rape" reproductive strategy prevails for subadult males (Galdikas, 1985b).

There has been some recent interest in the role of female mate choice in orang-utans, specifically by Schürmann & van Hooff (1986). They argue that female choice has been overlooked in most models of orang-utan reproductive behavior. According to Schürmann & van Hooff, concealed ovulation allows female orang-utans to choose their mates, providing additional selection for patterns of sexual dimorphism. The basis of their model is two principal observations. First, long calls (loud, long vocalizations) given exclusively by adult males (MacKinnon, 1974) are thought to both mediate male spacing and attract receptive females (Rodman, 1973; Galdikas, 1978, 1983; Schürmann, 1981), and second, females prefer mating with adult males.

While it does seem true that both nulliparous and parous females prefer to mate with adult males (Galdikas, 1985a,b; Schürmann & van Hooff, 1986) and that long calls regulate spacing between males through an approach–avoidance system based upon dominance relationships (Mitani, 1985a), the evidence supporting the usage of long calls to attract females is contradictory. Long calls are given both spontaneously and in response to sudden stimuli such as falling trees (Rodman, 1988). Pilo-erection and branchshaking often accompany a long call, which can last between 1–2 minutes (Mitani, 1985a), and a state of sexual excitement seems to accompany long calls as well (Schürmann, 1982).

Schürmann (1982) reported several observations in which a nulliparous female was attracted to an adult male giving long calls and once actually provoked the male into making a long call in order to get the male sexually aroused. The significance of such behavior can be questioned on several grounds. First, it seems that nulliparous females are both highly attracted to, and show proceptive behavior towards, adult males, as evidenced by nulliparous females always being the initiator during 47 observed copulations of young females with adult males at Ketambe in the Gunung Leuser Reserve in Sumatra (Schürmann & van Hoff, 1986). Second, the proposed heightened sexual arousal associated with a long call of an adult male may increase his willingness to copulate with a nulliparous female, which may explain the possible attraction of young females to long calls. Finally, whether these copulations with young females produce offspring is questionable given that some evidence exists for an extended period of adolescent sterility in female orang-utans (Rijksen, 1978).

Contradictory to Schürmann's findings of female mate choice, Mitani (1985a) found no evidence supporting the hypothesis that long calls are used to attract females in an experiment using replays of recorded long calls in Kutai Game Reserve in Borneo. In fact, a receptive adult female in consort with an adult male avoided the caller rather than being attracted to him after giving long calls. This lends support to Schürmann's (1982) observation that nulliparous and parous females may not behave similarly in sexual relationships. Mitani (1985a) concluded that long calls function to advertise an adult male's presence, deter other males from approaching his area and are adaptations for male–male competition.

The influence of mate choice has currently been given a secondary role by Rodman & Mitani (1987) in the production of both body size and sexual dimorphism in the orang-utan. Their conclusions are based primarily upon work done in Borneo where population densities are moderate compared to the higher densities in Sumatra where Schürmann & van Hoff (1986) have worked. Biases may be present as to which specific type of sexual selection, intra- or intersexual, may play a more significant role due to differences in population density. This may cause differences in behavior between respective study areas.

Until further studies are completed that compare geographical variations in reproductive behavior, Rodman & Mitani's (1987) proposed secondary role of female mate choice in producing sexual dimorphism in the orang-utan seems reasonable. Even if future studies do reveal that mate choice plays a significant role, one would expect that the selective forces would produce ontogenetic growth patterns similar to those produced by male–male competition, i.e., the traits important to males in intrasexual selection are most likely the traits females prefer and choose (C. van Schaik, pers. comm.). Therefore, mate choice cannot be completely ruled out as a viable explanation; however, it may not play as significant a role in the development of sexual dimorphism in orang-utans as does male–male competition.

Finally, let us examine the possible role of ecological selection acting in orang-utans to produce sexual dimorphism. Orang-utans are primarily fruit-eaters, but also at times rely heavily upon leaves, shoots, flowers, epiphytes, lianas, woodpith, bark, mineral-rich soils, ants, termites, bird eggs and even meat (MacKinnon, 1974; Galdikas, 1980; Sugardjito & Nurhuda, 1981). Ecological selection theory predicts that sex differences in morphology and/or behavior should be present primarily due to differences related to dietary habits, e.g., larger males should be eating more high-fibre, low-quality food than females. Some discussion of sex differences in diet is reported in the literature; however, most dietary differences are conflicting in the proportions of fruit, bark, insects and leaves eaten between the sexes. Rodman (1977) reported that an adult male spent less time feeding on fruit (58·6% of 144 hours *vs.* 67·1% of 227 hours) and termites (0·8% *vs.* 1·9%) and more on bark (16·5% *vs.* 4·9%) than three females during a 2 month study period. Contrary to Rodman's findings, Galdikas (1978, 1988) reported that adult males feed slightly more on fruits, considerably more on termites and less on both bark and young leaves than females during a 4 year study. Both studies, however, concur that males are less selective than females and that the number of feeding bouts per day does not significantly differ between the sexes. Galdikas (1988) suggested females are more selective and consume more types of foods per day than males, due to either nutritional requirements related to parental investment or to lower the intake of toxin levels due to the increased usage of bark.

From the above discussion of dietary differences, it is clear that no conclusions pertaining to which food types are eaten more often by respective sexes can be drawn from current analyses. However, most workers agree that some dietary differences are present because of the large degree of body size dimorphism in orang-utans. Some behavioral differences are also present. For example, adult males tend to live in the lower canopy and spend more time on the ground. Therefore, some degree of ecological selection is most likely acting in orang-utans. However, whether it is a primary selective force in producing body size dimorphism has been questioned. Rodman & Mitani (1987) argue that the known dietary differences are not large enough to suggest that the principal selective factor involved with body size differences is a reduction in feeding competition, but rather are a secondary result of increased male size due to male–male competition for access to females. However, the possibility that ecological selection currently contributes to the maintenance of the large degree of body size dimorphism that characterizes orang-utans cannot be excluded (C. van Schaik, pers. comm.).

Finally and most important, ecological selection cannot produce secondary sexual characteristics (male calls, cheek pads, large canine, throat sac, prognathism) according to Rodman & Mitani (1987), because such features are related to reproductive success rather than to reduced feeding competition. For example, the enlarged throat sac of adult males may help in long distance transmission of the long call, while the cheek pads may focus the sound in one direction. As previously mentioned, long calls are interpreted as an adaptation to male–male competition. Therefore, ecological selection may be acting to reduce feeding competition due to large differences in body size, but the known secondary sexual characteristics of the adult male orang-utan most likely are not the product of ecological selection.

The results of the morphometric analyses reported above provide good examples in support of the intrasexual selection theory currently proposed by behavioral ecologists. Univariate analyses reveal that males have an initially stronger degree of prognathism that continues through adult ages. Various dimensions (lower facial length, alveolar height, maximum cranial length, basion to prosthion) associated with prognathism show a departure from ontogenetic scaling in bivariate analyses. Sexual dimorphism in canine size may have a role in producing sexual differences in prognathism in such a way as to increase the anterior projection of the large canines in males. In this fashion, prognathism may be involved with the defense of a male's range and, therefore, male–male competition most likely is the underlying selective agent.

It is noteworthy that all departures from ontogenetic scaling occur in cranial proportions that are associated with secondary sexual characteristics. A pattern such as this might be expected in the production of secondary sexual characteristics in the orang-utan. Although the presence of some facial prognathism as well as both rudimentary cheek pads and laryngeal sacs are found in female orang-utans (Winkler, 1989), their obvious presence in adult males most likely cannot be explained simply as males growing longer in time following the same ontogenetic trajectory as females. Given the extreme size of both the laryngeal sac and cheek pads, one would expect some non-allometric shape differences to exist in these morphological features. The observed growth pattern seems to indicate that the target of selection in the case of the male orang-utan may be specific shape differences that relate to the development of secondary sexual characteristics. Since ecological selection may not be able to produce secondary sexual characteristics, we argue that male–male competition plays a strong selective role in producing the marked degree of sexual dimorphism in orang-utans. Because some of the dimensions related to food processing functions are ontogenetically scaled (palate width posterior to the canine, palate length), these proportional increases in size/shape relationships are probably a secondary result of increases in male body size. This would help explain some of the reported dietary and related niche differences between the sexes.

The results of the multivariate analyses corroborate the sexually dimorphic growth patterns revealed in univariate and bivariate analyses. Dimensions such as palate width posterior to the canine, palate length and biorbital width are clearly related to allometric size differences. However, other dimensions such as palate width at the canine, lower facial length and bimastoid width are clearly related to sexual selection. Differentiation of these processes is evident by age class 3 when the initial stages of development of the secondary sexual characteristics can be seen. Thus, we argue that the ontogenetic growth patterns revealed by the univariate, bivariate and multivariate analyses provide a strong example of morphometric data that corroborates the behavioral data in support of the sexual selection theory as the major explanation for the marked degree of sexual dimorphism in orang-utans.

Summary

We examine ontogenetic patterns of cranial sexual dimorphism in a large sample of Bornean orang-utans (*Pongo pygmaeus pygmaeus*) by means of univariate, bivariate and multivariate statistics. Univariate analyses reveal that males tend to be significantly larger per age class, starting by the mid-juvenile age class and continuing through the mature adult age class. A stronger initial development of prognathism in males is found. Bivariate analyses reveal two major growth patterns are present, ontogenetic scaling and acceleration. Ontogenetic scaling occurs in dimensions associated with orbit size, brain size and certain proportions of the dental arcade. The dimensions associated with the development of the secondary sexual characteristics seem to follow the accelerated growth pattern. Multivariate analyses reveal that the younger age classes seem to be separated by shape, principal component II, whereas the older age classes seem to be separated by size and shape variation as it is related to the allometry vector. The findings from the univariate and bivariate analyses are supported by the results of the multivariate analyses. The possible selective forces underlying the marked degree of sexual dimorphism in orang-utans are briefly examined. Currently, evidence from the behavioral ecology of wild orang-utans best supports the male–male competition theory rather than ecological selection as the primary selective force. We argue that our morphometric analyses corroborate the behavioral data in support of the intrasexual selection theory as the underlying selective regime primarily responsible for the marked degree of sexual dimorphism in orang-utans.

Acknowledgements

We thank Professor Dr G. Ziegelmayer for permission to study the SELENKA collection of orang-utan crania. We also thank Dr Matthew Ravosa and Anne Gomez for inviting us to participate in their symposium and to contribute to this special issue of the *Journal of Human Evolution*. Dr Brian Shea's comments on earlier manuscripts are greatly appreciated, as well as comments from three anonymous reviewers and Dr Carel van Schaik's insights that helped to improve this manuscript. Special thanks go to Dr Luci Kohn for her many helpful discussions and comments throughout the duration of this project. This research was supported by a Senior Professor Fulbright Award and a grant from the Graduate School, University of Wisconsin–Madison to W.L.

References

Airoldi, J. P. & Flury, B. K. (1988). An application of common principal component analysis to cranial morphometry of *Microtus californicus* and *M. ochrogaster* (Mammalia, Rodentia). *J. Zool., Lond.* **216,** 21–36.
Alberch, P., Gould, S. J., Oster, G. F. & Wake, D. B. (1979). Size and shape in ontogeny and phylogeny. *Paleobiology* **5,** 296–317.
Anderson, T. W. (1963). Asymptotic theory for principal component analysis. *Ann. Math. Stat.* **34,** 122–148.
Ashton, E. H. & Zuckerman, S. (1956). Cranial crests in the Anthropoidea. *Proc. Zool. Soc. Lond.* **126,** 581–634.
Baker, A. J. (1980). Morphometric differentiation in New Zealand populations of the house sparrow (*Passer domesticus*). *Evolution* **34,** 638–653.
Beynon, A. D., Dean, M. C. & Reid, D. J. (1991). Histological study of the chronology of the developing dentition in gorilla and orangutan. *Am. J. phys. Anthrop.* **86,** 189–203.
Biegert, J. (1957). Der Formwandel des Primatenschädels und seine Beziehungen zur ontogenetischen Entwicklung und den phylogenetischen Spezialisationen der Kopforgane. *Morph. Jb.* **98,** 77–199.
Bromage, T. G. (1985). Taung facial remodeling: a growth and development study. In (P. V. Tobias, Ed.) *Hominid Evolution: Past, Present, and Future*, pp. 239–246. New York: Alan R. Liss.
Bromage, T. G. (1989). Ontogeny of the early hominid face. *J. hum. Evol.* **18,** 751–773.
Campbell, N. A. & Atchley, W. R. (1981). The geometry of canonical variate analysis. *Syst. Zool.* **30,** 268–280.
Chatterjee, S. & Price, B. (1977). *Regression Analysis By Example.* New York: John Wiley & Sons.
Cochard, L. R. (1985). Ontogenetic allometry of the skull and dentition of the rhesus monkey (*Macaca mulatta*). In (W. L. Jungers, Ed.) *Size and Scaling in Primate Biology*, pp. 231–255. New York: Plenum Press.
Cock, A. G. (1966). Genetical aspects of metrical growth and form in animals. *Q. Rev. Biol.* **41,** 131–190.
Corner, B. D. & Richtsmeier, J. T. (1991). Morphometric analysis of craniofacial growth in *Cebus apella*. *Am. J. phys. Anthrop.* **84,** 323–342.
Dean, M. C. & Wood, B. A. (1981). Developing pongid dentition and its use for aging individual crania in comparative cross-sectional growth studies. *Folia primat.* **36,** 111–127.
Demment, M. W. (1983). Feeding ecology and the evolution of body size of baboons. *Afr. J. Ecol.* **21,** 219–233.
Eckhardt, R. B. (1975). The relative body weights of Bornean and Sumatran Orangutans. *Am. J. phys. Anthrop.* **42,** 439–450.
Fooden, J. & Izor, R. J. (1983). Growth curves, dental emergence norms, and supplementary morphological observations in known-age captive orangutans. *Am. J. Primat.* **5,** 285–301.
Galdikas, B. M. F. (1978). Orangutan adaptation at Tanjung Puting Reserve, Central Borneo. Ph.D. Dissertation, University of California at Los Angeles.
Galdikas, B. M. F. (1979). Orangutan adaptation at Tanjung Puting Reserve: mating and ecology. In (D. A. Hamburg & E. R. McCown, Eds) *The Great Apes*, pp. 195–234. Menlo Park, CA: Benjamin/Cummings Publishing Co.
Galdikas, B. M. F. (1980). Modern adaptations in orang-utans? *Nature* **291,** 266.
Galdikas, B. M. F. (1981). Orangutan reproduction in the wild. In (C. E. Graham, Ed.) *The Reproductive Biology of the Great Apes*, pp. 281–300. New York: Academic Press.
Galdikas, B. M. F. (1983). The orangutan long call and snag crashing at Tanjung Puting Reserve. *Primates* **24,** 371–384.
Galdikas, B. M. F. (1985*a*). Adult male sociality and reproductive tactics among orangutans at Tanjung Puting. *Folia primat.* **45,** 9–24.
Galdikas, B. M. F. (1985*b*). Subadult male orangutan sociality and reproductive behavior at Tanjung Puting. *Am. J. Primat.* **8,** 87–99.

Galdikas, B. M. F. (1988). Orangutan diet, range, and activity at Tanjung Puting, Central Borneo. *Int. J. Primatol.* **9,** 1–35.

Gavan, J. A. (1953). Growth and development of the chimpanzee: a longitudinal and comparative study. *Hum. Biol.* **25,** 93–143.

Gould, S. J. (1966). Allometry and size in ontogeny and phylogeny. *Biolog. Rev.* **41,** 587–640.

Gould, S. J. (1975a). On the scaling of tooth size in mammals. *Am. Zool.* **15,** 351–362.

Gould, S. J. (1975b). Allometry in primates, with emphasis on scaling and the evolution of the brain. In (F. S. Szalay, Ed.) *Approaches to Primate Paleobiology (Contr. Primatol. 5.)*, pp. 244–292. Basel: Karger.

Gould, S. J. (1977). *Ontogeny and Phylogeny*. Cambridge, MA: Harvard University Press.

Hamilton, S. & Johnston, R. F. (1978). Evolution in the house sparrow. VI. Variability and niche width. *Auk* **95,** 313–323.

Handford, P. (1983). Continental patterns of morphological variation in a South American sparrow. *Evolution* **37,** 920–930.

Harvey, P. H. & Mace, G. M. (1982). Comparisons between taxa and adaptive trends: problems of methodology. In (King's College Research Group, Eds) *Current Problems in Sociobiology*, pp. 343–361. Cambridge: Cambridge University Press.

Hrdlička, A. (1907). Anatomical observations on a collection of orang skulls from western Borneo. *Proc. U.S. Natn. Mus.* **31,** 539–568.

Humphries, J. M., Bookstein, F. L., Chernoff, B., Smith, G. R., Elder, R. L. & Poss, S. G. (1981). Multivariate discrimination by shape in relation to size. *Syst. Zool.* **30,** 291–308.

Hursh, T. M. (1975). A multivariate study of chimpanzee and gorilla crania. Ph.D. Dissertation, Harvard University.

Huxley, J. S. (1932). *Problems of Relative Growth*. London: Methuen.

Jolicoeur, P. (1963a). The multivariate generalization of the allometry equation. *Biometrics* **19,** 497–499.

Jolicoeur, P. (1963b). The degree of generality of robustness in *Martes americana*. *Growth* **27,** 1–27.

Jolicoeur, P. & Mosimann, J. E. (1960). Size and shape variation in the painted turtle. A principal component analysis. *Growth* **24,** 339–354.

Kay, R. F. (1982). Sexual dimorphism in Ramapithecinae. *Proc. natn. Acad. Sci. U.S.A.* **79,** 209–212.

Leigh, S. R. & Cheverud, J. M. (1991). Sexual dimorphism in the baboon facial skeleton. *Am. J. phys. Anthrop.* **84,** 193–208.

Leutenegger, W. & Larson, S. (1985). Sexual dimorphism in the postcranial skeleton of New World primates. *Folia primat.* **44,** 82–95.

Leutenegger, W. & Shell, B. (1987). Variability and sexual dimorphism in canine size of *Australopithecus* and extant hominoids. *J. hum. Evol.* **16,** 359–367.

Leutenegger, W. & Masterson, T. J. (1989a). The ontogeny of sexual dimorphism in the cranium of Bornean orang-utans (*Pongo pygmaeus pygmaeus*): I. Univariate analysis. *Z. Morph. Anthrop.* **78,** 1–14.

Leutenegger, W. & Masterson, T. J. (1989b). The ontogeny of sexual dimorphism in the cranium of Bornean orang-utans (*Pongo pygmaeus pygmaeus*): II. Allometry and heterochrony. *Z. Morph. Anthrop.* **78,** 15–24.

Livezey, B. C. & Humphrey, P. S. (1984). Sexual dimorphism in continental streamer-ducks. *Condor* **86,** 368–377.

MacKinnon, J. (1974). The behavior and ecology of wild orang-utans (*Pongo pygmaeus*). *Anim. Behav.* **22,** 3–74.

MacKinnon, J. (1979). Reproductive behavior in wild orangutan populations. In (D. A. Hamburg & E. R. McCown, Eds) *The Great Apes*, pp. 195–234. Menlo Park, CA: Benjamin/Cummings Publishing Co.

Martin, R. D. & Barbour, A. D. (1989). Aspects of line-fitting in bivariate allometric analyses. *Folia primat.* **53,** 65–81.

Masterson, T. J. & Leutenegger, W. (1990). The ontogeny of sexual dimorphism in the cranium of Bornean orang-utans (*Pongo pygmaeus pygmaeus*) as detected by principal components analysis. *Int. J. Primatol.* **11,** 517–539.

Mitani, J. C. (1985a). Sexual selection and adult male orangutan long calls. *Anim. Behav.* **33,** 272–283.

Mitani, J. C. (1985b). Mating behavior of male orangutans in the Kutai Reserve, East Kalimantan, Indonesia. *Anim. Behav.* **33,** 392–402.

Morrison, D. F. (1976). *Multivariate Statistical Methods*. New York: McGraw-Hill.

Mosimann, J. E. (1970). Size allometry: size and shape variables with characteristics of lognormal and gamma distributions. *J. Am. statist. Ass.* **65,** 930–945.

Oxnard, C. E. (1978). One biologist's view of morphometrics. *Ann. Rev. Ecol. Syst.* **9,** 219–241.

Pagel, M. D. & Harvey, P. H. (1988). Recent developments in the analysis of comparative data. *Q. Rev. Biol.* **63,** 413–440.

Ravosa, M. J. (1991a). Ontogenetic perspective on mechanical and nonmechanical models of primate circumorbital morphology. *Am. J. phys. Anthrop.* **85,** 95–112.

Ravosa, M. J. (1991b). The ontogeny of cranial sexual dimorphism in two old world monkeys: *Macaca fascicularis* (Cercopithecinae) and *Nasalis larvatus* (Colobinae). *Int. J. Primatol.* **12,** 403–426.

Reyment, R. A., Blackith, R. E. & Campbell, N. A. (1984). *Multivariate Morphometrics*, 2nd edn. New York: Academic Press.

Ricker, W. E. (1973). Linear regressions in fishery research. *J. Fish. Res. Bd. Can.* **30,** 409–434.

Rijksen, H. D. (1978). *A Field Study on Sumatran Orangutans (*Pongo pygmaeus abelii, *Lesson 1827). Ecology, Behavior, and Conservation*. Netherlands: Veenman and Zonen.
Rodman, P. S. (1973). Population composition and adaptive organization among orangutans of the Kutai Reserve. In (R. P. Micheal & J. H. Crook, Eds) *Comparative Ecology and Behavior of Primates*, pp. 171–209. London: Academic Press.
Rodman, P. S. (1977). Feeding behavior of orangutans in the Kutai Reserve, East Kalimantan. In (T. H. Clutton-Brock, Ed.) *Primate Ecology*, pp. 383–413. London: Academic Press.
Rodman, P. S. (1979). Individual activity patterns and the solitary nature of Orangutans. In (D. A. Hamburg & E. R. McCowan, Eds) *The Great Apes*, pp. 195–234. Menlo Park, CA: Benjamin/Cummings Publishing Co.
Rodman, P. S. (1984). Foraging and social systems of orangutans and chimpanzees. In (P. S. Rodman & J. G. H. Cant, Eds) *Adaptations for Foraging in Nonhuman Primates*, pp. 134–160. New York: Columbia University Press.
Rodman, P. S. (1988). Diversity and Consistency in ecology and behavior. In (J. H. Schwartz, Ed.) *Orang-utan Biology*, pp. 31–51. New York: Oxford University Press.
Rodman, P. S. & Mitani, J. C. (1987). Orangutans: sexual dimorphism in a solitary species. In (B. B. Smuts, D. L. Cheney, R. M. Seyfarth, R. W. Wrangham & T. T. Struhsaker, Eds) *Primate Societies*, pp. 146–154. Chicago: University of Chicago Press.
Röhrer-Ertl, O. (1984). Orang-Utan-Studien. *Hieronymus, Neuried*.
Röhrer-Ertl, O. (1988). Cranial growth. In (J. H. Schwartz, Ed.) *Orang-utan Biology*, pp. 201–223. New York: Oxford University Press.
Schultz, A. H. (1941). Growth and development of the orang-utan. *Contr. Embryol. Careg. Inst.* **29,** 57–110.
Schultz, A. H. (1962). Metric age changes and sex differences in primate skulls. *Z. Morph. Anthrop.* **52,** 239–255.
Schultz, A. H. (1969). *The Life of Primates*. New York: Universe Books.
Schürmann, C. L. (1981). Courtship and mating behavior of wild orangutans in Sumatra. In (A. B. Chiarelli & R. S. Corruccini, Eds) *Primate Behavior and Sociobiology*, pp. 130–135. Berlin: Springer-Verlag.
Schürmann, C. L. (1982). Mating behavior of wild orang-utans. In (L. E. M. de Boer, Ed.) *The Orang-utan: Its Biology and Conservation*, pp. 269–284. Netherlands: The Hague.
Schürmann, C. L. & van Hooff, A. R. A. M. (1986). Reproductive strategies of the orang-utan: new data and the reconsideration of existing sociosexual models. *Int. J. Primatol.* **7,** 265–287.
Selander, R. K. (1972). Sexual selection and sexual dimorphism in birds. In (B. G. Campbell, Ed.) *Sexual Selection and the Descent of Man, 1871–1971*. Chicago: Aldine.
Selenka, E. (1898). Rassen, Schädel und Bezahnung des Orangutan. *Stud. Entw. Schädelbau.* **1,** 1–99.
Shea, B. T. (1981). Relative growth of the limbs and trunk in the African apes. *Am. J. phys. Anthrop.* **56,** 179–201.
Shea, B. T. (1982). Growth and size allometry in the African Pongidae: cranial and postcranial analyses. Ph.D. Dissertation, Duke University.
Shea, B. T. (1983*a*). Size and diet in the evolution of African ape craniodental form. *Folia primat.* **40,** 32–68.
Shea, B. T. (1983*b*). Allometry and heterochrony in the African apes. *Am. J. phys. Anthrop.* **62,** 275–289.
Shea, B. T. (1985*a*). Ontogenetic allometry and scaling: a discussion based on the growth and form of the skull in African apes. In (W. L. Jungers, Ed.) *Size and Scaling in Primate Biology*, pp. 175–205. New York: Plenum Press.
Shea, B. T. (1985*b*). The ontogeny of sexual dimorphism in the African apes. *Am. J. Primat.* **8,** 183–188.
Shea, B. T. (1985*c*). Bivariate and multivariate growth allometry: statistical and biological considerations. *J. Zool., Lond.* **206,** 367–390.
Shea, B. T. (1986). Ontogenetic approaches to sexual dimorphism in anthropoids. *Hum. Evol.* **1,** 97–110.
Shea, B. T. (1989). Heterochrony in human evolution: the case for neoteny reconsidered. *Yearb. phys. Anthrop.* **32,** 69–101.
Simpson, S. W., Lovejoy, C. O. & Meindl, R. S. (1990). Hominoid dental maturation. *J. hum. Evol.* **19,** 285–297.
Smith, R. J. (1980). Rethinking allometry. *J. theor. Biol.* **87,** 97–111.
Smith, R. J. (1981). Interspecific scaling of maxillary canine size and shape in female primates: relationships to social structure and diet. *J. hum. Evol.* **10,** 165–173.
Snedecor, G. W. & Cochran, W. G. (1980). *Statistical Methods*, 7th edn. Ames, IA: The Iowa State University Press.
Somers, K. M. (1986). Multivariate allometry and removal of size with principal components analysis. *Syst. Zool.* **35,** 359–368.
Somers, K. M. (1989). Allometry, isometry and shape in principal components analysis. *Syst. Zool.* **38,** 169–173.
Sugardjito, J. & Nurhuda, N. (1981). Meat-eating behaviour in wild orang utans, *Pongo pygmaeus*. *Primates* **22,** 414–416.
Swindler, D. R. (1976). *Dentition of Living Primates*. New York: Academic Press.
Tanner, J. M. (1951). Some notes on the reporting of growth data. *Hum. Biol.* **23,** 93–159.
Tanner, J. M. (1962). *Growth at Adolescence*, 2nd edn. Oxford: Blackwell Scientific Publications.
Watts, E. S. & Gavan, J. A. (1982). Postnatal growth of nonhuman primates: the problem of the adolescent growth spurt. *Hum. Biol.* **54,** 53–70.
Werff Ten Bosch, J. J. Van Den. (1982). The physiology of reproduction of the orang-utan. In (L. E. M. de Boer, Ed.) *The Orang Utan. Its Biology and Conservation*, pp. 201–214. Netherlands: The Hague.

Winkler, L. A. (1987). Sexual dimorphism in the cranium of infant and juvenile orangutans. *Folia primat.* **49,** 117–126.
Winkler, L. A. (1989). Morphology and relationships of the orangutan fatty cheek pads. *Am. J. Primat.* **17,** 305–319.
Winkler, L. A., Conroy, G. C. & Vannier, M. W. (1988). Sexual dimorphism in exocranial and endocranial dimensions. In (J. H. Schwartz, Ed.) *Orang-utan Biology*, pp. 225–232. New York: Oxford University Press.
Winkler, L. A., Schwartz, J. H. & Swindler, D. R. (1991). Aspects of dental development in the orangutan prior to eruption of the permanent dentition. *Am. J. phys. Anthrop.* **86,** 255–271.
Wood, B. A. (1976). The nature and basis of sexual dimorphism in the primate skeleton. *J. Zool. Lond.* **180,** 15–34.

Appendix 1

All measurements and craniometric points, with the exception of orbital height, were defined according the Shea (1982, 1983a) and are as follows:

01 Palate width at canine: greatest palatal width along the alveolar ridge at the canines;
02 Palate width posterior to canine: greatest palatal width along the alveolar ridge posterior to the canines;
03 Palate length: staphylion to prosthion;
04 Lower facial length: mid point of basioccipital/basisphenoid suture to prosthion;
05 Alveolar height: nasospinale to prosthion;
06 Bimaxillary width: right zygomaxillare to left zygomaxillare;
07 Biorbital width: right frontomalare orbitale to left frontomalare orbitale;
08 Interorbital width: right interorbitale to left interorbitale;
09 Orbital width: left frontomalare orbitale to left interorbitale;
10 Orbital height: greatest perpendicular diameter of the left orbital entrance (Schultz, 1962);
11 Postorbital constriction: right infratemporale to left infratemporale;
12 Bizygomatic breadth: right zygion to left zygion;
13 Anterior basicranial width: most inferior point of the curvature of the medial border of the glenoid fossa on the right to the corresponding point on the left;
14 Basioccipital length: mid point basioccipital/basisphenoid suture to basion;
15 Biauricular width: right auriculare to left auriculare;
16 Bimastoid width: most lateral point on the mastoid process on the right to the corresponding point on the left;
17 Vault height: vertex to basion;
18 Maximum cranial length: opisthocranion to prosthion;
19 Minimum cranial length: opisthocranion to glabella;
20 Basicranial length: basion to nasion; and
21 Basion to prosthion.

Steven R. Leigh

Department of Anthropology,
Northwestern University, Evanston,
IL 60208 U.S.A.

Received 1 June 1991
Revision received 13 February
1992 and accepted 28 February
1992

Keywords: sexual dimorphism,
growth and development,
anthropoid primates.

Patterns of variation in the ontogeny of primate body size dimorphism

This study investigates variability in the ontogeny of sexual dimorphism in body weight in anthropoid primates. Specifically, the hypothesis that the ontogenetic bases of dimorphism vary in primates is tested with a large comparative sample. This sample allows both specification of the range of variability in patterns of dimorphic growth and careful examination of some correlates of ontogenetic variability.

The analysis is based on body weight growth in 45 species of captive primates. Growth curves are estimated for each sex in each species using a non-parametric regression technique (locally weighted least squares regression). These regressions facilitate estimation of the degree of differences in age at body weight maturation, the proportion of total adult dimorphism arising from sex differences in rate of growth, and the proportion of total dimorphism attributable to sex differences in duration of growth.

Primates show great diversity in how adult weight dimorphism develops. Although males tend to reach adult weight later than females, cases in which rate differences are exclusively responsible for body weight dimorphism are observed. More generally, sex differences in both rate and duration of growth occur together to produce dimorphism, but these differences may occur independently. Relationships between these processes and size appear complex, as do relations between these processes and dimorphism. These results suggest that primates have evolved a number of developmental pathways that lead to similar levels of adult dimorphism. It is expected that male patterns of growth respond primarily to sexual selection, but that female patterns of growth respond to natural selection.

Journal of Human Evolution (1992) **23**, 27–50

Introduction

Sexual dimorphism in body size is a common and well-recognized phenomenon in primates and in many other taxa (Ralls, 1977). Recent analyses of the evolution of primate sexual dimorphism delineate a number of correlations between body size dimorphism and socio-ecological variables. The ideas proposed to explain such variation in sexual dimorphism include sexual selection, allometric or size-related phenomena, ecological factors and phylogenetic inertia (see Cheverud *et al.*, 1985; Ely & Kurland, 1989; Gaulin & Sailer, 1984; Leutenegger & Cheverud, 1982, 1985; Rensch, 1959). Despite the high level of interest in the study of adult dimorphism, relatively few studies have provided detailed insight into the ontogeny of dimorphism. Moreover, ontogenetic studies of dimorphism have tended to be contradictory in their findings (see Watts, 1986).

The present study attempts to provide a basic understanding of the range of variability in dimorphic growth in primates. Specifically, this study expands previous research presented by Shea (1986) which shows that the ontogenetic bases of sexual dimorphism vary among primate species [see also Jarman (1983) and Georgiadis (1985) for terrestrial herbivores]. These studies indicate that there is interspecific diversity in the contribution of sex differences in rate of growth and duration of growth that lead to adult dimorphism. In other words, there are species-specific ways in which comparable levels of sexual dimorphism develop. Although these studies provide evidence for the presence of variability, the nature of this variation remains inadequately specified because relatively few species have been investigated, particularly for primates. Definition of the variability in dimorphic growth processes provides a foundation for more detailed explanatory analyses (see Leigh, 1992).

0047–2484/92/070027 + 24 $03.00/0 © 1992 Academic Press Limited

Ontogeny and dimorphism
Ontogeny is important in gaining an understanding of dimorphism because similar levels of body size dimorphism could, theoretically, be produced through quite disparate developmental processes. If the developmental processes that ultimately account for dimorphism vary substantially among primate species, then the presence of variable processes requires explanation. Previous researchers (Jarman, 1983; Fedigan, 1982; Ralls, 1977; Shea, 1985, 1986) have addressed the possibility that ontogenetic bases vary among species and have discussed the significance of actual or potential variation. However, the patterns of variation in the ontogenetic processes that result in primate body size dimorphism are not well documented.

The presence of marked variation in the ontogenetic processes that lead to adult body size dimorphism would imply that, in some instances, similar levels of adult dimorphism reflect differing evolutionary causes. For example, equivalent levels of adult dimorphism might be observed in two species even though rate differences in growth produced dimorphism in one species while duration differences resulted in dimorphism in another species. An attempt to explain the presence or degree of dimorphism in an interspecific analysis without knowledge about the underlying ontogenetic trajectories risks evaluation of a variable (adult dimorphism) that may be a consequence of distinct ontogenetic processes. Adult dimorphism can be seen as a result of such processes, reflecting the presence of such processes but not their nature. This problem is recognized by Shea, who suggested that interspecific variability in the developmental processes that produce dimorphism may be the outcome of "fundamentally different evolutionary pathways" (1986:98). This general theoretical approach follows Gould (1977) quite closely in anticipating that the evolution of widely divergent ontogenetic processes may produce similar adult morphologies. This perspective may imply that analyses of adult dimorphism have yielded inconsistent results and proven controversial (Ely & Kurland, 1989), because they focus on the consequences of dimorphic growth and not on the actual processes that generate dimorphism.

A number of previous studies investigated the relations between ontogeny and dimorphism. In a pioneering analysis, Wiley showed that bimaturism [which he defined as "a sexual difference in the age at onset of breeding" (1974:241)] is positively correlated with polygyny, large body size and body size dimorphism in grouse. According to Wiley, bimaturism can evolve when delays in male breeding confer reproductive advantages to males. In grouse, bimaturism most likely evolves because young males cannot compete effectively with older birds for mates. Moreover, females may tend to choose older males, limiting reproductive opportunities for young males. As a result of these (and other) factors, bimaturism evolves.

Jarman (1983) modified and enhanced Wiley's concept of bimaturism. Jarman defined bimaturism as a sex difference in duration of growth (usually body weight growth) and evaluated associations between bimaturism and dimorphism in *Bovidae* (antelope, sheep, cattle, etc.), *Cervidae* (deer) and *Macropodidae* (kangaroos, wallabies, rat kangaroos, etc.). In these taxa, sex differences in rate of growth may occur, but dimorphism always seems to be associated with bimaturism. Georgiadis (1985) offered comparable results for an analysis of body weight growth and sexual dimorphism in African ruminants. Georgiadis found that "no males attain asymptotic weight before conspecific females" (1985:79) and that bimaturism is much more important in the production of dimorphism than rate differences.

These studies of terrestrial herbivores would suggest that variability in body size dimorphism is a consequence of variable ontogenetic processes. As pointed out by Jarman, there are species-specific ways in which dimorphism develops (1983:513). Variation in patterns of ontogeny that result in dimorphism are seen as evolutionary responses to a number of factors which may include territoriality, mate competition and resource distribution plus interactions between these and other factors. In general, however, Jarman expects bimaturism to evolve when:

> "Males that eventually breed will be those that, as young males, adopted strategies of survival rather than competing dangerously for matings against the older males" (1983:487).

It should be noted that Jarman's findings were based primarily on evaluation of literature data. Moreover, the growth data used include body weight for some species but skeletal dimensions for others [such as skull weight in dall sheep (1983:493) and mandibular length in caribou (1983:496)]. Jarman's cross sectional growth curves were often based on somewhat limited data, and growth curves fitted by eye. Consequently, Jarman's results may be rather tentative. Georgiadis' (1985) analysis, which is technically more rigorous because it was based on regression analysis of body weight growth, does not conflict with Jarman's general assessment.

Shea (1986) approached the problem of ontogenetic variability and dimorphism in primates through a heterochronic analysis. In comparisons involving groups of closely related species, he found that variation in levels of bimaturism (defined as a sex difference in age at body weight maturity) was positively correlated with size and dimorphism among certain primate species. These species include macaques, baboons and African apes. In contrast, Shea found that in some West African forest monkeys (guenons), dimorphism apparently resulted mainly from sexual rate differences in growth.

Although Shea's study was rather general, it points to considerable diversity among groups of closely related primate species in the role that rate and duration differences in ontogeny play in the production of dimorphism. In some cases, similar levels of dimorphism may arise through rate differences, but in others, sex differences in the duration of the growth period (bimaturism) seem to account for dimorphism.

Growth studies of linear skeletal dimensions in primates may suggest less diversity in the ontogenetic processes that lead to dimorphism than is indicated by Shea's study of weight growth. It should be noted that such studies may not be directly comparable to analyses based on body weight primarily because analyses of skeletal growth describe growth in a single morphological system. In contrast, analyses of weight simultaneously incorporate growth of all morphological components. While the skeletal system contributes to mass and weight, the relation between this system and others is not well known in non-human primates. Nevertheless, in a discussion of detailed longitudinal studies of long bone growth based on rhesus macaques, chimpanzees and humans, Watts suggested that skeletal "sexual dimorphism in the tempo of maturation appears to have been conserved" among primate species (1986:162). This implies that interspecific variation in rate differences in growth are largely, but as Watts cautions, not exclusively, responsible for production of variable levels of skeletal length dimorphism in primates. This view contrasts with an earlier study by Gavan & Swindler (1966) who indicated that intraspecific differences in sitting height dimorphism among these taxa are largely the outcome of variation in bimaturism. Gavan & Swindler detected statistically significant sex differences in adolescent chimpanzee growth rates (1966: 184–185) and in rhesus growth rates. However, Gavan & Swindler (1966) attributed the differences in chimpanzee growth rates to small sample sizes rather than to developmental processes, possibly explaining the disparity between their findings and those of Watts (1986).

Review of these investigations illustrates that, although there is strong evidence for diversity in the developmental processes that result in sexual size dimorphism, there is some controversy as to the nature and degree of variation. In general, detailed comparative knowledge about the ontogenetic processes that result in dimorphism is lacking (in both weight and length). Because of this lack of information, the present study attempts to delineate the range of variability in processes of dimorphic growth in anthropoid primates. Exploration of this variability provides a foundation for further explanatory analyses of primate sexual dimorphism. Variability in the ontogenetic processes (rate and duration of growth) that result in dimorphism is anticipated in primates (Shea, 1986). The degree of bimaturism is also expected to vary. It is further expected that bimaturism is often size-related within groups of closely related primate species, as hypothesized by Shea (1986:107). An additional objective of this study is to provide basic descriptions of body weight growth in several primate species that reflect variability in patterns of dimorphic growth.

Materials and methods

Materials

The data consist of chronological age (years) and body weight measurements (kg) for captive primates. Forty-five anthropoid species are included in the analysis (Table 1).

Chronological age and body weight measurements were obtained from veterinary records for 2637 monkeys and apes. These animals are maintained at United States Regional Primate Research Centers and European and United States zoological parks and research laboratories. Additional data gathered from the veterinary records include species, subspecies, sex, individual identification number or name, health status, reproductive status, general body condition and health history. Animals were observed and photographed whenever possible.

Most of the data were obtained during routine physical examinations and tuberculosis tests. Data from animals requiring veterinary care as a result of trauma (such as lacerations and fractures) are also utilized, provided weights were obtained shortly after occurrence of trauma. Weights from pregnant, chronically ill, dehydrated or congenitally abnormal animals are excluded. A small number of animals identified in health records as obese (usually based on palpation by veterinary staff) are omitted from the analysis. No animals were designated as obese prior to attainment of adult weight. Nearly all animals are of known chronological age, accurate to the day of birth. Data from animals of unknown date of birth are utilized but only those data points collected after the animals had been in captivity for at least 10 years. For gorillas and orang-utans, only weights from animals in captivity for at least 15 years were used from wild-caught individuals.

The use of weight as a size variable requires careful scrutiny. Specifically, it is important to consider whether or not reliable evolutionary inferences can be based on an analysis of weight. Weight may mask differences in body composition that may be important in interspecific analyses. In addition, it is well known that weight is easily affected by environmental factors. As a result, weight can fluctuate markedly, especially after maturity.

These liabilities are balanced by several factors. First, nearly all previous literature on primate body size dimorphism focuses on weight (Cheverud *et al.*, 1985; Clutton-Brock, 1985; Clutton-Brock & Harvey, 1977; Clutton-Brock *et al.*, 1977; Gaulin & Sailer, 1984; Leutenegger & Cheverud, 1982, 1985). The reason for investigation of weight is obviously related to sexual selection theory, particularly hypotheses about the role of intermale

Table 1 Species for which growth data were collected and key for plot abbreviations

Species	Plot abbreviation	Individuals Male	Female
Cebuella pygmaea	CPYG	29	36
Callithrix jacchus	CJAC	33	39
Callimico goeldi	CLGOE	42	47
Saguinus fuscicollis	SFUS	18	19
Saguinus geoffroyi	SGEO	9	10
Saguinus imperator	SIMP	11	14
Saguinus oedipus	SOED	46	18
Leontopithecus r. rosalia	LROS	26	31
Saimiri sciureus	SSCI	32	28
Cebus apella	CAPEL	26	28
Callicebus moloch	CMOL	30	23
Aotus trivirgatus	ATRI	25	23
Ateles geoffroyi	AGEO	29	30
Alouatta caraya	ACAR	27	26
Allenopithecus nigroviridis	ANIG	6	10
Cercopithecus aethiops	CAET	30	30
Cercopithecus diana	CDIA	25	26
Cercopithecus neglectus	CNEG	29	23
Cercopithecus mitis	CMIT	27	37
Erythrocebus patas	EPAT	41	52
Miopithecus talapoin	MTAL	16	27
Cercocebus torquatus atys	CTOR	38	71
Macaca arctoides	MARC	52	58
Macaca cyclopis	MCYL	25	20
Macaca fascicularis	MFAS	13	13
Macaca mulatta	MMUL	52	58
Macaca nemestrina	MNEM	39	64
Macaca nigra	MNIG	27	31
Macaca radiata	MRAD	28	26
Macaca silenus	MSIL	39	41
Papio cynocephalus	PCYN	33	53
Papio (Mandrillus) leucophaeus	PLEU	5	10
Papio (Mandrillus) sphinx	PSPH	49	59
Colobus guereza	CGUE	46	49
Presbytis cristata	PCRI	9	19
Presbytis entellus	PENT	29	24
Presbytis obscura	POBS	19	17
Pygathrix nemaeus	PNEM	13	12
Hylobates lar	HLAR	24	25
Hylobates syndactylus	HSYN	19	21
Pongo pygmaeus	PPYG	42	41
Pan paniscus	PPAN	13	23
Pan troglodytes	PTRO	22	23
Gorilla gorilla	GGOR	59	50

Only those species listed in Tables 2 and 3 are included in plots. Because longitudinal data are pooled, regressions are based on larger numbers of data points than individuals. Results for species with fewer than about 20 individuals for each sex should be treated with caution.

competition in the evolution of dimorphism. Sexual selection should favor heavier males in cases where male–male competition (most directly, combat) influences mating success (Darwin, 1871). When sexual selection does produce larger males, it can be expected to do so similarly among species, probably through increased size of the musculo-skeletal system. It should be noted that this expectation does not obviate a role for natural selection in the determination of male body size. Second, and more generally, weight is recognized as a key variable in ecological, physiological, morphological, allometric and life history studies because it is an accurate approximation of mass (Jungers, 1985:350–351). As such, weight is inextricably linked to nearly all aspects of an organism's biology. Thus, weight offers an appropriate variable for many types of biological investigations, including the present study.

It is also necessary to consider the value of captive data in addressing evolutionary hypotheses. For the present analysis, two assumptions facilitate evolutionary inferences. First, it is assumed that captivity affects both sexes similarly. This study examines sex differences, and it is unlikely that captivity has differential effects by sex. Therefore, inferences about patterns of sexually differentiated growth in captive animals should reflect similar patterns in non-captive animals. Second, it is assumed that captivity affects all species similarly. Clearly, neither assumption can be tested with data currently available, and future research should focus on these assumptions.

There are notable advantages to using captive data. First, nearly all ages are exact and all weights are actual (see above). This is a very important feature because all previous analyses of adult body size dimorphism and some studies of ontogeny and dimorphism are based on developmentally aged samples. Limited data gathered for the present study suggest that dental maturity may not correspond to cessation of body weight growth. Dental eruption often appears complete (sometimes substantially) before attainment of adult weight, particularly in males (Leigh, 1992). As a result, estimates of adult weight based on dentally adult individuals may not reflect weight of animals after body weight growth cessation.

Second, the majority of the data are mixed-longitudinal (Cock, 1966), allowing some knowledge about individual variation in growth. Although the focus of this study is on comparisons among species, repeated measures (longitudinal data) are considered for all species.

Third, all the animals in this study have received good nutrition and veterinary care, minimizing disruptions to growth trajectories resulting from illness or parasites. Fourth, veterinary records, daily logs, interviews with animal care staff and direct observation allow some familiarity with a number of the animals studied. It should be added that data from zoos and regional primate research centers seem comparable, with no apparent differences in growth or in adult weight.

Methods

All data for each species are plotted with age as the independent variable and weight as the dependent variable. The data are treated cross-sectionally (or as pooled time-series data; Dielman, 1989). Evaluation of longitudinal data in this manner loses information about individual variation in growth. However, this approach can be employed because cross-sectional regression coefficients approximate the mean of longitudinal regression coefficients (Cock, 1966; Tanner, 1978) and because groups (e.g., males *vs*. females) are compared in this study. Tanner (1978) and Eveleth & Tanner (1990) recommend cross-sectional data in group comparisons. Finally, comparisons of results based on analyses of longitudinal and cross-sectional data drawn from the same samples suggest that both types of data supply adequate information for the present study (Leigh, 1992).

Non-parametric locally weighted least squares regression (lowess) is used to describe growth in these species. This method is employed because it makes no assumptions about the form of growth curves, an important consideration when numerous species are compared. According to Cleveland & Devlin (1988), lowess regression is analogous to calculation of moving averages in time series analysis.

Lowess regression fits small portions of a bivariate distribution of points through several steps (Efron & Tibshirani, 1991). Initially, a small "window" of points is selected in which there is a single "target" point. A weighting function is centered on this target point, with the highest weight at the target point. Weights at either side of the target point decrease, falling to zero at the edges of the window. The window and weighting function determine the position of a linear regression centered on the target point and calculated for all the points in the window. Data near the target point have a large influence on the position of the linear regression because weights are highest at this point. Iterative shifting of the target point and window results in a solid line through the entire bivariate distribution.

Varying the size of the window controls the linearity of the fit to the bivariate distribution. For example, a window that includes all data points would fit a linear weighted least squares line to the data. Conversely, extremely small windows could conceivably fit all points with unique x and y values. Efron & Tibshirani (1991) provided a non-technical description of this method, while Cleveland (1979) and Cleveland & Devlin (1988) detailed its mathematical foundations. All lowess regressions in the present analysis were performed using Systat packaged programs (Wilkinson, 1990).

Lowess regressions facilitate visual estimates of the two variables studied in this analysis (Figure 1; see also Cheverud *et al.*, 1992). These variables are the degree of bimaturism (in years), the proportion of dimorphism due to growth rate differences (pr) and the proportion of dimorphism resulting from body weight bimaturism (pt). Bimaturism is defined in the present study as a sex difference in age at cessation of body weight growth. Bimaturism is measured by subtracting age at female growth cessation from age at male growth cessation. The proportion of dimorphism due to difference in rate of growth is estimated by measuring the percentage of total dimorphism observed at the age of female growth cessation (Figure 1). In other words, the rate proportion is estimated by dividing r by the total dimorphism (D). The proportion due to difference in age at growth cessation is simply one minus the proportion due to rate. These two proportions measure the effects of rate differences and bimaturism, respectively, on dimorphism.

Values for pr and pt are calculated only for those species in which dimorphism exceeds 1·25. This procedure is necessary because pr and pt do not reflect varying degrees of dimorphism; even minute levels of dimorphism could be partitioned into rate and duration compounds by this method. For example, a mean sex difference of 1 g can be parcelled into pr and pt as easily as 100 kg of dimorphism. Thus, pr and pt are "dimorphism free" measures of ontogenetic sex differences. This property may preclude extensive use of pr and pt values in statistical hypothesis tests directed toward explaining variation in dimorphism.

Exclusion of non-dimorphic species helps ensure that the phenomenon of dimorphism is not confounded with monomorphism because of arithmetic artifacts. Limiting the analysis in this manner removes all callitrichids, *Aotus trivirgatus*, *Callicebus moloch*, *Presbytis obscura*, *Presbytis cristata* and hylobatids, resulting in a sample of 31 species. There are no species in the sample of 45 species with estimates of dimorphism between 1·15–1·25.

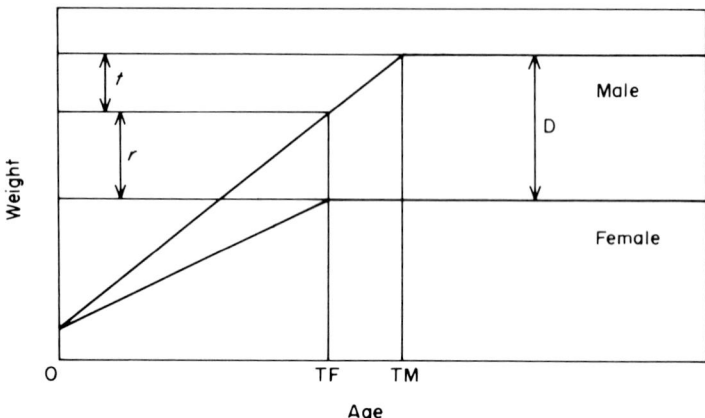

Figure 1. Idealized growth curves illustrating measurement of proportional contributions to dimorphism. Ontogenetic components of total adult dimorphism (D) are estimated by r, resulting from rate differences and t resulting from duration differences. Division of r by D gives the proportion of total dimorphism due to sex differences in rate of growth (pr). Division of t by D provides an estimate of proportion of total dimorphism arising from sex differences in duration of growth (pt).

Derivation of accurate estimates of bimaturism, pr and pt hinges on the ability to estimate ages at growth cessation. These estimates are obtained visually from lowess plots and accuracy to the nearest 3 months is attempted. Specifically, an inflection separating a post-natal growth rate from a flatter adult regression is taken as the age at growth cessation (see Figure 1). A straight-edge ruler is placed perpendicular to the x-axis on a printed copy of the lowess regression. An inflection defining age at female growth cessation is designated and male weight at this age is determined by intersection of the male lowess curve and the straight-edge (see Figure 1). Male age at growth cessation is estimated in a similar manner. Adult weights are calculated by averaging all data points at ages in excess of 1 year past the estimated age at growth cessation.

This method is subjective. In some cases, female growth decelerates over a long period of time, making a precise estimate of age at growth cessation difficult. This problem is, however, fairly infrequent. Longitudinal data were inspected in situations where estimation of growth cessation was problematic and these estimates were generally consistent with cross-sectional estimates.

Intraobserver error is evaluated for the percentage of dimorphism resulting from rate differences in growth (pr). Means for two trials over the sample of 31 species for which pr (and therefore, pt) are calculated are nearly identical (trial 1 pr mean=0·517, trial 2 pr mean=0·544). A t-test comparing these means suggests no significant differences ($t=0·352$, $P<0·73$). Thus, estimates seem consistent, at least within a single observer.

Dimorphism is measured as the mean of male adult weights divided by the mean of female adult weights. Female weight is taken as a measure of species size.

Results

Variation in bimaturism

Bimaturism is very commonly observed in primate species and there appears to be considerable diversity in the degree of bimaturism among primate species (Table 2). The length of the

Table 2   **Ages at growth cessation, bimaturism (measured as male–female) and duration of female growth as a percentage of male growth period**

Species	Age at growth cessation			Percent bimaturism
	Male	Female	Bimaturism	
S. fuscicollis	2·00	2·00	0·00	1·00
S. geoffroyi	1·00	1·00	0·00	1·00
S. imperator	1·50	1·50	0·00	1·00
S. oedipus	2·00	2·00	0·00	1·00
L. r. rosalia	2·00	2·50	−0·50	1·25
C. goeldi	1·50	2·00	−0·50	1·33
C. jacchus	2·25	2·25	0·00	1·00
C. pygmaea	1·50	1·50	0·00	1·00
S. sciureus	4·50	3·50	1·00	0·78
C. apella	7·50	4·50	3·00	0·60
C. moloch	3·50	3·50	0·00	1·00
A. trivirgatus	2·80	3·20	−0·50	1·36
A. geoffroyi	6·50	5·50	1·00	0·85
A. caraya	7·50	4·50	3·00	0·60
A. nigroviridis	5·50	4·50	1·00	0·82
C. aethiops	5·50	4·00	1·50	0·72
C. diana	8·50	6·50	2·00	0·76
C. mitis	7·50	9·00	−1·50	1·20
C. neglectus	5·50	5·25	0·25	0·95
E. patas	6·00	4·50	1·50	0·75
M. talapoin	7·50	5·50	2·00	0·73
C. t. atys	7·00	5·75	1·25	0·82
M. arctoides	9·75	8·00	1·75	0·82
M. cyclopis	8·50	8·50	0·00	1·00
M. fascicularis	8·50	6·50	2·00	0·76
M. fuscata	11·00	8·00	3·00	0·72
M. mulatta	9·50	5·50	4·00	0·58
M. nemestrina	8·50	6·75	1·75	0·79
M. nigra	8·50	8·50	0·00	1·00
M. radiata	8·75	6·50	2·25	0·74
M. silenus	7·50	4·50	3·00	0·60
P. cynocephalus	9·00	7·50	1·50	0·83
P. (M.) leucophaeus	9·00	9·00	0·00	1·00
P. (M.) sphinx	9·50	7·75	1·75	0·82
C. guereza	6·00	5·00	1·00	0·83
P. cristata	4·50	4·50	0·00	1·00
P. entellus	6·25	5·50	0·75	0·88
P. obscura	5·50	5·50	0·00	1·00
P. nemaeus	4·50	4·50	0·00	1·00
H. lar	6·25	6·25	0·00	1·00
H. syndactylus	7·50	7·50	0·00	1·00
P. pygmaeus[1]				
P. paniscus	10·50	8·50	2·00	0·81
P. troglodytes	12·00	11·50	0·50	0·96
G. gorilla	12·50	8·50	4·00	0·68

[1] See text for discussion.

female growth period averages about 90% of the male growth period. For the 31 dimorphic species, the average length of the growth period for females in comparison to males decreases to about 80%.

Variability in bimaturism occurs within groups of closely related species (Table 2). An increase in bimaturism is observed when squirrel monkeys (*Saimiri sciureus*) are compared to black-capped capuchins (*Cebus apella*). Comparable variability is observed in atelines [*Ateles geoffroyi* (spider monkeys) and *Alouatta caraya* (black howler monkeys)]. It should be noted that results for howler monkeys are somewhat tentative.

Bimaturism is highly variable in guenons, ranging from 3·5 years in diana monkeys (*Cercopithecus diana*) to −1·5 years in blue monkeys (*Cercopithecus mitis*). Negative bimaturism (defined as attainment of male adult weight prior to attainment of female adult weight) is an unexpected result, particularly for a species as dimorphic as *C. mitis* (see below). Minimal levels of negative bimaturism are also noted (but with differing causes and consequences) for some non-dimorphic species (Table 2, see below).

In macaques, the best-represented clade in this study, there is extreme variation in bimaturism. *Macaca mulatta* (rhesus macaque) shows approximately 4 years of bimaturism, a value that is high both in comparison to some other macaques and to anthropoids in general. Two macaque species [*Macaca cyclopis* (Taiwan macaques) and *Macaca nigra* (Sulawesi black "ape")] exhibit dimorphism in the absence of bimaturism.

Although underrepresented in this data base, baboons (including *Mandrillus* species) seem to show rather limited bimaturism. Data for *Papio* (*Mandrillus*) *leucophaeus* are limited. Similarly, bimaturism is restricted in colobines, but this taxon is also not well sampled.

Variability in bimaturism is pronounced in hominoids. About 2 years of bimaturism characterize bonobos (*Pan paniscus*), but there is little bimaturism in common chimpanzees (*Pan troglodytes*). Bimaturism is expressed in gorillas (*Gorilla gorilla*) and female gorillas grow for about the same amount of time as female bonobos. Orang-utans (*Pongo pygmaeus*) are highly unusual in that male body weight growth may not cease (i.e., unlimited male growth).

A scatterplot of bimaturism against size for the dimorphic species suggests no clear relations between these variables (Figure 2), at least at a broad level. Comparisons among groups of closely related species are presented in subsequent sections.

Variability in the effects of rate and duration differences in growth
Proportions pr and pt measure the contributions of rate differences and bimaturism to dimorphism, respectively. Analysis of 31 dimorphic species in this sample suggests that differences in rate of growth and differences in duration of growth are both very important in production of body size dimorphism (Table 3). On average, rate differences in growth (pr) account for about 54% of the total dimorphism while duration differences (pt) account for the remaining 46% of the total dimorphism.

There appears to be considerable variation in the effects of rate and duration differences on dimorphism (Table 3). In addition, comparable levels of dimorphism are clearly the outcome of variation in both rate and duration of growth (Table 4). In some cases, rate differences account for virtually all the observed dimorphism (pr = 1·00). In other instances, bimaturism almost exclusively results in dimorphism. Thus, summary of pr and pt by averages is somewhat misleading because of the presence of extreme values.

The proportion of dimorphism resulting from rate differences (and consequently, the proportion due to bimaturism) appears independent of species size when all 31 species are considered (Figure 3). Comparison of closely-related species may suggest more complexity in

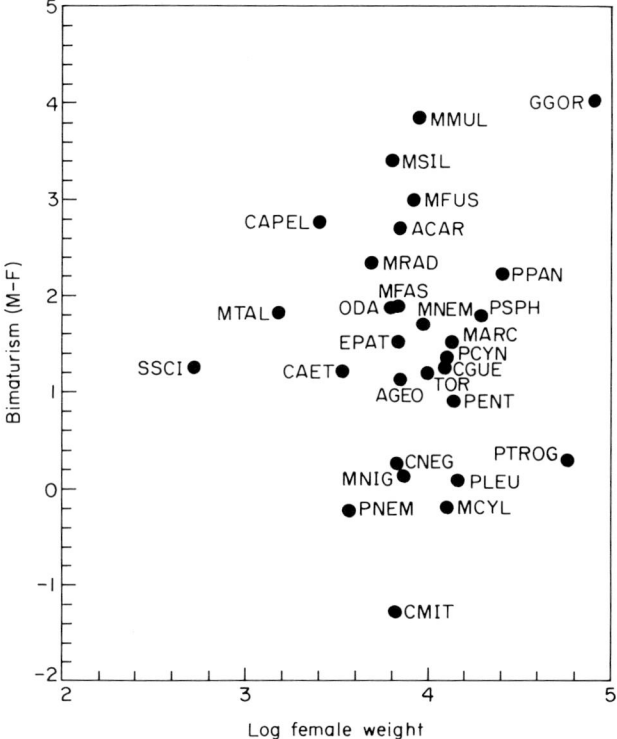

Figure 2. Scatterplot of bimaturism (sex difference in age at body weight growth cessation) *vs.* size (log female adult weight).

the relations between proportional ontogenetic components of dimorphism, size and adult dimorphism. Small sample sizes complicate statistical assessment of this possibility. In addition, variation in pr and pt is independent of variation in dimorphism. This property may limit the strength of assessments of relations between these variables (Figure 4).

In *Cebidae*, represented by *Cebus apella* and *Saimiri sciureus*, increased differences in duration of growth correspond to increased size and dimorphism. The reverse appears to be true for guenons, with bimaturism in *Miopithecus talapoin* (talapoins) and *Cercopithecus aethiops* (vervets) giving way to increased rate differentiation in progressively larger and more dimorphic *C. diana*, *C. mitis* and *Cercopithecus neglectus* (DeBrazza's guenon). In other words, rate differentiation appears to be correlated with increased species size and dimorphism in guenons.

Macaques show great diversity in values of pr and pt. Associations between size, dimorphism and either rate or duration differences do not seem to show clear patterns.

In *Papio cynocephalus papio* (guinea or red baboons), rate differences seem most important in the production of dimorphism [see Coelho, 1985; Glassman *et al.*, 1984, for *Papio cynocephalus anubis* (olive baboons)] and the same appears generally true for *Mandrillus* species. A group composed of mangabeys (represented here by *Cercocebus torquatus atys*) and baboons might suggest that increases in size and adult dimorphism correspond to increases in the importance of rate differences. The lack of additional papionins complicates interpretation of these results.

Table 3 Proportional components to dimorphism in anthropoid primates

Species	pr[1]	pt[2]	Dimorphism[3]	Female weight (kg)	Female growth cessation age (years)[4]
S. sciureus	0·62	0·38	1·44	0·67	3·50
C. apella	0·21	0·79	1·91	3·20	4·50
A. geoffroyi	0·70	0·30	1·34	6·46	5·50
A. caraya	0·74	0·26	1·86	3·77	4·50
A. nigroviridis	0·83	0·17	1·71	4·66	4·50
C. aethiops	0·08	0·92	1·53	3·45	4·00
C. diana	0·44	0·56	1·82	4·94	6·50
C. mitis	1·00	0·00	1·84	5·59	9·00
C. neglectus	0·86	0·14	1·89	5·66	5·25
E. patas	0·80	0·20	1·89	6·69	4·50
M talapoin	0·00	1·00	1·32	1·23	5·50
C. t. atys	0·61	0·39	1·54	7·82	5·75
M. arctoides	0·68	0·32	1·47	12·1	8·00
M. cyclopis	0·93	0·07	1·36	11·2	8·50
M. fascicularis	0·13	0·87	1·53	5·15	6·50
M. fuscata	0·27	0·73	1·47	11·0	8·00
M. mulatta	0·17	0·83	1·60	8·37	5·50
M. nemestrina	0·50	0·50	1·93	7·50	6·50
M. nigra	1·00	0·00	1·75	8·88	8·50
M. radiata	0·13	0·87	1·79	5·90	6·50
M. silenus	0·06	0·94	1·63	5·80	4·50
P. cynocephalus	0·61	0·39	1·65	14·3	7·50
P. (M.) leucophaeus	1·00	0·00	2·14	15·3	9·00
P. (M.) sphinx	0·60	0·40	2·17	16·4	7·75
C. guereza	0·89	0·11	1·33	9·26	5·00
P. entellus	0·60	0·40	1·49	13·5	5·50
P. nemaeus	1·00	0·00	1·53	6·86	4·50
P. pygmaeus	0·50	0·50	2·04	60·9	15·0
P. paniscus	0·31	0·69	1·38	33·9	8·50
P. troglodytes	0·73	0·27	1·27	45·0	11·5
G. gorilla	0·39	0·61	1·93	80·8	8·50

[1]pt represents the percentage of adult dimorphism attributable to sex differences in duration of growth. [2]pr is a measure of the percentage of adult dimorphism attributable to sex differences in rate of growth. [3]Dimorphism is measured by the ratio of (male adult weight/female adult weight). [4]Ages at body weight growth cessation are estimated from lowess regression (see text for further discussion).

The colobines are probably insufficient for adequate comparisons. Effects of rate differences do, however, consistently exceed effects of duration differences in these species.

The hominoids exhibit diversity in patterns of dimorphic growth, illustrating quite clearly that there are species-specific ways in which dimorphism develops. Gorillas and bonobos show a tendency for duration differences to account for adult dimorphism, but common chimpanzees exhibit a large rate component. Longitudinal and cross-sectional data suggest that growth in male orang-utans may be unlimited, making estimates of pr and pt somewhat arbitary for orang-utans.

Examples of ontogenetic variability

Evaluation of pr and pt values offers an overview of ontogenetic processes that lead to dimorphism. However, additional variation can be illustrated by comparisons of body weight growth curves among selected species. For example, rate differences bring about

Table 4 **Proportional components to dimorphism in anthropoid primates ranked according to level of dimorphism**

Species	pr[1]	pt[2]	Dimorphism[3]
P. troglodytes	0·73	0·27	1·27
M. talapoin	0·00	1·00	1·32
C. guereza	0·89	0·11	1·33
A. geoffroyi	0·70	0·30	1·34
M. cyclopis	0·93	0·07	1·36
P. paniscus	0·31	0·69	1·38
S. sciureus	0·62	0·38	1·44
M. arctoides	0·68	0·32	1·47
M. fuscata	0·28	0·73	1·47
P. entellus	0·60	0·40	1·49
C. aethiops	0·08	0·92	1·53
M. fascicularis	0·13	0·87	1·53
P. nemaeus	1·00	0·00	1·53
C. t. atys	0·61	0·39	1·54
M. mulatta	0·17	0·83	1·60
M. silenus	0·06	0·94	1·63
P. cynocephalus	0·61	0·39	1·65
A. nigroviridis	0·83	0·17	1·71
M. nigra	1·00	0·00	1·75
M. radiata	0·13	0·87	1·79
C. diana	0·44	0·56	1·82
C. mitis	1·00	0·00	1·84
A. caraya	0·74	0·26	1·86
C. neglectus	0·86	0·14	1·89
E. patas	0·80	0·20	1·89
C. apella	0·21	0·79	1·91
M. nemestrina	0·50	0·50	1·92
G. gorilla	0·39	0·61	1·93
P. pygmaeus	0·50	0·50	2·04
P. (M.) leucophaeus	1·00	0·00	2·14
P. (M.) sphinx	0·60	0·40	2·17

[1]pt represents the percentage of adult dimorphism attributable to sex differences in duration of growth. [2]pr is a measure of the percentage of adult dimorphism attributable to sex differences in rate of growth. [3]Dimorphism is measured as male adult weight divided by female adult weight.

adult dimorphism in spider monkeys (Figure 5). Dimorphism in this species can be seen largely as a result of female growth deceleration and slight bimaturism. A contrasting pattern is illustrated by *C. mitis* (Figure 6) in which rate differences result in dimorphism due to pronounced growth spurts in males and to an extended female growth period. Female growth appears to have two distinct phases prior to attainment of adult weight, a pattern not seen in other species. Males reach adult body weight prior to females. This assessment is strongly supported by detailed longitudinal data (Leigh, 1992).

A pattern of significant rate differentiation is seen in guinea (red) baboons (Figure 7). Female growth is prolonged and decelerates over an extended period of time. Although males grow for a longer period of time than females, growth rate differences seem mainly responsible for the observed dimorphism. The longer duration of male growth accounts for a small degree of additional dimorphism because the male growth rate is slow during this period (cf. *A. geoffroyi*; Figure 5).

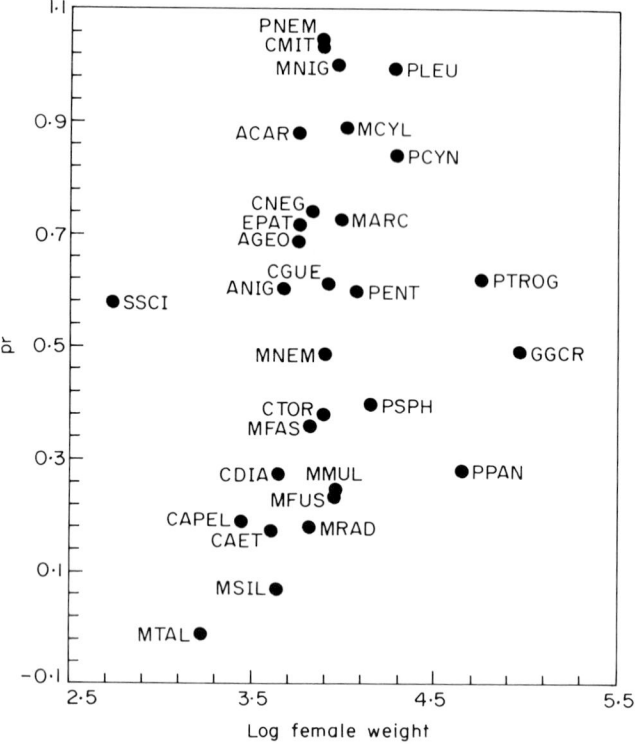

Figure 3. Scatterplot of proportion of total dimorphism resulting from sex differences in rate of growth vs. size (log female adult weight).

Taxa in which bimaturism almost exclusively results in dimorphism include *Cebus apella* (black-capped capuchins; Figure 8) and *Macaca silenus* (lion-tailed macaques; Figure 9). Roughly equal contributions of bimaturism and rate differences seem to account for dimorphism in *Macaca nemestrina* (pig-tailed macaques; Figure 10).

Monomorphic species follow the general pattern of decelerating growth seen for *A. trivirgatus* (Figure 11). The decelerating pattern is also observed in large-bodied monomorphic species such as siamangs (*Hylobates syndactylus*). There may be slight differences in rates of growth in monomorphic species, but these are balanced by countervening differences in duration of growth (Figure 11). More specifically, males of some monomorphic species may grow slightly faster than conspecific females. However, male growth duration is shorter than female growth duration in some monomorphic species (negative bimaturism) in these cases. This pattern characterizes *A. trivirgatus*, *Callimico goeldi* and *Leontopithecus rosalia rosalia*, but cannot be unambiguously detected in other monomorphic species.

Discussion

The results of this study allow clarification of a number of issues, centering on the variability of bimaturism in primates, variability in the effects of rate differences and duration differences on dimorphism and correlations between these variables, size and dimorphism. Finally, patterns of variation in growth are considered.

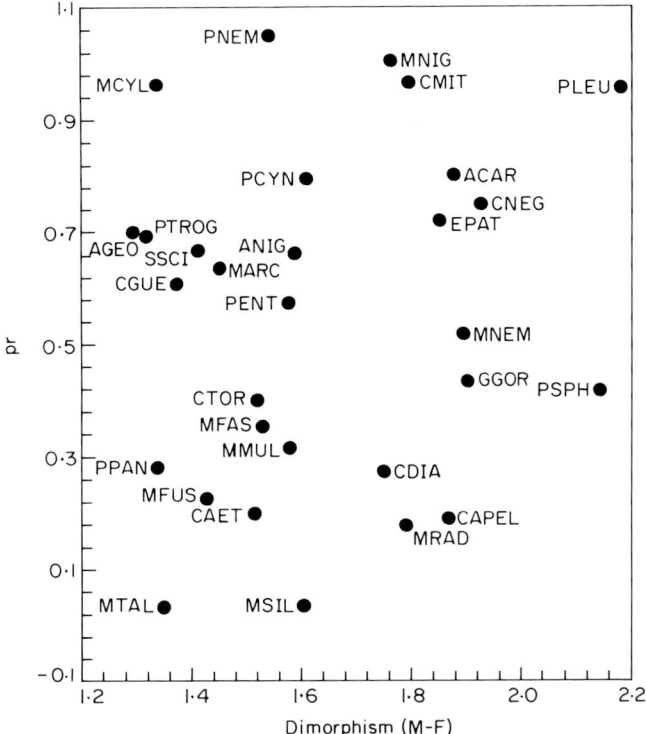

Figure 4. Scatterplot of proportion of total dimorphism resulting from sex differences in rate of growth *vs.* dimorphism (male adult weight/female adult weight).

The expectation that the ontogenetic bases of adult dimorphism vary interspecifically (Georgiadis, 1985; Jarman, 1983; Shea, 1986) is well substantiated by the present study. The presence of bimaturism is very common, and it is highly variable in its degree. This is also illustrated by numerous cases in which sex differences in rate of growth account for the vast majority of adult dimorphism. Putting aside variation in types of rate differences, this result would suggest that there are potentially very marked differences in the ways that primate species attain adult dimorphism. Moreover, the presence of dimorphism in association with negative bimaturism (*C. mitis*) or in the absence of bimaturism [*M. cyclopis*, *M. nigra*, *P. (Mandrillus) leucophaeus*, *Pygathrix nemaeus*] is an unexpected result. Of these cases, that of *C. mitis* is best documented. Additional data would be helpful in further verifying the absence of bimaturism in these other species [especially in *P. (Mandrillus) leucophaeus*].

This study indicates that bimaturism and sexual rate differences may be independent of one another. In other words, the absence of bimaturism leads either to monomorphism or dimorphism. Furthermore, the lack of bimaturism does not imply either the presence or absence of a sex difference in rate of growth. Bimaturism (positive bimaturism) always leads to some level of dimorphism. Combinations of these two processes are most important in producing adult dimorphism, but it is essential to realize they can impact dimorphism independently.

The present analysis permits a more detailed understanding of the variability in ontogenetic processes that lead to dimorphism than previous analyses (Shea, 1986; Watts, 1986).

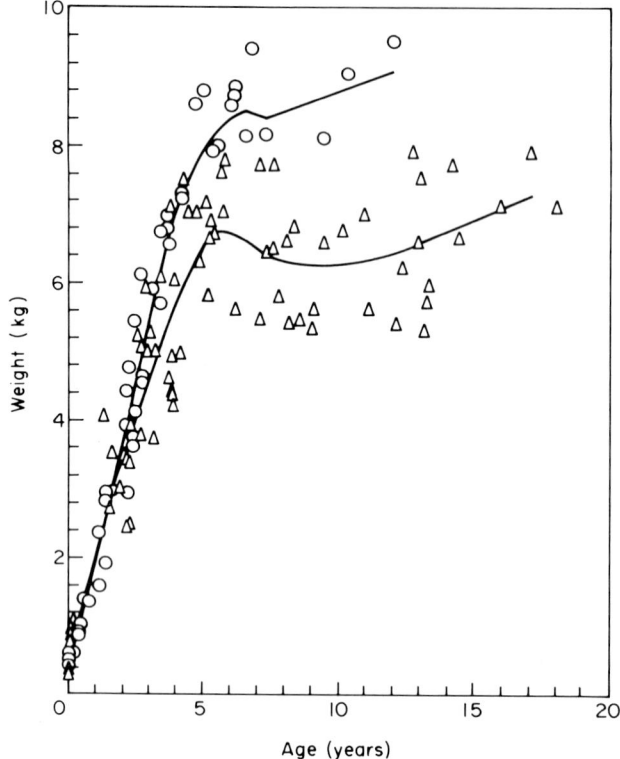

Figure 5. Lowess estimated growth curves for *Ateles geoffroyi*. Persistent deceleration of female growth seems to account for a large portion of the observed dimorphism. Male growth decelerates and ceases later than female growth.

Shea argues that body weight bimaturism is a common feature of primate growth and suggests that there may be great variation in the role that bimaturism plays (1986:108). The present study confirms Shea's point through investigation of a much more complete and better documented data base. High levels of dimorphism can be attained in the absence of major differences in duration of growth, a result that parallels Shea's (1986) earlier result for guenons.

Reconsideration of Watts' (1986) assessment of ontogenetic inputs into dimorphism may be required if findings for weight growth are generalizable to growth in length. Specifically, her finding of conservative (perhaps nearly constant) interspecific levels of skeletal bimaturism contrasts with the results of the present study. Limited variation in bimaturism would imply that dimorphism is produced mainly through rate differences in growth. The diversity observed in the present analysis contrasts with this idea. Differences in conclusions between Watts and the present study may be due either to greater species diversity in the present study or to differences in the variables analysed.

Negative bimaturism is observed for some non-dimorphic species and at least one strongly dimorphic species (*C. mitis*). Males and females reach adult body weight at the same time in several dimorphic species. Unlimited growth in *P. pygmaeus* implies extreme body weight bimaturism. A similar pattern is reported by Jolicoeur (1985) for male elephant seals

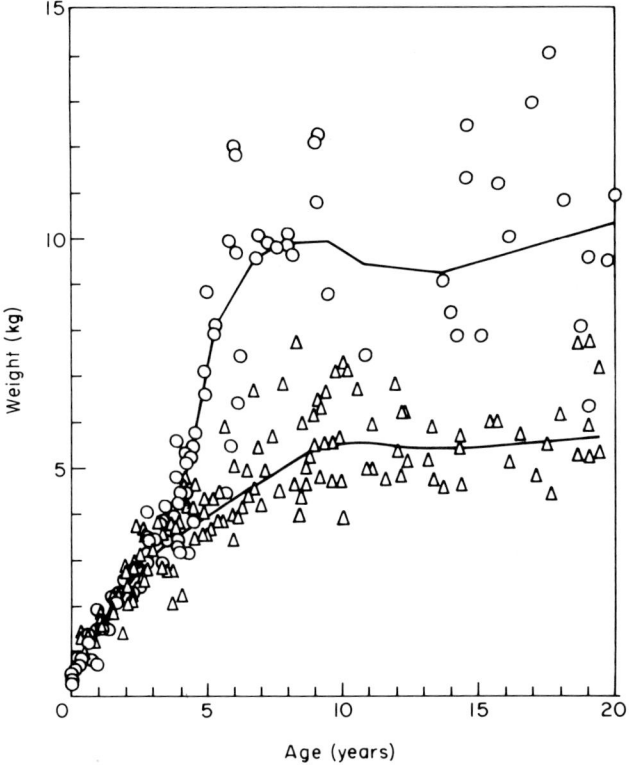

Figure 6. Lowess estimated growth curves for *Cercopithecus mitis*. A marked increase in the rate of male growth seems to account for dimorphism in this species.

(*Mirounga leonina*) and by Jarman (1983) for several species of large terrestrial herbivores. Thus, the present study indicates that there is variation in bimaturism and especially in the effects of bimaturism on dimorphism.

The diversity in the ways that macaques attain dimorphism requires special attention. The variation in this genus is not paralleled by other genera and might imply that better representation of other genera would show comparable diversity. In any case, it is clear that macaques have apparently evolved a number of different pathways to dimorphism.

It should be noted that *M. mulatta* (rhesus) shows a very high level of bimaturism, which is not readily apparent from some previous studies of weight growth in this species. Previous studies often did not include sufficient data for older males to allow accurate estimation of growth cessation. For example, van Wagenen & Catchpole (1956) report male weights up to 8 years of age, Gavan & Hutchinson (1973) up to 7·5 years and Kirk (1972) up to 8 years. Given that *M. mulatta* seems derived in terms of a high level of body weight bimaturism, it is extremely interesting that, in terms of length growth bimaturism, this species shows a pattern comparable to other species (see Watts, 1986).

The present analysis does not unambiguously support the hypothesis that bimaturism is consistently correlated with size within groups of closely related species (cf. Shea, 1986: 107); a more complex situation is revealed. For example, within groups of closely related cercopithecoids (guenons, mangabeys and baboons), size, dimorphism and differences in

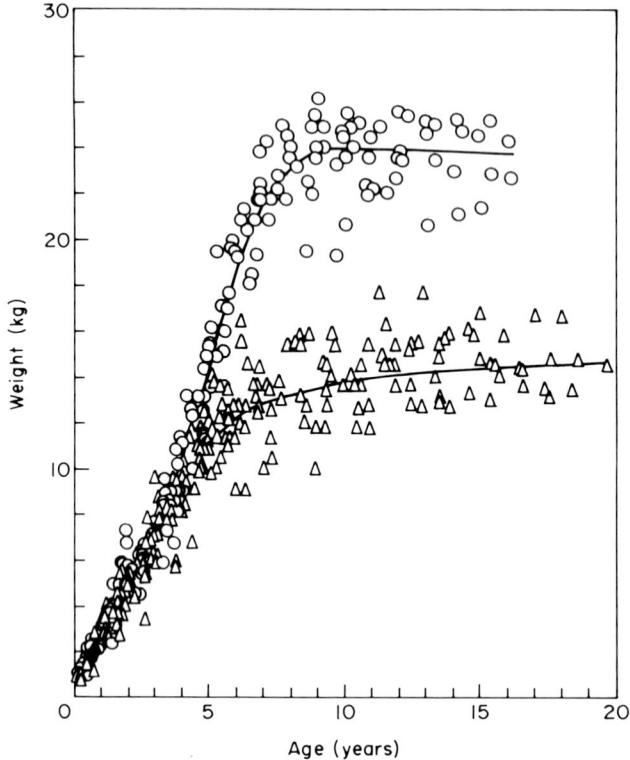

Figure 7. Lowess estimated growth curves for *Papio cynocephalus papio*. An increase in male growth rate plus prolonged female growth deceleration account for the majority of the dimorphism in this taxon.

rate of growth may be positively associated. There are no consistent associations of this kind within *Macaca*. Bimaturism shows associations with size in Cebidae (*S. sciureus* to *C. apella*) and a positive association between dimorphism, size and bimaturism could be suggested for hominoids. The lack of consistency in size-relatedness of these variables may imply that other ecological or social variables play major roles in determining the presence and degree of rate differences and bimaturism in primates. Research currently in progress addresses such issues (Leigh, 1992).

Variation by sex among species in growth rate and duration illustrates the complexity of developmental processes that result in dimorphism. For example, males of many species often exhibit marked growth spurts. These growth spurts frequently have large effects on adult dimorphism. Generally, the effects of growth spurts on dimorphism seem more substantial than the effects of male growth prolongation because late male growth rates are often very slow.

Male growth spurts tend to define rate-dimorphic species (e.g. *C. mitis*, *P. c. papio*). In these species, the male growth rate increases, quickly separating male and female growth trajectories. However, in duration-differentiated species, male growth rates tend to be stable, with little or no increase in growth rate (e.g., *M. silenus* and *C. apella*). Consequently, two basic patterns of male growth can be distinguished, involving either a growth spurt or prolonged period of growth.

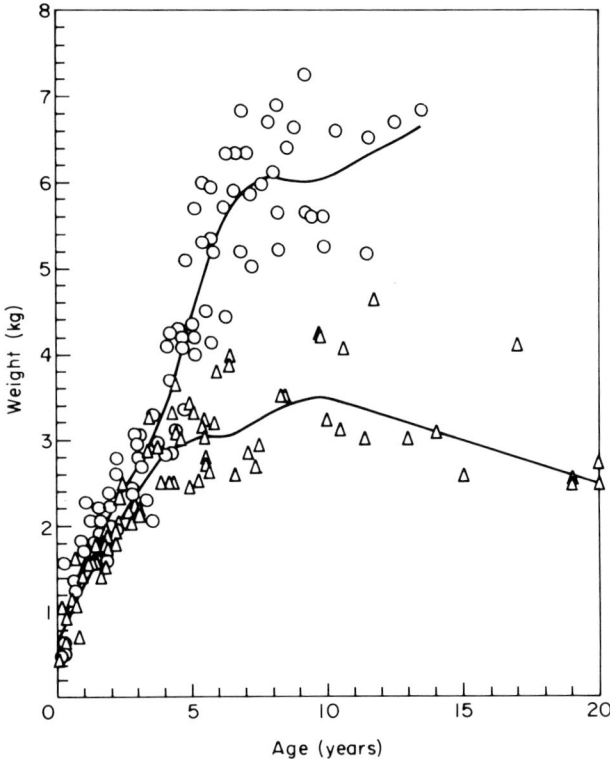

Figure 8. Lowess estimated growth curves for *Cebus apella*. Prolonged male growth is primarily responsible for the level of dimorphism observed in this species.

It is possible that these two contrasting male developmental trajectories have differing evolutionary bases. Dimorphism in some species may be largely the outcome of evolutionary factors that increase male growth rate. In other species, factors that prolong male growth duration may be seen as causes of dimorphism. Thus, an explanation of adult dimorphism would require identification of these factors and a model of how they influence ontogeny. Focus on adult dimorphism confounds these potentially separate underlying growth processes.

It is generally assumed that evolutionary modification of male ontogenetic trajectories is the primary factor in producing variation in dimorphism. Although variation in male growth rate and duration is obviously important, the present study indicates that variation in female growth rate and duration are key to the ontogeny of dimorphism (Fedigan, 1982; Shea, 1986; Willner & Martin, 1985). For example, prolongation of female growth in *C. mitis* and *P. c. papio* (see above) can be seen as a mechanism that, *ceteris paribus*, limits dimorphism in these species. In other words, cessation of female growth in these species at younger ages would produce extremely high levels of dimorphism. Prolongation of female growth in *C. mitis* is particularly important in producing rate-based dimorphism. In contrast, the pattern of fairly abrupt cessation of female growth in some species (e.g., *C. apella*, *M. silenus*) would tend to increase levels of adult dimorphism. Compared to closely related species, ages at growth cessation in female *Erythrocebus patas* (Rowell, 1977) and in *G. gorilla* (Shea, 1986) may be

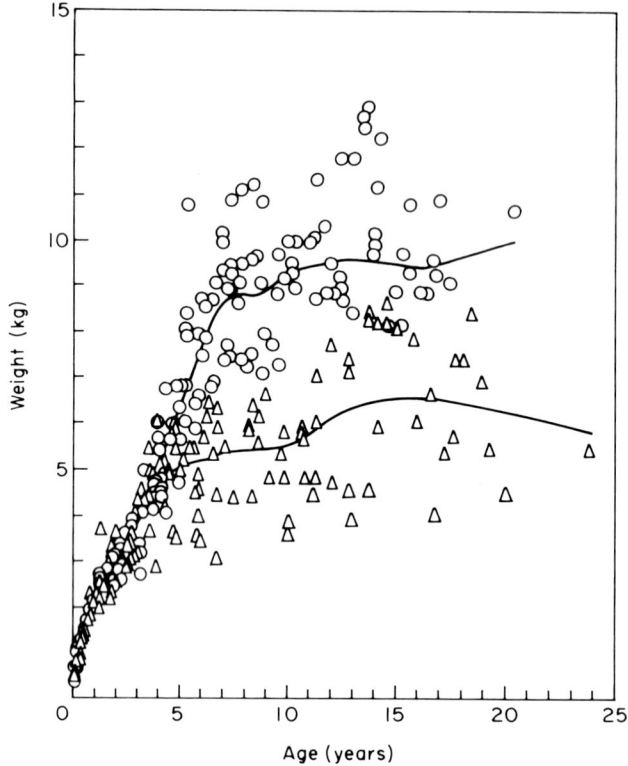

Figure 9. Lowess estimated growth curves for *Macaca silenus*. Duration differences in growth produce the majority of dimorphism in this species.

early. However, the situation for African apes is not straightforward because female *P. paniscus* and *G. gorilla* share approximately equal ages at growth cessation. This would suggest that *P. troglodytes* exhibits a derived pattern of prolonged female growth relative to bonobos and gorillas. In general, however, these results suggest that high levels of dimorphism in these species are related to factors that influence female growth rate and duration (see Shea, 1986). As with diversity in male growth, the "female component" of dimorphism is unrecognizable in studies of adult dimorphism, but can be detected in comparative ontogenetic analyses (Leigh, 1992). Moreover, the effects of female growth on dimorphism may be largely responsible for the lack of consensus as to the evolutionary correlates of dimorphism in primates (see Gaulin & Sailer, 1984; Leutenegger & Cheverud, 1982, 1985; Ely & Kurland, 1989).

It is possible that variation in male growth is ultimately most strongly influenced by sexual selection, which should increase dimorphism. The presence of variation in male ontogenetic pathways may reflect differences in kinds of sexual selection or variability in responses to sexual selection. The expectation of major effects of sexual selection does not deny a role for natural selection. However, this prediction differs both from Watts' (1986:162) suggestion that natural selection substantially affects male growth and from Shea's (1986:108) proposition of a more general role for natural selection.

Variation in female growth may be primarily the outcome of natural selection. Natural selection on female growth rate and duration could serve to increase or decrease dimorphism.

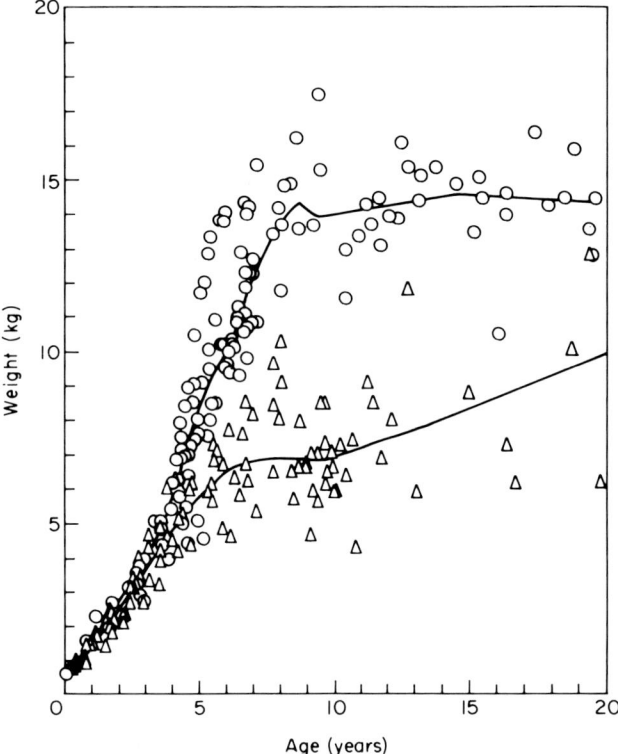

Figure 10. Lowess estimated growth curves for *Macaca nemestrina*. Differences in rate and duration of growth produce similar quantities of dimorphism in this species.

Analyses of body weight growth allow some insight into the processes that influence female growth and, therefore, dimorphism.

It is the interplay of these (and other) factors that produces a given level of dimorphism. Although investigation of the evolutionary bases of the sex differences in growth detailed here are beyond the scope of the present study, it is important to recognize that diversity in ontogenetic bases of dimorphism is characteristic of primates. An ontogenetic perspective lends the opportunity to address, in much more detail than is available by comparison of adult endpoints, the causal factors that are involved in the evolution of dimorphism. Such studies must attempt to explain variation in rate and duration of growth in each sex. Investigation of sex differences in these parameters should lead to a fuller understanding of primate dimorphism.

Summary and conclusions

This analysis investigates the ontogeny of body weight dimorphism in 45 species of anthropoid primates. The study attempts to supplement previous analyses of ontogenetic diversity in primates (Shea, 1986) through analysis of an extensive sample of captive primates. Previous studies of primates (Shea, 1986) and terrestrial herbivores (Georgiadis, 1985; Jarman, 1983) found variation in the ontogenetic bases of sexual dimorphism.

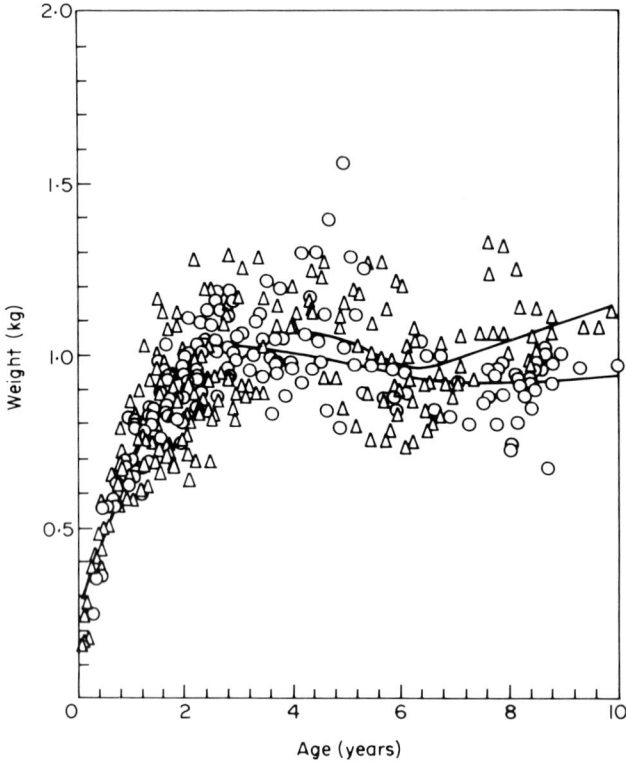

Figure 11. Lowess estimated growth curves for *Aotus trivirgatus*. Males seem to grow slightly faster in this species, but cease growth earlier than females. The male curve lies below the female curve between the ages of 4 and 10 years.

The contributions of sex differences in rate and duration of weight growth are evaluated in this study. In addition, the relations between size and bimaturism and between varying levels of dimorphism and bimaturism are investigated. These problems are addressed using estimates of sex differences in growth derived from non-parametric regression.

The results of this analysis suggest that there is considerable diversity in the ontogenetic processes that lead to body weight dimorphism in anthropoid primates. Bimaturism varies considerably within groups of closely related species. In addition, rate differences are important in producing adult dimorphism, but their contributions to dimorphism vary substantially among groups of closely related species. In some cases, this variation is correlated with size and dimorphism, but additional studies that explore evolutionary factors that underlie these correlations must be undertaken. The large degree of variability in ontogenetic processes that result in adult dimorphism in primates may imply that analyses of adult dimorphism are of limited value in explaining the evolution of body size dimorphism.

Acknowledgments

This research was supported by the generosity of many individuals and institutions. In particular, assistance, access to records and helpful discussions were provided by Dr Anne

Baker (Brookfield Zoo); Dr Tom Meehan and Ms Pat Sass (Lincoln Park Zoo); Dr Andy Teare and Mr Jan Rafert (Milwaukee County Zoo); Ms Ingrid Portan (St. Louis Zoo); Dr Amy Shima and Ms Kay Munduate (San Diego Zoo and San Diego Wild Animal Park); Dr Doug Armstrong and Ms Sarah Junior (Henry Doorly Zoo, Omaha); Dr Tim Reichard (Toledo Zoo); Dr Albert Lewandowski (Cleveland Zoo); Dr Richard Cambre (Denver Zoo); Dr Mike Burton (Cheyenne Mountain Zoo, Colorado Springs, CO); Drs Paul Calle, Danny Wharton, Fred Koontz and Robert Cook (Bronx Zoo); Dr Lynne Kramer (Columbus Zoo); Dr Mark Campbell (Cincinatti Zoo); Dr Julian Duvall (Indianapolis Zoo); Dr William Russell (Tulsa Zoo); Drs Debra Forthman, Beth Stevens, Rita Macmanamon and Ms Cindy Thorstad (Atlanta Zoo); Dr Roy Burns (Louisville Zoo); Dr Mark Peckham, Dr Joe Flanagan and Ms Barbara Lester (Houston Zoo); Ms Donna Todd and Dr Rodney Walker (Jackson Zoo); Drs Mark Stetter and Susan Wells (Audubon Park Zoo, New Orleans); Mr Jake Yelverton (Louisiana Purchase Gardens and Zoo, Monroe, LA); Ms Linda Sanders (Baton Rouge Zoo); Mr Will Sugg, Drs Edwin Gould and Benjamin Beck (National Zoological Park); Ms Connie Sweet (Santa Ana Zoo); Dr Gary Kuehn and Mr Victor Bolanos (Los Angeles Zoo); Dr Les Shobert (North Carolina Zoo); Ms Greta Macmillan (Knoxville Zoo); Dr Barbara Baker and Ms Karen Lindquist (Pittsburg Zoo); Dr A. J. Smith (Santa Barbara Zoo); Mr Donald Moore (Syracuse Zoo); Dr Andrew Hendrickx (California/Davis Regional Primate Research Center); Dr Jim Else, Dr Harold McClure, Dr Jeremy Dahl and Ms Sue Setzekorn (Yerkes Regional Primate Research Center); Dr Margaret Clarke (Delta Regional Primate Research Center); Dr Suzanne Iliff-Sizemore (Oregon Regional Primate Research Center); Dr Curtis Port and Ms Mary Dal Corrobbo (G. D. Searle and Company); Dr L. R. Cochard (Northwestern); Dr M. Inoue (Himeji Gakuin Women's College, Japan); Dr Freeland Dunker, Ms Gail Hedberg and Dr Myron Sulak (San Francisco Zoo); Dr Andrew Petto (New England Regional Primate Research Center); Dr Phyllis Dolhinow and Mr Elsworth Ray (University of California, Berkeley); Dr Neil Bemment (Paignton Zoo); Dr Volker Sommer (University of Gottingen); Mr Robert Evans (San Antonio Zoo); and Dr Nate Flessness (International Species Information System). Drs Brian Shea, Larry Cochard, Jim Cheverud, Matt Ravosa, Jill Bullington and Ms Anne Gomez provided helpful comments on this and earlier versions of this paper. Dr Elizabeth Watts provided perceptive advice and editorial comments on this manuscript. Dr Jim Cheverud's method for calculating proportional contributions to dimorphism is appreciated. Partial financial support for this project was provided by Sigma Xi, Northwestern University and the Wenner-Gren Foundation.

References

Cheverud, J. M., Dow, M. M. & Leutenegger, W. L. (1985). The quantitative assessment of phylogenetic constraints in comparative analyses: sexual dimorphism in body weight among primates. *Evolution* **39,** 1335–1351.
Cheverud, J. M., Dittus, W. P. J. & Wilson, P. (1992). Primate population studies at Polonnaruwa. I. Somatometric growth in a natural population of toque macaques (*Macaca sinica*). *J. hum. Evol.* **23,** 51–77.
Clutton-Brock, T. H. (1985) Size and sexual dimorphism in primates. In (W. L. Jungers, Ed.) *Size and Scaling in Primate Biology*, pp. 51–60. New York: Plenum Press.
Clutton-Brock, T. H. & Harvey, P. H. (1977). Primate ecology and social organization. *J. Zool. Lond.* **183,** 1–39.
Clutton-Brock, T. H., Harvey, P. H. & Rudder, B. (1977). Sexual dimorphism, socionomic sex ratio, and body weight in primates. *Nature* **263,** 797–799.
Cleveland, W. S. (1979). Robust locally weighted regression and smoothing scatterplots. *J. Am. stat. Assoc.* **74,** 829–836.
Cleveland, W. S. & Devlin, S. J. (1988). Locally weighted regression: an approach to regression analysis by local fitting. *J. Am. stat. Assoc.* **83,** 596–610.

Cock, A. G. (1966). Genetical aspects of growth and form in animals. *Q. Rev. Biol.* **41,** 131–190.
Coelho, A. M. (1985). Baboon dimorphism: growth in weight, length, and adiposity from birth to eight years of age. In (E. S. Watts, Ed.) *Nonhuman Primate Models for Human Growth and Development*, pp. 125–159. New York: Alan R. Liss.
Darwin, C. R. (1871). *The Descent of Man, and Selection in Relation to Sex.* Princeton: Princeton University Press (Facsimile of 1st Edition).
Dielman, T. E. (1989). *Pooled Cross-sectional and Time Series Data Analysis.* New York: Marcel Dekker.
Efron, B. & Tibshirani, R. (1991). Statistical data analysis in the computer age. *Science* **253,** 390–395.
Ely, J. & Kurland, J. A. (1989). Spatial autocorrelation, phylogenetic constraints, and the causes of sexual dimorphism in primates. *Int. J. Primatol.* **10,** 151–171.
Eveleth, P. B. & Tanner, J. M. (1990). *World-wide Variation in Human Growth.* Cambridge: Cambridge University Press.
Fedigan, L. M. (1982). *Primate Paradigms: Sex Roles and Social Bonds.* Montreal: Eden Press.
Gaulin, S. J. C. & Sailer, L. D. (1984). Sexual dimorphism in weight among the primates: the relative importance of allometry and sexual selection. *Int. J. Primatol.* **5,** 515–535.
Gavan, J. A. & Hutchinson, T. C. (1973). The problem of age estimation: a study using the rhesus monkey (*Macaca mulatta*). *Am. J. phys. Anthrop.* **38,** 69–82.
Gavan, J. A. & Swindler, D. R. (1966). Growth rates and phylogeny in primates. *Am. J. phys. Anthrop.* **24,** 181–190.
Georgiadis, N. (1985). Growth patterns, sexual dimorphism and reproduction in African ruminants. *Afr. J. Ecol.* **23,** 75–87.
Glassman, D. M., Coelho, A. M., Carey, K. D. & Bramblett, C. A. (1984). Weight growth in savannah baboons: a longitudinal study from birth to adulthood. *Growth* **48,** 425–433.
Gould, S. J. (1977). *Ontogeny and Phylogeny.* Cambridge: Belknap.
Jarman, P. (1983). Mating system and sexual dimorphism in large terrestrial mammalian herbivores. *Biol. Rev.* **58,** 485–520.
Jolicouer, P. (1985). A flexible 3-parameter curve for limited or unlimited somatic growth. *Growth* **49,** 271–281.
Jungers, W. L. (1985). Body size and scaling of limb proportions in primates. In (W. L. Jungers, Ed.) *Size and Scaling in Primate Biology*, pp. 345–382. New York: Plenum Press.
Kirk, J. H. (1972). Growth of maturing *Macaca mulatta*. *Lab. Anim. Sci.* **22,** 573–575.
Leigh, S. R. (1992). Ontogeny and sexual dimorphism in anthropoid primates. Ph.D. Dissertation, Northwestern University.
Leutenegger, W. L. & Cheverud, J. M. (1982). Correlates of sexual dimorphism in primates: ecological and size variables. *Int. J. Primatol.* **3,** 387–402.
Leutenegger, W. L. & Cheverud, J. M. (1985). Sexual dimorphism in primates: the effects of size. In (W. L. Jungers, Ed.) *Size and Scaling in Primate Biology*, pp. 33–50. New York: Plenum Press.
Ralls, K. (1977) Sexual dimorphism in mammals: avian models and unanswered questions. *Am. Nat.* **111,** 918–938.
Rensch, B. (1959). *Evolution above the Species Level.* New York: Columbia University Press.
Rowell, T. E. (1977). Variation in age at puberty in monkeys. *Folia primat.* **27,** 284–296.
Shea, B. T. (1985). The ontogeny of sexual dimorphism in the African Apes. *Am. J. phys. Anthrop.* **8,** 183–188.
Shea, B. T. (1986). Ontogenetic approaches to sexual dimorphism in anthropoids. *Hum. Evol.* **1,** 97–110.
Tanner, J. M. (1978). *Fetus into Man.* Cambridge: Harvard University Press.
van Wagenen, G. & Catchpole, H. R. (1956). Physical growth of the rhesus monkey (*Macaca mulatta*). *Am. J. phys. Anthrop.* **14,** 245–273.
Watts, E. S. (1986). Evolution of the human growth curve. In (F. Falkner & J. M. Tanner, Eds) *Human Growth*, Vol. 1, 2nd edn, pp. 153–166. New York: Plenum Press.
Wiley, R. H. (1974). Evolution of social organization and life history patterns among grouse. *Q. Rev. Biol.* **49,** 201–227.
Wilkinson, L. (1990). *Systat.* Evanston, IL: Systat.
Willner, L. A. & Martin, R. D. (1985). Some basic principles of mammalian sexual dimorphism. In (J. Ghesquiere, R. D. Martin & R. Newcombe, Eds) *Human Sexual Dimorphism*, pp. 1–42. London: Taylor and Francis.

James M. Cheverud* & Paul Wilson
Department of Anatomy & Neurobiology, Washington University School of Medicine, St. Louis, Missouri 63110, U.S.A.

Wolfgang P. J. Dittus*
Department of Zoological Research, National Zoological Park, Smithsonian Institution, Washington DC 20008, U.S.A. and Institute of Fundamental Studies, Kandy, Sri Lanka

Received 21 June 1991
Revision received 22 October 1991 and accepted 3 January 1992

Keywords: growth, toque macaque, body measurements, Polonnaruwa, sexual dimorphism, *Macaca sinica*.

Primate population studies at Polonnaruwa. III. Somatometric growth in a natural population of toque macaques (*Macaca sinica*)

Growth studies of non-human primates in natural populations are rare due to difficulties in obtaining measurements on animals of known age. We report a cross-sectional analysis of growth in the natural population of toque macaques (*Macaca sinica*) from Polonnaruwa, Sri Lanka, including the timing of growth and physical maturation and the developmental basis of sexual dimorphism.

Twenty-eight measurements were collected from 274 macaques aged 2 weeks to 34 years. Non-parametric spline curves were fit to the distribution of measurement values over age separately by sex. The spline curves suggest four age intervals in which linear regressions of trait on age and analysis of covariance of measurement value with age, sex, and their interaction, can be used to parametrically test specific hypotheses about growth and sexual dimorphism.

Most growth had ceased for both sexes in the adult phase. Prior to this, females had two and males three distinct growth phases. There was little sexual dimorphism in growth rates among infants and young juveniles (birth to 2·5 years). The age limits for subsequent growth periods differed by sex and trait. Thus, skeletal limb growth ceased in juvenile females by about 5·5 years, but continued into an additional subadult phase in males reaching completion by about 7·5 years, whereas muscle mass and weight reached maturity by about 8 and 12 years in females and males, respectively. An adolescent growth spurt was evident only for body weight in males, but cannot be ruled out for other traits. These macaques grow for one and a half to twice as long as macaques from other species in laboratory colonies. Much of the sexual dimorphism among adults arises from bimaturism, although sex differences in growth rates among older juveniles are also responsible for a significant portion of adult dimorphism.

Journal of Human Evolution (1992) **23,** 51–77

Introduction

Non-human primate growth has been studied from a comparative evolutionary perspective and as a model for human growth (Watts, 1985*a*). Nearly all of our information on non-human primate growth is derived from laboratory populations due to the need for known-age animals in growth studies. These studies have established that many non-human primates, like humans, display a biphasic growth pattern with distinct early and late growth periods (Laird, 1967; Watts, 1985*a*; Bogin, 1988) and have suggested the possibility that some non-human primates display an adolescent growth spurt in specific morphological features (Watts & Gavan, 1982; Watts, 1985*a*; Bogin, 1988). The timing of these growth periods is known primarily from small samples of captive specimens in a few species, including common chimpanzees, macaques and baboons (Watts & Gavan, 1982; Coelho, 1985; Glassman *et al.*, 1984; Coelho *et al.*, 1984). It has become clear from field studies that rates of maturation of primates in their natural habitats are considerably slower than in their well-fed counterparts under captive or semi-captive management (e.g., Dittus, 1975; Dittus & Thorington, 1981; Altmann *et al.*, 1977; Altmann & Alberts, 1987; Phillips-Conroy & Jolly, 1988). Hence, the timing of growth periods under laboratory conditions is of questionable utility for purposes of

*Authors to whom requests for reprints should be addressed.

evolutionary analyses. Unfortunately, information from captive populations is often all that is available (Leigh, 1991).

We report the results of a cross-sectional analysis of growth in a series of somatometric features in the natural population of toque macaques (*Macaca sinica*) from Polonnaruwa, Sri Lanka (Dittus, 1977a, 1988). The population has undergone an intensive but largely non-invasive study of demography, behavior and ecology for nearly 25 years, providing a rich source of basic data which can be combined with somatometric data to study growth. Earlier studies of development in this natural population, based only on body weight, indicated significantly faster growth rates among juvenile males than females aged 0·1 to 5·5 years old, a growth spurt in adolescent males, and continued growth in males and females after 10 and 7 years old, respectively (Dittus, 1977a). The aim of this study was to investigate, in a more comprehensive way, the nature and timing of infant and juvenile growth periods and the timing of somatic maturation in males and females.

In addition to providing basic data on growth in a natural population of non-human primates, we investigate the developmental basis of sexual dimorphism in body measurements. Adult sexual dimorphism can arise from two sources, sexual bimaturism and sex differences in growth rates (Shea, 1986; Gavan & Swindler, 1966). Sexual bimaturism occurs when the sexes reach maturity, or complete growth, at different ages. These two developmental processes may reflect different evolutionary histories and adaptations (Shea, 1986) and differ in their relative importance in a comparative perspective (Leigh, 1991), but are not known in detail for wild populations.

Human sexual dimorphism for weight arises from a combination of these underlying developmental processes, with males growing for a longer period than females and displaying a more exaggerated adolescent growth spurt (Tanner, 1962). Shea (1986) briefly reviewed non-human primate sexual dimorphism, and provided examples in which each developmental process takes precedence in producing sex differences. We examine the extent to which sex differences in toque macaques are due to a bimaturism and rate differences for a variety of somatic characters.

Materials and methods

Population

The toque macaques living at the Nature Sanctuary and Archaeological Reserve at Polonnaruwa, Sri Lanka, have been the subjects of a long-term study by W. Dittus. The population has been observed continuously from September 1968 to May 1972 and from March 1975 through 1991. Periodic observations were made from May 1972 to March 1975. The natural dry evergreen forest inhabited by the macaques and many aspects of their demography, ecology and behavior have been described earlier (Dittus, 1977a, 1977b, 1988).

The population contains approximately 600 individuals in 23 social groups. All animals were individually identified using methods described in Dittus & Thorington (1981). Groups were censused once a month and during the birth season individual females were observed every few days to obtain accurate birth dates. Birth, death and emigration were recorded for each animal. Chronological age at trapping was obtained from known birth dates, although ages of animals born before 1968 were estimated given their level of morphological development as determined in 1968 (Dittus, 1988). At this time, only a few of the oldest animals have estimated ages.

Measurements

Over the last few years, the Polonnaruwa population of macaques have been systematically trapped by social group (and released unharmed) in order to collect genetic, morphological and a variety of biomedical data on the population. Trapping was done following the birth season. To date, twenty-eight measurements, including weight and the length, width and circumference of the head, trunk and extremities, have been collected from 274 animals, 124 males and 150 females, ranging in age from 2 weeks to 34 years (see Appendix 1 for trait definitions). In the trapping process, individuals were held in cages and liberally fed for varying durations prior to being weighed and measured. Variation in weight and other measurements (such as abdominal circumference) due to feeding was therefore likely to be greater than if all animals had been weighed early in the morning prior to feeding. A genetic analysis of these measurements is described in Cheverud & Dittus (1992).

W. Dittus measured approximately half the social groups and research assistants measured the others. All observers practised measurements on several test animals in order to establish both intra- and inter-observer reliability. This involved learning to make minor adjustments in the placement of calipers at specific locations on various anatomical structures used as anchors for measurement. Difficult landmarks, such as the distal margins of rounded condyles, were typically measured several times on a given animal until a consistent measure was obtained. Skeletal lengths and widths had the least error and the effect of error in measuring over articulated segments (such as crown-rump length) was diminished when the lengths were over extended portions of the body. Most error probably involved abdominal circumferences where gut contents could have contributed to the variation. The observations were screened for outliers after data collection by visually inspecting plots of trait score against age separately by sex. Outliers were set to missing values (0·2% of the data) before the analyses reported here.

Age at trapping was obtained from the demographic records. The data analysed here are cross-sectional in nature and thus, may smooth out some of the variability apparent in longitudinal growth studies. It can be particularly difficult to delineate an adolescent growth spurt of small size, as may exist in macaques (Watts & Gavan, 1982; Watts, 1985*b*), with cross-sectional data due to variation in the timing of the spurt among individual animals (Boas, 1892; Tanner, 1951). Also, the age distribution of the animals tended to be clumped in yearly intervals due to the consistent birth season of the macaques, making judgements about cessation of growth limited to yearly intervals. This constrains the precision with which growth events can be located in time.

Spline curves

Growth was analysed first by fitting a spline curve to the distribution of trait values by age, separately by sex using the algorithm and computer program described by Schluter (1988). The logic of this approach follows, and is drawn largely from Schluter (1988). In order to estimate a growth function, f, from a collection of individual measurements, X_i, and age, A_i we select the function which is most likely given the data at hand:

$$\log(f) = \Sigma \log(X_i; A_i, f)$$

where $\log(f)$ is the total log likelihood of the function, f, and the term $\log(X_i; A_i, f)$ is the $\log_e$ probability that the measurement equals X at age A given this particular function f. The summation is over all individuals. Regrettably, the function which maximizes this likelihood is any function which passes through each of the individual data points. The function would be very rough, bouncing up and down to pass through each data point. We do not expect a regular process such as growth to, on average, follow a complex zig-zagging path. Thus, it is usual to include a penalty for roughness in choosing the appropriate function. This is accomplished by maximizing the penalized log likelihood:

$$\log(f) = \Sigma \log(X_i; A_i, f) - n\alpha J(f)$$

where n is the sample size, α is a non-negative constant refered to as the smoothing parameter and $J(f)$ measures the roughness of the function f as its summed squared curvature. Therefore, the more complex the function the greater its penalty and the lower its overall likelihood $[\log(f)]$.

If we assume that measurement values are normally distributed at each age and have constant variance, the function that maximizes the penalized log likelihood at any given α value is a cubic spline. A cubic spline is a function incorporating $n+1$ cubic polynomials spliced seamlessly at each of the n data points. The value of the smoothing parameter α controls the extent of the penalty for roughness and determines the form of the function f. When α is low, the penalty for roughness is low and the function chosen will be complex while when α is high the function approaches a straight line. So we must have some criterion for choosing a suitable level for the smoothing parameter. A common choice for α is to pick one with maximal predictive power for the data at hand using cross-validation. For any specified α and individual i, let f_i be the function that maximizes the penalized log likelihood for the data excluding individual i. Then the score of the excluded individual i (X_i) is estimated using f_i. This is done n times, excluding each individual in turn and using the same level of α to obtain the sum of the squared differences between observed and predicted values for the excluded individuals. The whole procedure is repeated with a series of different α values and the α finally chosen is the one which minimizes the squared difference between the predicted and observed values for excluded individuals.

A spline curve is essentially a running average placed through the middle of the bivariate data distribution. The number of points included in this "running average" varies with the local density of the data. The spline provides a local estimate of the regression surface in contrast to parametric models which provide a global fit to the data and can thus be extremely inaccurate in any given local region. Standard parametric growth curves are special cases of cubic splines and will result from the analysis if they are appropriate for the data.

Spline curves represent a non-parametric approach to growth analysis. No specific model (or growth curve equation) is fit to the data. This has several advantages for growth analysis. First, the shape of the growth curve is free to vary as the data vary, instead of being constrained to follow a particular mathematical model. Laird (1967) found it impossible to fit the entire growth period of macaques to a single mathematical function and instead recommended piecing together separate functions for the infant and juvenile growth periods. As noted by Watts & Gavan (1982), choice of any given equation requires prior knowledge of the growth dynamics. Also, growth curve coefficients obtained from fitting particular models are often quite difficult to interpret in terms of growth rate *vs.* maturational dimorphism. The mathematical models used to fit growth curves require a negative correlation between estimates of growth rate and maturation age. Thus, growth curve models do not allow us to separately evaluate the relative importance of growth rate dimorphism and maturational dimorphism to adult sex differences.

A disadvantage of the non-parametric spline curve approach to growth analysis is that there are no growth curve coefficients available to compare across the sexes, making quantitative statements about growth difficult to test statistically. For this reason, we first utilize the spline curves and then perform parametric analyses based on the spline curve results. The method also assumes normally distributed residuals of uniform variance across ages. For the data reported here, these conditions are generally true, although heteroscedasticity (increasing variance of residuals with increasing age) is evident for some traits (such as weight). In these cases, the low variance ages should have been weighted more strongly than the high variance ages in constructing the spline, although the overall curve is not likely to be severely affected.

The first step in the analysis is the inspection of the spline curves themselves. This inspection suggests growth dynamics for each of the traits. A judgement was made concerning critical ages at which growth rates appeared to change using both the numerical and graphical results of the spline curve analysis. Both the original spline curves (measurement plotted against age) and pseudo velocity curves derived from the splines (change in measurement divided by change in age plotted against age; Coelho, 1985) were generated and subjectively assessed. The form of the spline curves of the raw data can be used to suggest specific hypotheses about growth rates and maturation age which can be tested with standard statistical models. Spline curves, as with any running average technique, are least reliable at the edges of the bivariate distribution and are also less reliable in regions with relatively little data. Peculiarities of growth curves should not be overinterpreted, especially in regions of sparse data and at the earliest and latest ages at the beginning and end of the curve.

In order to roughly measure the contribution of rate differences and bimaturism to adult sexual dimorphism, the expected measurement difference between the sexes based on the spline curves was recorded for males and females at the age at which female growth was judged to be complete (typically at 5·5 or 8 years, depending on the measurement) and at the age at which male growth was judged to be complete (typically at 7·5 or 12 years, depending on the measurement). The difference in trait values at the female maturation age was then expressed as a percentage of the difference at the male maturation age. This proportional difference is the proportion of adult sexual dimorphism due to dimorphism in growth rates. One minus this percentage was considered as the proportion of adult sexual dimorphism due to continued subadult male growth beyond the age of female maturation, or adult sexual dimorphism due to bimaturism. Determination of a precise age at which growth ceases is somewhat subjective and can be subject to error. Error may be due to age clumping and the cross-sectional nature of the data. Age clumping limits resolution while cross-sectional data tends to smooth over sharp distinctions seen in longitudinal data. Furthermore, growth may slow over several years before finally ceasing, making it difficult to judge a precise age for growth cessation. In these circumstances we tended to choose an age for growth cessation after which growth did not occur, erring on the side of later ages of maturation. Thus, the ages given are likely to be upper bounds for individual growth.

Parametric analyses
Specific hypotheses about the sexual dimorphism of growth were tested using linear regression and analysis of covariance (Sokal & Rohlf, 1981), with the following analysis of covariance model:

$$y_{ijk} = a + sex_i + age_j + (sex \times age)_{ij} + e_{ijk}$$

where y is the trait value, a is a constant, *sex* represents the effect of male *vs.* female, *age* is the linear effect of age in months, *(sex × age)* is the effect of the interaction between sex and age on trait values and e is the residual. A significant interaction effect indicates that the growth rate is significantly different between the sexes over the age interval considered. This is the particular significance test of interest for our hypotheses. A significant effect of age indicates that growth takes place during the interval while a significant effect of sex indicates that the sexes are significantly different over the interval considered. The significance tests for sex differences and age (growth) should only be considered when the interaction effect is not significant. The significance tests for growth, irrespective of sex, are not presented because of the predicted heterogeneity of growth across the sexes. Instead, the sex-specific slopes can be tested for statistical significance using their associated standard errors.

Inspection of the spline curves suggests that formal tests of sex differences in growth rates be obtained for four different age periods. First, linear regressions are calculated separately by sex for the infant and early juvenile period, from birth to 2·5 years. This is the age by which the first growth period is complete, although it is likely to be a maximum age of completion for individual animals due to the cross-sectional nature of the data. While growth rates may decrease somewhat over this interval, our current data do not allow resolution of the detailed dynamics. A second regression is calculated for the later juvenile period from 2·5 years to the age of growth cessation in females (typically 5·5 or 8 years, depending on the trait). The third regression covers the subadult period for males and young adult period for females, after female growth has ceased (typically 5·5 to 7·5 years or 8 to 12 years, depending on the trait). Finally, a regression was fit to the adult data (typically after 5·5 or 8 years in females and 7·5 or 12 years in males, depending on the trait). Thus, for females, animals included in the third regression are a subset of those used for the regression against adult age. Analyses of covariance were performed for these four age intervals to test for significant sex differences in growth rates.

Results

Age and sex-specific means and standard deviations for the 28 traits are presented in Appendix 2. After 7 years of age, 2-year age intervals were used because of the small samples available for single years. All measurements are significantly different between the sexes in adults as determined by *t*-tests using animals 13 years and older (see Appendix 2). Females have completed growth for all traits by 8 years, while males continue growing to 12 years.

Spline curves
Representative spline curves for both sexes are presented in Figures 1–10. In general, for each trait and both sexes, there is an early period of fast growth (infant and early juvenile period) followed by an extended period of slower growth which continues until growth ceases (late juvenile period in females and late juvenile plus subadult period in males), and finally, a period during which growth has stopped (adult period). The infant and early juvenile growth period extends from birth to about 2·0–2·5 years. The age of final cessation of growth varied both by sex and measurement.

The postcranial traits fall into two basic categories with regard to the timing of their growth periods as judged from the splines. Linear measurements of the extremities (including arm, forearm, hand, thigh, leg, foot and tail length and foot and hand width) and ventral

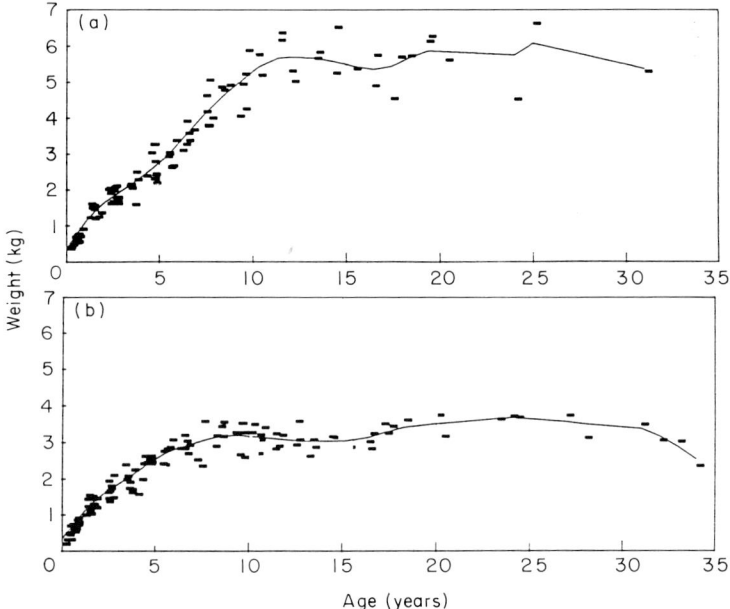

Figure 1. Growth curves for body weights in (a) males and (b) females. A spline curve is fit separately for each sex. Growth is complete at 8 years in females and 12 years in males.

trunk length cease growing at about 5·5 years in females and 7·5 years in males. Weight, most measurements of the trunk (including crown–rump length, occipital to tail base, abdominal and thoracic circumferences, and shoulder and hip widths) and circumferences of the extremities (including arm, thigh, and calf circumference) cease growing at about 8 years in females and 12 years in males. Aged animals (20–25 years) tended to have smaller limb circumferences and weights than younger adults, perhaps indicating a loss of muscle tissue at advanced ages. Examples of these growth patterns are shown in Figures 1–6.

Growth in head measurements typically does not follow the patterns observed for the postcranium. Instead, several of the measurements showed unique growth patterns. These traits showed relatively large variability around the spline curves because of the relatively small amount of absolute growth experienced postnatally. Variability around the spline curves hinders interpretation and leads to a general lack of statistical significance in analyses of covariance. Head length and breadth and biorbital and bizygomatic widths follow a growth pattern similar to that found for weight, trunk and limb circumferences, although male growth seems to continue to nearly 15 years of age for head length (see Figure 7) and breadth. Jaw width and upper facial height cease to grow in females at 5 and 7·5 years, respectively, with male growth ceasing at about 10 years (see Figure 8). Lower facial height and nasal width did not show obvious dimorphism in age at maturity in the spline curves, growth slowed in both sexes by 10 and 15 years of age, respectively, in these traits (see Figures 9 and 10). Even so, significant growth in male lower facial height continues throughout adulthood.

Inspection of the pseudo-velocity curves indicated evidence for an adolescent growth spurt in male body weights (see Figure 11). In males, growth rate for weight starts accelerating at about 5 years, peaks at about 6 years, and then decelerates until growth is complete at 12

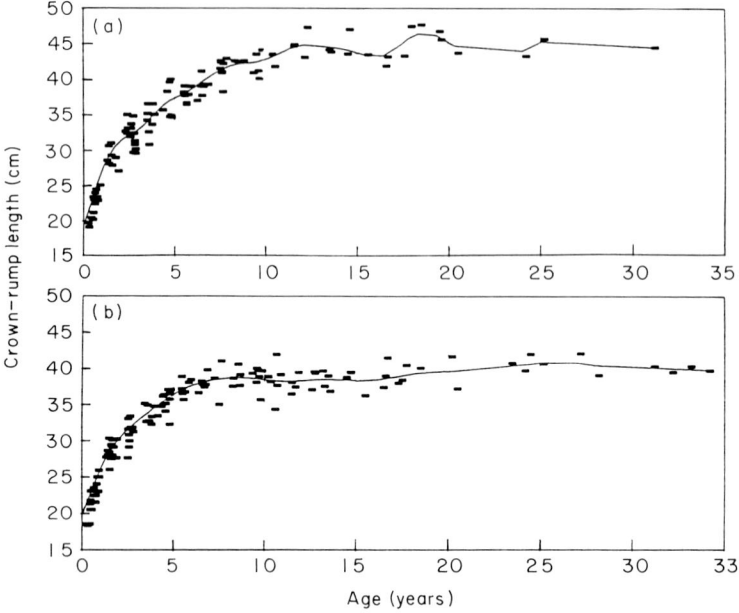

Figure 2. Growth curves for crown–rump length in (a) males and (b) females. Growth is complete at 8 years in females and 12 years in males.

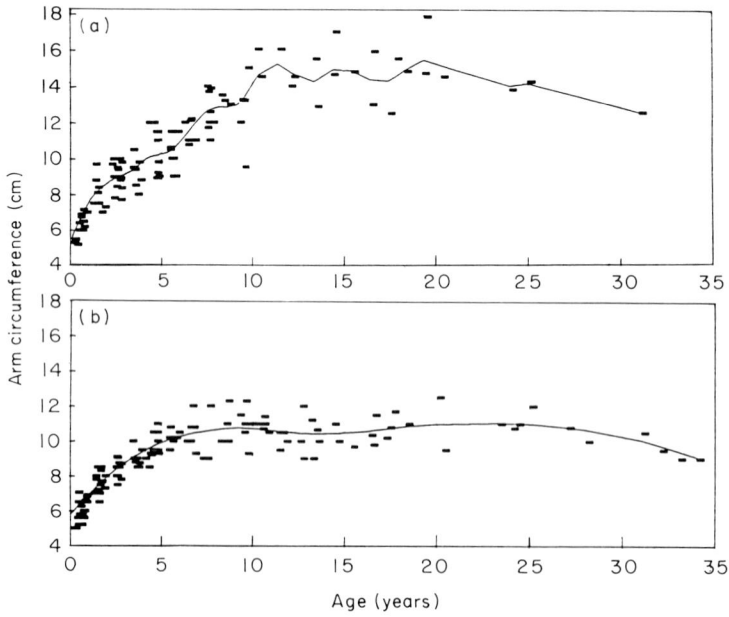

Figure 3. Growth curves for arm circumference in (a) males and (b) females. Growth is complete at 8 years in females and 12 years in males.

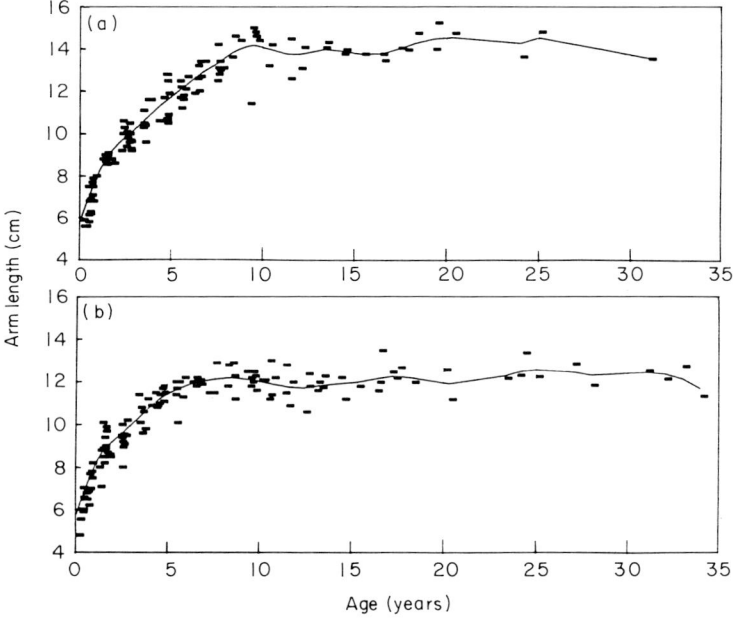

Figure 4. Growth curves for arm length in (a) males and (b) females. Growth is complete at 5·5 years in females and 7·5 years in males.

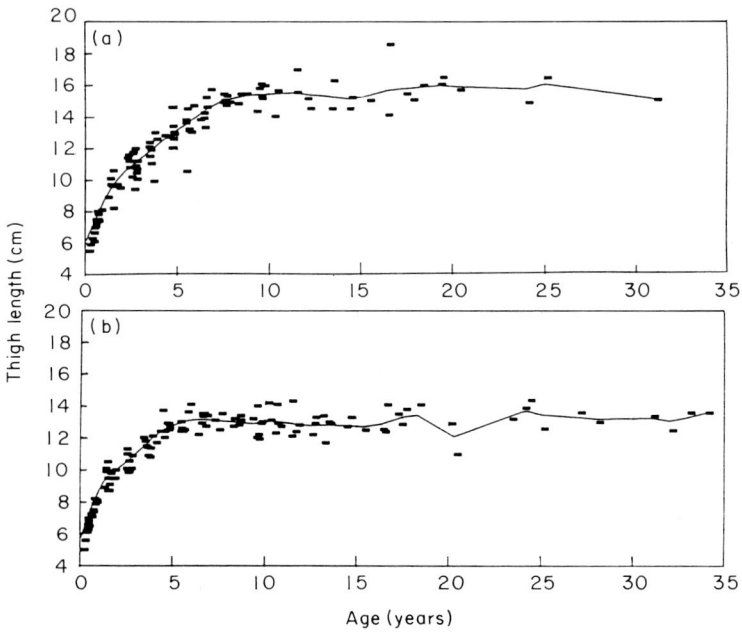

Figure 5. Growth curves for thigh length in (a) males and (b) females. Growth is complete at 5·5 years in females and 7·5 years in males.

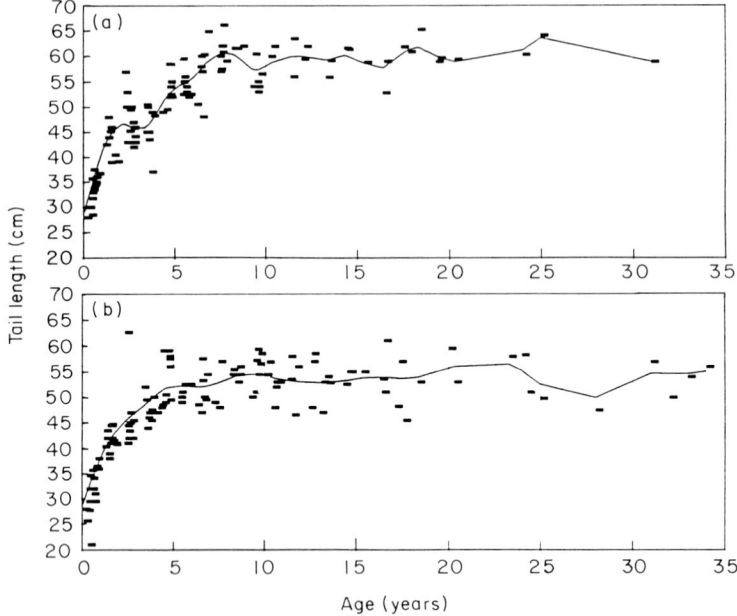

Figure 6. Growth curves for tail length in (a) males and (b) females. Growth is complete at 5·5 years in females and 7·5 years in males.

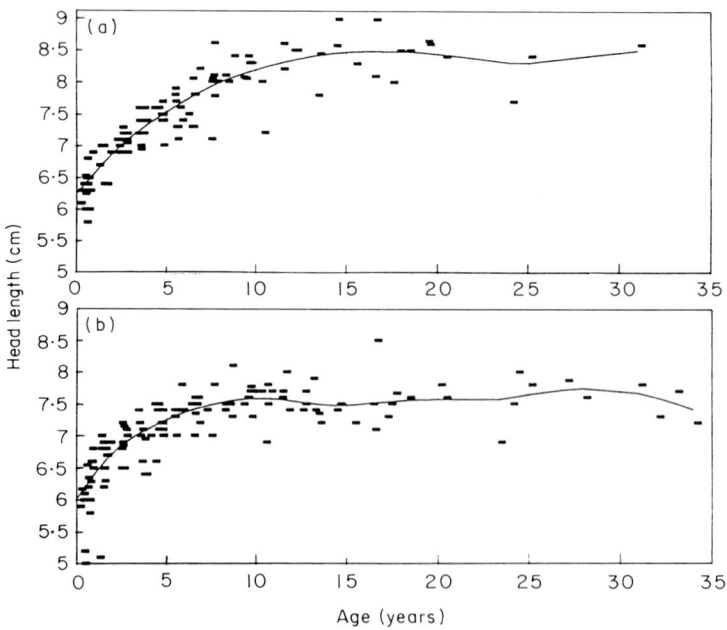

Figure 7. Growth curves for head length in (a) males and (b) females. Growth is complete at 8 years in females and 15 years in males.

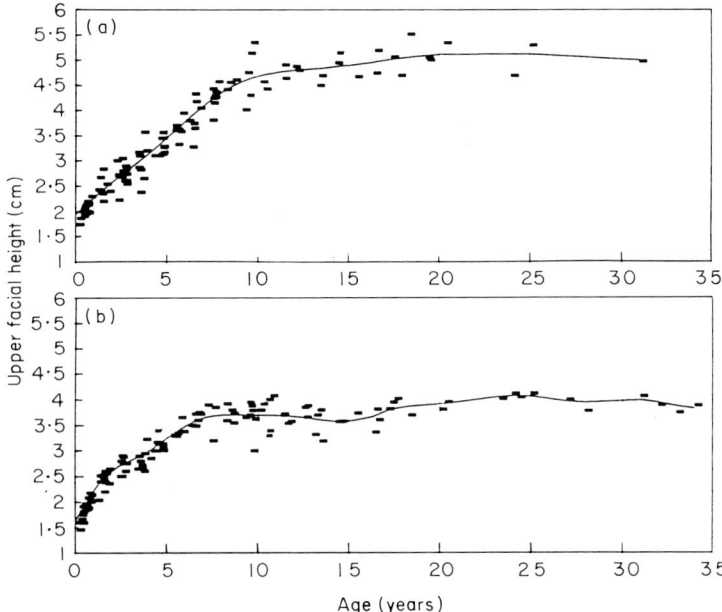

Figure 8. Growth curves for upper facial height in (a) males and (b) females. Growth is complete at 7·5 years in females and 10 years in males.

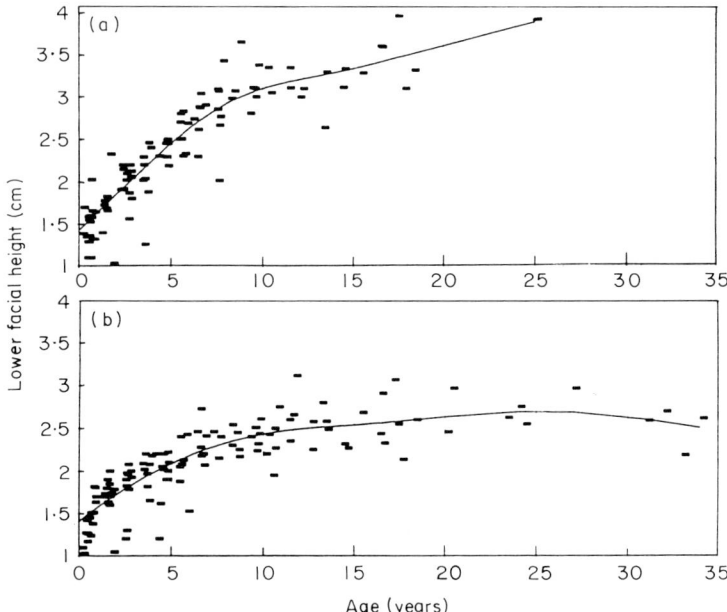

Figure 9. Growth curves for lower facial height in (a) males and (b) females. Growth is complete at 10 years in females and in males.

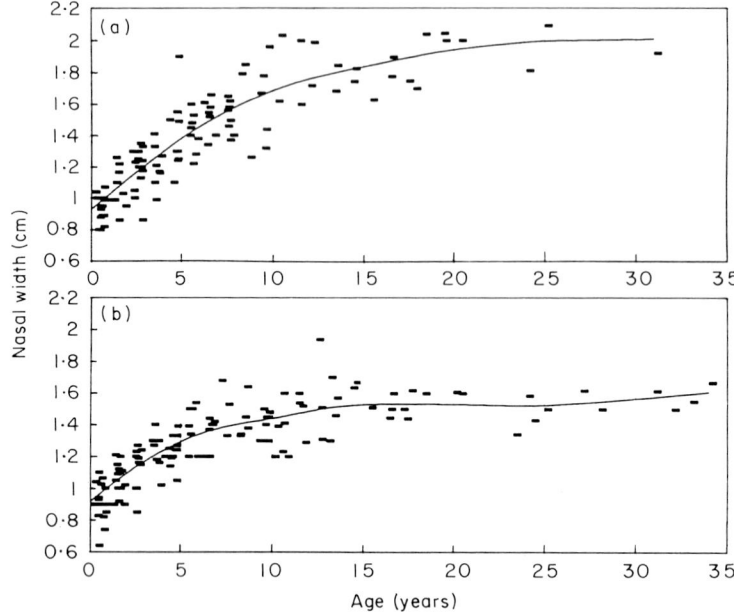

Figure 10. Growth curves for nasal width in (a) males and (b) females. Growth is complete at 15 years in females and in males.

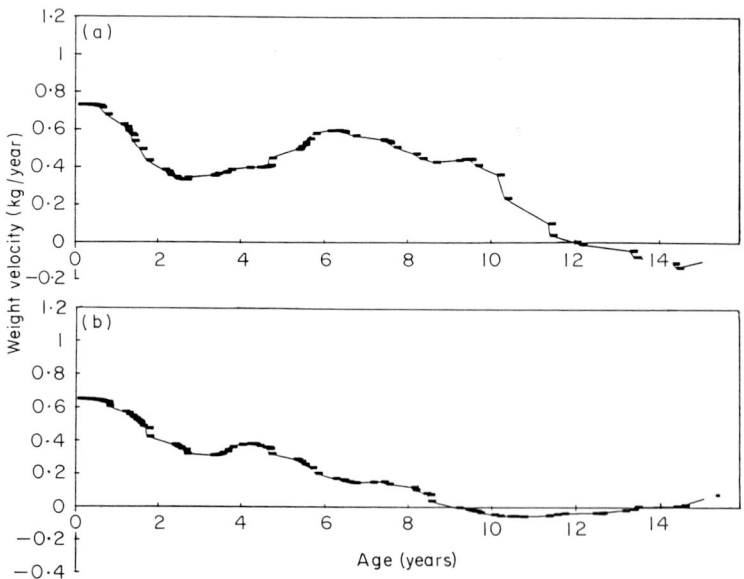

Figure 11. Pseudo-velocity curves for weight in (a) males and (b) females. Male curve displays an adolescent growth spurt. While the female curve shows a slight increase in velocity over a restricted age range starting at an age of 4, the increase in velocity is too small and the length of the increased growth period too short for this to be taken as a strong indication of a growth spurt.

Table 1   **Percent of adult dimorphism due to differences in growth rates and age at maturity**

Trait	Female age at maturity (year)	Male age at maturity (year)	Growth rate (%)	Age at maturity (%)
Weight	8	12	49	51
Crown–rump length	8	12	50	50
Occipital to tail base	8	12	38	62
Abdominal circumference	8	12	49	51
Thoracic circumference	8	12	55	45
Arm circumference	8	12	51	49
Calf circumference	8	12	34	66
Thigh circumference	8	12	39	61
Trunk length (ventral)	5·5	7·5	12	88
Shoulder width	8	12	30	70
Arm length	5·5	7·5	36	64
Forearm length	5·5	7·5	18	82
Hand length	5·5	7·5	61	39
Hand width	5·5	7·5	32	68
Hip width	8	12	58	42
Thigh length	5·5	7·5	30	70
Leg length	5·5	7·5	25	75
Foot length	5·5	7·5	47	53
Foot width	5·5	7·5	61	39
Tail length	5·5	7·5	32	68
Head length	8	15	45	55
Head breadth	8	15	62	38
Biorbital width	8	12	56	44
Bizygomatic width	8	12	49	51
Jaw width	5	10	4	96
Upper facial height	7·5	10	46	54
Lower facial height	10	10	100	0
Nasal width	15	15	100	0

years. While there is a slight increase in velocity at 4 years of age in females, this increase is not large or persistent enough to warrant positive identification as a growth spurt. Pseudo-velocity curves for the other traits did not provide solid evidence for a growth spurt, although the presence of a spurt cannot be ruled out on the basis of the present cross-sectional data.

Bimaturism vs. *growth rate dimorphism*
The proportion of adult sexual dimorphism due to growth rate dimorphism (during the infant and juvenile phase) and bimaturism (continued growth of subadult males) as judged from the spline curves is presented in Table 1. For example, the weight predicted from the spline curve at 8 years of age (when female growth ceases) is 3·12 kg in females and 4·40 kg in males, a difference of 1·28 kg. At 12 years (when male growth ceases), the predicted values are 3·07 and 5·69 kg for females and males, respectively, or a difference of 2·62 kg. The ratio of the difference at 8 years to the difference at 12 years measures the proportion of difference due to dimorphism in juvenile growth rates, 49% in this case. Overall, about 40% of adult dimorphism is due to differences in growth rates and 60% is due to differences in maturational age. Bimaturism is particularly important for postcranial body lengths (70% of

Table 2 Linear regressions in male and female toque macaques over limited age ranges representing the infant and early juvenile, later juvenile, subadult male and adult growth periods

Trait	Age[1]	Sex	n	Intercept (S.E.)	Growth rate (S.E.)	Probability of growth rate dimorphism[2]
Weight	<2.5	M	28	0.340 (0.050)	0.760 (0.059)	
		F	39	0.326 (0.044)	0.649 (0.043)	0.021
	<8	M	54	0.630 (0.124)	0.450 (0.124)	
		F	47	0.992 (0.124)	0.291 (0.026)	0.000
	<12	M	14	1.581 (0.379)	0.369 (0.140)	
		F	22	3.633 (0.571)	−0.049 (0.058)	0.002
	>12	M	20	5.235 (0.629)	0.022 (0.033)	
	>8	F	58	3.066 (0.110)	0.008 (0.006)	0.664
Crown–rump length	<2.5	M	36	20.367 (0.465)	5.740 (0.354)	
		F	49	20.629 (0.450)	4.789 (0.316)	0.025
	<8	M	51	26.860 (0.155)	1.958 (0.155)	
		F	41	28.537 (0.827)	1.470 (0.164)	0.024
	<12	M	14	37.468 (3.392)	0.536 (0.346)	
		F	22	40.918 (3.297)	−0.254 (0.335)	0.060
	>12	M	17	44.793 (2.862)	−0.001 (0.168)	
	>8	F	55	37.335 (0.548)	0.095 (0.030)	0.315
Occipital to tail base	<2.5	M	36	15.940 (0.568)	5.303 (0.433)	
		F	49	16.264 (0.486)	4.606 (0.341)	0.102
	<8	M	50	20.387 (0.895)	1.953 (0.168)	
		F	41	22.634 (1.413)	1.490 (0.280)	0.073
	<12	M	14	26.777 (4.784)	0.939 (0.487)	
		F	21	34.465 (4.829)	−0.185 (0.489)	0.063
	>12	M	16	41.979 (4.425)	−0.279 (0.260)	
	>8	F	54	31.612 (0.730)	0.079 (0.040)	0.106
Abdominal circumference	<2.5	M	36	12.122 (0.440)	4.428 (0.336)	
		F	49	11.768 (0.456)	4.587 (0.320)	0.368
	<8	M	52	18.875 (0.744)	1.127 (0.141)	
		F	41	18.244 (1.039)	1.033 (0.206)	0.353
	<12	M	14	16.492 (6.089)	1.452 (0.620)	
		F	22	22.807 (5.117)	0.387 (0.519)	0.099
	>12	M	18	32.428 (2.642)	−0.065 (0.156)	
	>8	F	55	25.439 (0.867)	0.151 (0.047)	0.077
Thoracic circumference	<2.5	M	36	15.252 (0.342)	3.676 (0.260)	
		F	49	14.815 (0.308)	3.323 (0.216)	0.148
	<8	M	51	18.509 (0.669)	1.593 (0.126)	
		F	41	19.201 (0.722)	1.236 (0.143)	0.040
	<12	M	14	24.606 (3.216)	0.908 (0.328)	
		F	22	30.734 (2.021)	−0.219 (0.205)	0.002
	>12	M	18	34.717 (2.358)	−0.036 (0.140)	
	>8	F	54	27.941 (0.475)	0.052 (0.026)	0.936
Arm circumference	<2.5	M	36	5.777 (0.213)	1.474 (0.162)	
		F	49	5.612 (0.160)	1.274 (0.112)	0.150
	<8	M	52	6.746 (0.462)	0.751 (0.087)	
		F	41	7.567 (0.442)	0.438 (0.087)	0.011
	<12	M	14	6.016 (3.721)	0.792 (0.379)	
		F	22	11.982 (1.518)	−0.128 (0.154)	0.007
	>12	M	18	14.097 (1.949)	0.035 (0.115)	
	>8	F	55	10.999 (0.330)	−0.024 (0.018)	0.542
Calf circumference	<2.5	M	36	5.924 (0.201)	1.432 (0.153)	
		F	49	5.541 (0.168)	1.303 (0.118)	0.071
	<8	M	52	6.949 (0.442)	0.616 (0.084)	
		F	41	7.513 (0.478)	0.381 (0.095)	0.041
	<12	M	14	5.265 (1.868)	0.677 (0.190)	
		F	22	13.135 (0.993)	−0.281 (0.101)	0.001

Table 2 Continued.

Trait	Age[1]	Sex	n	Intercept (S.E.)	Growth rate (S.E.)	Probability of growth rate dimorphism[2]
Thigh circumference	>12	M	17	13·007 (1·093)	−0·026 (0·064)	
	>8	F	54	10·470 (0·256)	−0·025 (0·014)	0·767
	<2·5	M	28	7·456 (0·495)	2·957 (0·360)	
		F	27	7·291 (0·510)	2·798 (0·355)	0·377
	<8	M	40	10·457 (0·842)	1·100 (0·151)	
		F	31	11·590 (0·842)	0·640 (0·170)	0·033
	<12	M	13	10·797 (3·755)	0·951 (0·384)	
		F	13	21·189 (3·724)	−0·503 (0·387)	0·007
	>12	M	12	23·116 (3·753)	−0·184 (0·217)	
	>8	F	35	16·561 (0·793)	−0·042 (0·042)	0·459
Trunk length (ventral)	<2·5	M	36	12·661 (0·410)	4·778 (0·313)	
		F	49	13·546 (0·397)	3·955 (0·279)	0·027
	<5·5	M	32	17·038 (1·736)	1·944 (0·437)	
		F	28	16·564 (1·343)	2·048 (0·321)	0·433
	<7·5	M	16	18·229 (4·380)	1·605 (0·655)	
		F	13	23·330 (5·483)	0·641 (0·837)	0·184
	>7·5	M	35	31·442 (0·752)	0·048 (0·049)	
	>5·5	F	73	27·495 (0·359)	0·062 (0·022)	0·431
Shoulder width	<2·5	M	36	4·789 (0·198)	1·297 (0·151)	
		F	49	5·089 (0·177)	0·822 (0·124)	0·009
	<8	M	52	5·778 (0·343)	0·483 (0·065)	
		F	41	6·308 (0·560)	0·279 (0·111)	0·056
	<12	M	15	2·651 (1·309)	0·802 (0·134)	
		F	22	8·156 (1·372)	0·031 (0·139)	0·001
	>12	M	18	13·552 (1·624)	−0·147 (0·096)	
	>8	F	55	8·411 (0·254)	0·016 (0·014)	0·129
Arm length	<2·5	M	36	6·102 (0·169)	1·734 (0·129)	
		F	49	6·247 (0·201)	1·462 (0·141)	0·087
	<5·5	M	31	7·751 (0·455)	0·791 (0·114)	
		F	28	8·194 (0·574)	0·647 (0·137)	0·218
	<7·5	M	18	10·084 (1·399)	0·418 (0·206)	
		F	13	10·788 (1·478)	0·180 (0·226)	0·240
	>7·5	M	35	13·642 (0·450)	0·029 (0·033)	
	>5·5	F	73	11·822 (0·143)	0·018 (0·009)	0·879
Forearm length	<2·5	M	36	6·357 (0·169)	1·709 (0·129)	
		F	49	6·279 (0·173)	1·504 (0·122)	0·129
	<5·5	M	31	7·322 (0·455)	0·937 (0·114)	
		F	28	8·148 (0·602)	0·732 (0·144)	0·140
	<7·5	M	18	9·957 (2·051)	0·488 (0·302)	
		F	13	14·025 (1·240)	−0·322 (0·189)	0·038
	>7·5	M	35	14·047 (0·535)	−0·013 (0·039)	
	>5·5	F	73	11·842 (0·144)	0·011 (0·009)	0·750
Hand length	<2·5	M	36	5·155 (0·131)	0·916 (0·100)	
		F	49	5·073 (0·134)	0·900 (0·094)	0·454
	<5·5	M	32	5·407 (0·265)	0·581 (0·067)	
		F	28	6·536 (0·397)	0·231 (0·095)	0·002
	<7·5	M	18	7·243 (1·256)	0·217 (0·185)	
		F	13	7·381 (1·309)	0·048 (0·200)	0·286
	>7·5	M	35	8·531 (0·336)	0·024 (0·024)	
	>5·5	F	73	7·595 (0·110)	0·009 (0·007)	0·302
Hand width	<2·5	M	36	1·666 (0·038)	0·239 (0·029)	
		F	48	1·559 (0·044)	0·208 (0·031)	0·245
	<5·5	M	32	1·988 (0·120)	0·075 (0·030)	
		F	27	1·985 (0·171)	0·046 (0·041)	0·289

Table 2 Continued.

Trait	Age[1]	Sex	n	Intercept (S.E.)	Growth rate (S.E.)	Probability of growth rate dimorphism[2]
	<7.5	M	18	1.656 (0.394)	0.138 (0.058)	
		F	13	1.635 (0.833)	0.097 (0.127)	0.375
	>7.5	M	35	2.583 (0.108)	0.006 (0.008)	
	>5.5	F	71	2.252 (0.049)	0.005 (0.003)	0.144
Hip width	<2.5	M	35	4.318 (0.130)	1.173 (0.102)	
		F	49	4.294 (0.113)	1.009 (0.079)	0.102
	<8	M	52	5.112 (0.322)	0.562 (0.036)	
		F	40	5.737 (0.288)	0.421 (0.057)	0.066
	<12	M	15	8.468 (1.356)	0.164 (0.139)	
		F	22	9.629 (0.606)	−0.099 (0.062)	0.030
	>12	M	18	10.860 (0.634)	−0.022 (0.038)	
	>8	F	53	8.346 (0.119)	0.028 (0.007)	0.671
Thigh length	<2.5	M	36	6.200 (0.193)	2.237 (0.147)	
		F	49	6.460 (0.189)	1.818 (0.133)	0.020
	<5.5	M	32	8.247 (0.658)	0.973 (0.058)	
		F	27	8.969 (0.524)	0.743 (0.125)	0.164
	<7.5	M	18	10.127 (1.765)	0.657 (0.260)	
		F	13	15.256 (1.860)	−0.314 (0.284)	0.014
	>7.5	M	35	14.997 (0.539)	0.031 (0.039)	
	>5.5	F	73	12.895 (0.175)	0.011 (0.011)	0.912
Leg length	<2.5	M	36	6.831 (0.212)	1.870 (0.161)	
		F	49	6.814 (0.246)	1.772 (0.173)	0.345
	<5.5	M	32	8.549 (0.504)	0.881 (0.127)	
		F	28	9.758 (0.884)	0.597 (0.211)	0.129
	<7.5	M	18	10.560 (2.473)	0.612 (0.365)	
		F	13	12.568 (3.097)	0.126 (0.473)	0.220
	>7.5	M	35	15.341 (0.539)	−0.027 (0.039)	
	>5.5	F	73	12.829 (0.187)	0.018 (0.011)	0.198
Foot length	<2.5	M	35	7.341 (0.147)	1.422 (0.112)	
		F	49	7.276 (0.131)	1.252 (0.092)	0.121
	<5.5	M	32	8.543 (0.355)	0.647 (0.089)	
		F	28	8.467 (0.583)	0.538 (0.139)	0.252
	<7.5	M	18	8.768 (1.206)	0.553 (0.178)	
		F	13	10.667 (1.709)	0.055 (0.261)	0.063
	>7.5	M	35	12.283 (0.266)	0.031 (0.019)	
	>5.5	F	73	11.105 (0.127)	−0.003 (0.008)	0.122
Foot width	<2.5	M	35	1.556 (0.044)	0.288 (0.033)	
		F	48	1.610 (0.049)	0.202 (0.034)	0.040
	<5.5	M	32	1.723 (0.106)	0.154 (0.027)	
		F	28	2.114 (0.190)	0.034 (0.045)	0.011
	<7.5	M	18	1.649 (0.365)	0.137 (0.054)	
		F	13	2.866 (0.639)	−0.085 (0.097)	0.022
	>7.5	M	34	2.590 (0.103)	0.007 (0.007)	
	>5.5	F	72	2.274 (0.041)	0.007 (0.003)	0.953
Tail length	<2.5	M	36	30.603 (1.005)	8.313 (0.766)	
		F	47	29.933 (1.110)	7.255 (0.797)	0.176
	<5.5	M	32	33.226 (2.186)	4.089 (0.550)	
		F	27	37.080 (3.951)	3.188 (0.957)	0.200
	<7.5	M	18	32.475 (10.00)	3.752 (1.474)	
		F	13	51.405 (13.95)	0.050 (2.130)	0.083
	>7.5	M	35	58.557 (1.900)	0.074 (0.137)	
	>5.5	F	71	52.544 (0.977)	0.074 (0.061)	0.269
Head length	<2.5	M	35	6.160 (0.094)	0.404 (0.072)	
		F	49	5.963 (0.111)	0.420 (0.078)	0.443

Table 2 Continued.

Trait	Age[1]	Sex	n	Intercept (S.E.)	Growth rate (S.E.)	Probability of growth rate dimorphism[2]
	<8	M	52	6·621 (0·126)	0·176 (0·024)	
		F	41	6·591 (0·168)	0·124 (0·033)	0·104
	<15	M	21	7·342 (0·438)	0·083 (0·039)	
		F	34	7·830 (0·227)	−0·026 (0·020)	0·004
	>15	M	14	8·492 (0·474)	−0·004 (0·023)	
	>8	F	60	7·521 (0·096)	0·002 (0·005)	0·377
Head Breadth	<2·5	M	36	5·120 (0·065)	0·323 (0·049)	
		F	49	4·889 (0·072)	0·329 (0·051)	0·474
	<8	M	52	5·257 (0·121)	0·198 (0·023)	
		F	41	5·327 (0·180)	0·107 (0·036)	0·016
	<15	M	21	6·378 (0·333)	0·080 (0·030)	
		F	33	6·037 (0·198)	0·005 (0·018)	0·013
	>15	M	14	7·552 (0·509)	−0·014 (0·024)	
	>8	F	58	6·283 (0·097)	−0·016 (0·005)	0·350
Biorbital width	<2·5	M	36	3·714 (0·055)	0·452 (0·042)	
		F	48	3·632 (0·054)	0·429 (0·037)	0·341
	<8	M	52	4·002 (0·111)	0·235 (0·021)	
		F	41	4·341 (0·130)	0·131 (0·026)	0·001
	<12	M	15	4·548 (0·528)	0·182 (0·054)	
		F	22	5·482 (0·409)	−0·014 (0·042)	0·003
	>12	M	18	6·514 (0·457)	−0·002 (0·027)	
	>8	F	55	5·227 (0·074)	0·013 (0·004)	0·652
Bizygomatic width	<2·5	M	36	3·709 (0·136)	0·780 (0·104)	
		F	49	3·723 (0·125)	0·661 (0·088)	0·192
	<8	M	52	4·520 (0·211)	0·275 (0·040)	
		F	41	4·564 (0·248)	0·224 (0·049)	0·222
	<12	M	15	2·941 (1·038)	0·448 (0·106)	
		F	22	5·139 (0·692)	0·105 (0·070)	0·004
	>12	M	18	7·481 (0·519)	0·023 (0·031)	
	>8	F	55	6·117 (0·126)	0·013 (0·007)	0·953
Jaw width	<2·5	M	36	2·290 (0·151)	0·337 (0·115)	
		F	49	2·229 (0·112)	0·405 (0·078)	0·308
	<5	M	20	2·269 (0·629)	0·309 (0·197)	
		F	28	2·701 (0·590)	0·145 (0·141)	0·357
	<10	M	34	2·746 (0·595)	0·144 (0·082)	
		F	31	3·228 (0·493)	0·030 (0·065)	0·140
	>10	M	25	4·225 (0·290)	0·004 (0·016)	
	>5	F	58	3·367 (0·138)	0·007 (0·009)	0·496
Upper facial height	<2·5	M	36	1·905 (0·050)	0·375 (0·038)	
		F	49	1·691 (0·042)	0·472 (0·030)	0·046
	<7·5	M	48	1·866 (0·117)	0·314 (0·023)	
		F	41	2·039 (0·099)	0·237 (0·020)	0·011
	<10	M	11	2·767 (1·351)	0·314 (0·023)	
		F	13	4·106 (1·015)	−0·046 (0·112)	0·095
	>10	M	26	4·442 (0·189)	0·028 (0·011)	
	>7·5	F	60	3·512 (0·080)	0·015 (0·005)	0·182
Lower facial height	<2·5	M	36	1·366 (0·071)	0·286 (0·054)	
		F	49	1·332 (0·061)	0·227 (0·043)	0·196
	<10	M	60	1·468 (0·105)	0·188 (0·018)	
		F	54	1·675 (0·094)	0·080 (0·015)	0·001
	>10	M	18	2·476 (0·266)	0·056 (0·018)	
		F	43	2·439 (0·114)	0·006 (0·006)	0·012
Nasal width	<2·5	M	36	0·881 (0·035)	0·138 (0·027)	
		F	48	0·894 (0·032)	0·105 (0·022)	0·176

Table 2 Continued.

Trait	Age[1]	Sex	n	Intercept (S.E.)	Growth rate (S.E.)	Probability of growth rate dimorphism[2]
	<15	M	72	1·072 (0·050)	0·060 (0·007)	
		F	75	1·132 (0·039)	0·031 (0·005)	0·001
	>15	M	14	1·608 (0·234)	0·015 (0·011)	
		F	26	1·414 (1·107)	0·005 (0·005)	0·190

[1] The maximum age in the regression, the minimum age being the maximum for the previous regression model. The last of the regressions are for adults and include all animals above the specified age. Age in years.
[2] Probabilities are for growth rate differences during the specified age interval (growth rate dimorphism).

adult dimorphism due to bimaturism, not including hands and feet) while it is less important for weight and postcranial circumferences and widths (54% of adult dimorphism due to bimaturism). Growth of the head displayed a variety of results from no bimaturism (lower facial height and nasal width) to nearly complete bimaturism (jaw width).

Parametric analyses
The regression equations, appropriate standard errors and probabilities of growth rate differences by sex are presented for each trait in Table 2 for separate regressions during the infant and early juvenile, late juvenile, subadult male or young adult female and adult periods.

Comparing growth rates among the ages. In general, growth rates declined with age, being highest during the infant and early juvenile period. Growth during the late juvenile period was usually about half the rate observed during the earlier period for males and about one-third the rate observed during the earlier period for females. Males grew less rapidly during the subadult than during the late juvenile period. In keeping with our selection of age categories based on the splines, female age-mates to subadult males (young adult females) had stopped growing altogether. Males showed several exceptions to the trend for lower growth rates with increasing age. Abdominal, arm and calf circumferences, shoulder width, hand width and bizygomatic width had slightly higher growth rates in the subadult phase than in the preceding late juvenile growth period.

Comparing growth rates across the sexes. Growth rates during the infant–early juvenile period were only rarely significantly different between the sexes (7 of 28 traits). Even so, all of the 20 postcranial measurements bar one (abdominal circumference) grew faster in males than in females, and six of these were statistically significant. One-half of the eight cranial measurements indicated faster growth in males, the other half in females, but only upper facial height grew significantly faster in females. These results indicate that growth rate differences by sex may develop early in life, especially for postcranial traits, but that these early rate differences are relatively small and contribute only a small proportion to adult dimorphism.

During the late juvenile period males nearly always grew at a faster rate than females (only ventral trunk length grew slightly faster in females than in males) and this was statistically significant for 13 of 28 measurements at the 5% level. The differences in growth rate also tended to be larger than in the earlier period. These results indicate significant sex differences in growth rate during the juvenile period for many measurements.

As expected given our delineation of age groups based on inspection of the splines, the largest differences in growth rates occurred in the subadult male and early adult female period, when female growth had stopped and male growth continued. During the young adult period, females did not grow for any measurement. For measurements with positive (but not statistically significant) female growth rates, males grew more than twice as fast as females (hand width is the only exception to this trend). Fourteen of 28 measurements showed significant sex differences in growth rates at the 5% level and 20 were significant at the 10% level, even though this growth rate comparison had the smallest sample sizes of the four tested. The measurements not showing a significant difference at the 10% level displayed significant male growth but not significant female growth (except for hand length, in which neither sex showed significant growth). Differences in growth rate during the subadult male–young adult female period are due to differences in the age of physical maturation in males and females. Adult sexual dimorphism arises, in part, from these differences in maturational age.

Discussion

Growth patterns and timing
Not surprisingly, we found distinct infant and juvenile growth periods for wild toque macaques. This corresponds to earlier results for rhesus macaques from Laird (1967) and Watts & Gavan (1982), for baboons (Glassman *et al.*, 1984; Coelho, 1985), chimpanzees (Laird, 1967; Watts & Gavan, 1982) and humans (Tanner, 1962; Bogin, 1988). The first period lasted from birth to, at most, 2·5 years. This is similar to the age of 23 months given by Laird (1967) for the end of early growth patterns in rhesus macaques using longitudinal data. The early period did not appear to differ in length among the various measurements studied. There is some potential for heterogeneity in growth rates within this age range which cannot be adequately resolved with this data, but this awaits further study.

The period of later juvenile growth differed among the measurements, with measures related to muscle mass (weight, dorsal trunk lengths, trunk breadths, extremity circumferences) growing several years longer than linear skeletal measures of limb size. Skeletal measurements complete growth by about 5·5 years in females and 7·5 years in males, while muscle mass measures continue growing to about 8 years in females and 12 years in males. The age of growth cessation found here for skeletal measures agrees reasonably well with the ages of epiphyseal fusion reported for the rhesus macaques from Cayo Santiago (Cheverud, 1981).

Somatometric measurements of the Cayo Santiago rhesus macaques (Turnquist & Kessler, 1989) provide very similar results to those found here for toque macaques. While not a laboratory colony, the rhesus macaques of Cayo Santiago are liberally provisioned and thus cannot be considered as a feral population in terms of somatic growth. The animals in this population are likely to be larger than animals from a feral population of rhesus macaques. Even so, Turnquist & Kessler (1989) found the same pattern of growth difference between skeletal and muscle mass traits, skeletal traits (arm, forearm, hand, thigh, leg and foot lengths) ceasing growth between 4–6 years in females and 6–10 years in males, while traits reflecting muscle mass (weight, crown–rump length, arm and thigh circumference) appear to continue to grow until 10–14 years in both sexes. While the timing of growth is similar in the Cayo rhesus macaques and the Sri Lankan toque macaques, the Cayo rhesus macaques grow to about twice the body weight in the same period of time. The size differences between members of these two populations are largely due to differences in growth rates, not to ages of maturation or size at birth. Whether the same results would be obtained for a comparison of feral rhesus and toque macaques depends on whether the free-ranging but provisioned macaques on Cayo Santiago mature at the same age as a feral rhesus macaque population.

Baboons also seem to grow on a schedule similar to that found for macaques (Sigg et al., 1982), suggesting that much of the size variation among papionins may be due to differences in growth rates. These interspecific comparisons are in contrast to comparisons between sexes within species where bimaturism is the most important cause of size differences.

The results of growth studies for laboratory populations of non-human primates differ importantly from those reported here. Growth is greatly accelerated in laboratory populations. Data from laboratory-reared rhesus macaques suggest that growth in weight is complete by about 6 years in females and 8 years in males (Kirk, 1972). This is 2–4 years earlier than in the provisioned free-ranging rhesus macaques of Cayo Santiago or the feral toque macaques of Polonnaruwa. Laboratory-reared baboons complete growth of weight and crown-rump length by about 6–7 years (Coelho, 1985). Again, this is much earlier than reported for feral baboons, in which females complete weight growth at about 6–7 years, while males continue growing to about 12 years (Sigg et al., 1982). Altmann & Alberts (1987) found that the growth rate for weight reported for captive baboons was double the rate they observed in wild animals. Dental eruption is also greatly accelerated in captive baboons (Phillips-Conroy & Jolly, 1988; Kahumbu & Eley, 1991).

There is no good evidence for an adolescent growth spurt in the data reported here, except for male body weight. This is not surprising given the cross-sectional nature of the data, age clumping in the sample due to natural birth peaks and small size of the spurt expected for macaques (Watts & Gavan, 1982). Thus, we cannot rule out the possibility of growth spurts for other traits or for female weight based on our present data. The magnitude of the juvenile growth spurt for skeletal dimensions suggested by earlier studies of non-human primates is so small that its existence has been appropriately questioned (Bogin, 1988). However, the evidence seems more secure for weight and muscle mass measurements (Bogin, 1988), especially in male baboons (Coelho, 1985). We have a similar result in this study for male body weight. However, longitudinal growth studies of this population will be needed to confirm this finding.

The limb circumferences and weight also display a tendency to decrease at advanced ages, perhaps indicating the effects of muscle wastage and decline in condition at this period of life (see Figure 3). The same trend appears in the data from the free-ranging colony of rhesus macaques on Cayo Santiago (Turnquist & Kessler, 1989). However, longitudinal data would be necessary to confirm this decline among aged animals since secular trends could also produce this result.

Developmental bases of sexual dimorphism
The developmental bases of adult sexual dimorphism in these body characters include both rate and maturation age dimorphism. Bimaturism dominates, especially for body lengths. The average proportion of adult dimorphism due to bimaturism for postcranial body lengths is 70% (if the hands and feet are not included). This proportion is likely to be underestimated due to the difficulty in identifying a precise age at which growth stops. The evolution of

sexual dimorphism for body lengths appears to be due, primarily, to the prolongation of male growth during the subadult period, or conversely, the contraction of the female juvenile growth period, with relatively little modification of growth rates. The difference in growth periods for males and females corresponds to differences in epiphyseal fusion and is accomplished by delayed epiphyseal fusion in males relative to females, especially at the elbow and knee joints (Cheverud, 1981).

In contrast, sexual dimorphism in postcranial circumferences and widths are due to rate and maturation dimorphism in nearly equal proportions (54% of adult dimorphism due to maturational differences). Most of the significant differences among older juveniles concern rates of growth for widths, circumferences and weight. This tallies with an earlier estimate indicating that juvenile males gain weight faster than their female peers (Dittus, 1977a). Even so, the proportion due to growth rate dimorphism is likely overestimated due to the difficulties in precisely assessing when growth stops. This bias may be particularly strong for females and postcranial circumferences and widths due to the relatively long period of declining growth rates in female muscle mass measurements. Thus, taking female morphology as primitive and male as derived, sexual dimorphism in body widths and circumferences, largely reflecting muscular development, have evolved through a combination of increased juvenile male growth rates and prolonged subadult male growth.

Measurements of the head display a variety of developmental bases for adult dimorphism. For some measurements adult dimorphism is produced solely by rate differences (lower facial height, nasal width) while for others adult dimorphism is almost entirely caused by differences in age at maturity (jaw width). In general, head measurements tend to differ more in growth rates between the sexes than postcranial measurements. In a cross-sectional growth study of rhesus macaque crania, Cheverud & Richtsmeier (1986) found that the face grows nearly twice as fast in males as it does in females during the juvenile growth period (about 3–7 years), while differences in age of maturation appear minor. Sirianni (1985) found that facial measurements grow approximately two to three times faster in juvenile male pigtailed macaques (*Macaca nemestrina*) than in juvenile females. These results correspond to our findings for lower facial height, nasal width and other facial measurements.

The developmental bases of sexual dimorphism vary considerably with the kind of measurement considered. Postcranial skeletal dimensions show a considerable degree of bimaturism while sexual dimorphism in muscle mass and cranial features are due to both rate differences and bimaturism.

Summary and conclusions

Somatic growth in the wild toque macaques from Polonnaruwa, Sri Lanka, is similar in pattern to that reported for laboratory populations of macaques and baboons, but growth is prolonged by several years. The period of infant and early juvenile growth ends at about 2·5 years, followed by a later juvenile growth period which varies in length by sex and type of measurement. Weight and measurements influenced by muscle mass complete growth by 8 years in females and 12 years in males, while skeletal lengths of the extremities complete growth by 5·5 years in females and 7·5 years in males. An adolescent growth spurt was detected for male body weight.

Sexual dimorphism is due primarily to bimaturism, males growing for a prolonged period relative to females, although growth rate dimorphism during the later juvenile period when both sexes are still growing also contributes significantly to adult dimorphism. Bimaturism was particularly important for skeletal length measurements. During the infant and early juvenile period, growth rate estimates for postcranial traits were slightly, but consistently, faster in males than in females.

Preliminary information comparing toque macaque growth to other macaques and baboons indicates that much of the variation among adult papionins may be due to differences in infant and juvenile growth rates with relatively little contribution from variation in age at maturity. However, due to acceleration of growth likely in laboratory and provisioned colonies, more data from feral macaque and baboon populations is required to make secure interspecific comparisons.

Acknowledgements

We thank the Office of the President of the Republic of Sri Lanka for research permission and especially C. Ponnamperuma, Director, Institute of Fundamental Studies and members of the Department of Zoology, University of Peradeniya. F. Bayart, E. Berkeley, S. Goonatillake and S. Freit assisted with the data collection. T. Diaz and D. Pernikoff helped supervise the field work, while N. Basnayeke, V. Coomaraswamy and S. Nathanael helped with data summaries. D. Kleiman lent administrative assistance and D. Melnick collaborated in the research. We also thank L. Kohn, S. Leigh and J. Phillips-Conroy for their comments on an earlier version of this manuscript. The field program by W. Dittus was supported by NSF grant BNS-8609665, the Harry Frank Guggenheim Foundation, Smithsonian Institution Scholarly Studies Program and the Friends of the National Zoo. Analysis was supported by NSF grant BSR-8906041 to J. Cheverud.

References

Altmann, J. & Alberts, S. (1987). Body mass and growth rates in a wild primate population. *Oecologia* **72**, 15–20.
Altmann, J., Altmann, S. A., Hausfater G. & McCluskey G. A. (1977). Life history of yellow baboons: physical development, reproductive parameters, and infant mortality. *Primates* **18**, 315–330.
Boas, F. (1892). The growth of children. II. *Science* **19**, 281–282.
Bogin, B. (1988). *Patterns of Human Growth*. Cambridge: Cambridge University Press.
Cheverud J. M. (1981). Epiphyseal union and dental eruption in *Macaca mulatta*. *Am. J. phys. Anthrop.* **56**, 157–167.
Cheverud, J. M. & Dittus, W. P. J. (1992). Primate population studies at Polonnaruwa. II. Heritability of body measurements in a natural population of toque macaques (*Macaca sinica*). *Am. J. Primat.* **27**, (in press).
Cheverud, J. M. & Richtsmeier, J. T. (1986). Finite element scaling applied to sexual dimorphism in rhesus macaque (*Macaca mulatta*) facial growth. *Syst. Zool.* **35**, 381–399.
Coelho, A. M. (1985). Baboon dimorphism: growth in weight, length, and adiposity from birth to 8 years of age. In (E. S. Watts, Ed.) *Nonhuman Primate Models for Human Growth and Development*, pp. 125–160. New York: Alan R. Liss.
Coelho, A. M., Glassman, D. M. & Bramblett, C. A. (1984). The relation of adiposity and body size to chronological age in olive baboons. *Growth* **48**, 445–454.
Dittus, W. P. J. (1975). Population dynamics of the toque monkey, *Macaca sinica*. In (R. H. Tuttle, Ed.) *Socioecology and Psychology of Primates*, pp. 125–152. The Hague: Mouton Publishers.
Dittus, W. P. J. (1977a). The social regulation of population density and age-sex distribution in the toque macaque. *Behaviour* **63**, 281–322.
Dittus, W. P. J. (1977b). The ecology of a semi-evergreen forest community in Sri Lanka. *Biotropica* **9**, 268–286.
Dittus, W. P. J. (1988). Group fission among wild toque macaques as a consequence of female resource competition and environmental stress. *Anim. Behav.* **36**, 1626–1645.
Dittus, W. P. J. & Thorington, R. W. (1981). Techniques for aging and sexing primates. In (Subcommittee on Conservation of Natural Primate Populations, Eds) *Techniques for the Study of Primate Population Ecology*, pp. 81–131. Washington, DC: National Academy of Sciences Press.
Gavan, J. A. & Swindler, D. R. (1966). Growth rates and phylogeny in primates. *Am. J. phys. Anthrop.* **24**, 181–190.
Glassman, D. M., Coelho, A. J., Carey, K. D. & Bramblett, C. A. (1984). Weight growth in savannah baboons: a longitudinal study from birth to adulthood. *Growth* **48**, 425–433.

Kahumbu, P. & Eley, R. M. (1991). Teeth emergence in wild olive baboons in Kenya and formulation of a dental schedule for aging wild baboon populations. *Am. J. Primat.* **23,** 1–9.
Kirk, J. H. (1972). Growth of maturing *Macaca mulatta*. *Lab. Anim. Sci.* **22,** 573–575.
Laird, A. K. (1967). Evolution of the human growth curve. *Growth* **31,** 345–355.
Leigh, S. (1991). The ontogeny of body size dimorphism in anthropoid primates. *Am. J. phys. Anthrop.* (suppl.) **12,** 113.
Phillips-Conroy, J. E. & Jolly, C. J. (1988). Dental eruption schedules of wild and captive baboons. *Am. J. Primat.* **15,** 17–29.
Schluter, D. (1988). Estimating the form of natural selection on a quantitative trait. *Evolution* **42,** 849–861.
Shea, B. T. (1986). Ontogenetic approaches to sexual dimorphism in anthropoids. *Hum. Evol.* **1,** 97–110.
Sigg, H., Stolba, A., Abegglen, J.-J. & Dasser, V. (1982). Life history of hamadryas baboons: physical development, infant mortality, reproductive parameters and family relationships. *Primates* **23,** 473–487.
Sirianni, J. E. (1985). Nonhuman primates as models for human craniofacial growth. In (E. S. Watts, Ed.) *Nonhuman Primated Models for Human Growth and Development*, pp. 95–124. New York: Alan R. Liss.
Sokal, R. R. & Rohlf, F. J. (1981). *Biometry*. San Francisco: W. H. Freeman & Co.
Tanner, J. M. (1951). Some notes on the reporting of growth data. *Hum. Biol.* **23,** 93–159.
Tanner, J. M. (1962). *Growth and Adolescence*. Oxford: Blackwell Scientific Publications.
Turnquist, J. E. & Kessler, M. J. (1989). Free-ranging Cayo Santiago rhesus monkeys (*Macaca mulatta*): I. Body size, proportion, and allometry. *Am. J. Primat.* **19,** 1–13.
Watts, E. S. (1985*a*). Introduction. In (E. S. Watts, Ed.) *Nonhuman Primate Models for Human Growth and Development*, pp. 1–8. New York: Alan R. Liss.
Watts, E. S. (1985*b*). Adolescent growth and development of monkeys, apes and humans. In (E. S. Watts, Ed.) *Nonhuman Primate Models for Human Growth and Development*, pp. 41–66. New York: Alan R. Liss.
Watts, E. S. & Gavan, J. A. (1982). Postnatal growth of nonhuman primates: the problem of the adolescent spurt. *Hum. Biol.* **54,** 53–70.

Appendix 1

Body measurements collected from the toque macaques at Polonnaruwa. Measurements were recorded in cm except for weight which was recorded in kg.
Weight: weight in kg at trapping (scale).
Crown–Rump length (CRNRMP): distance between the vertex (top of the head) and the caudal tip of the ischial tuberosities (caliper).
Occiput to tail base (OCCTBS): dorsal trunk length from the base of the skull to the base of the tail (caliper).
Trunk length (ventral) (TRL): distance between suprasternale (the upper border of the sternal notch) and symphysion (the midsagittal point along the cranial border of the pubic symphysis) (caliper).
Abdominal circumference (ABDCIRC): circumference of the abdomen measured above the pelvis (tape measure).
Thoracic circumference (THORCIRC): circumference of the thoracic cavity measured at the level of the breast (tape measure).
Biacromial width (shoulder width) (BAW): distance between right and left acromion (most lateral point of the upper extremity) (caliper).
Bitrochanteric width (hip width) (BTW): distance between the right and left trochanterion laterales (most lateral point on the greater trochanters of the femur) measured with legs held together in flexed position (caliper).
Arm circumference (BICPCIRC): maximum circumference on the arm (tape measure).
Arm length (ARL): distance from acromion to radiale [the most lateral and proximal point of the radial head (at elbow)] (caliper).
Forearm length (FAL): distance from radiale to stylion (the most distal point on the radial styloid process) (caliper).
Hand length (HNL): distance from stylion to chirodactylion (the most distal tip of the third digit) (caliper).
Hand width (HNW): distance from metacarpale mediale (head of the second metacarpal) to metacarpale laterale (head of the fifth metacarpal) (caliper).
Calf circumference (CLFCIRC): maximum circumference of the calf (tape measure).
Thigh circumference (THGHCIRC): maximum circumference of the thigh (tape measure).
Thigh length (THL): distance from trochanterion laterale to femorale (most distal point of the femur's lateral condyle at knee) (caliper).
Leg length (LEL): distance between tibiale (most proximal point on the medial condyle of the tibia at the knee) and sphyrion (the most distal point on the medial malleolus of the tibia at the ankle) (caliper).
Foot length (FTL): distance from pternion (heel) to pododactylion (most distal point on the third digit) (caliper).
Foot width (FTW): distance from metatarsale mediale (medial point of the head of the second metatarsal) to metatarsale laterale (the lateral side of the fifth metatarsal head) (caliper).
Tail length (TL): distance from the base of the tail to the tip of the most caudal vertebra (tape measure).
Head length (HDL): distance from glabella (most anterior point on the frontal bone) to inion (most posterior point on the occipital) (caliper).

Head breadth (HDB): maximum transverse distance between right and left parietal bones (caliper).
Biorbital width (BOW): distance between the lateral margins of the orbits (caliper).
Bizygomatic width (BZW): maximum distance between the lateral surfaces of the zygomatic arches (caliper).
Bigonial width (BGW): maximum distance between the right and left angles (gonia) of the mandible (caliper).
Upper facial height (UFH): distance between nasion (depression below brow ridges) and intradentale superior (gingiva between upper medial incisors) (caliper).
Lower facial height (LFH): distance between the most inferior point of the mandibular symphysis and intradentale inferior (gingiva between lower medial incisors) (caliper).
Nasal width (NSW): maximum distance between the lateral margins of the nostrils (caliper).

Appendix 2

Age and sex specific means (and standard deviations) for somatometric traits in the toque macaques from Polonnaruwa, Sri Lanka.

Age	Crown–rump length M	F	Occipital to tail base M	F	Abdominal circumference M	F	Thoracic circumference M	F	Arm circumference M	F
0–1	22.36 (1.69)	22.24 (2.18)	17.74 (1.94)	17.67 (2.02)	13.63 (1.63)	13.43 (2.23)	16.48 (1.14)	16.10 (1.67)	6.28 (0.65)	6.07 (0.67)
1–2	29.20 (1.75)	28.54 (1.14)	24.28 (2.05)	24.15 (1.22)	19.31 (1.92)	19.13 (1.80)	21.07 (1.09)	19.91 (1.10)	8.04 (0.84)	7.61 (0.57)
2–3	31.98 (1.79)	31.12 (1.86)	26.03 (2.51)	26.05 (1.97)	21.39 (1.64)	21.77 (1.55)	23.12 (1.60)	22.33 (1.06)	8.95 (0.74)	8.48 (0.62)
3–4	33.89 (2.43)	33.40 (1.16)	27.66 (1.80)	27.33 (1.98)	23.03 (1.29)	21.42 (1.37)	23.51 (0.84)	23.21 (0.81)	9.17 (0.82)	8.98 (0.45)
4–5	36.77 (2.35)	35.46 (1.51)	29.40 (2.48)	29.97 (3.38)	24.69 (1.61)	23.33 (1.23)	25.57 (1.27)	25.09 (0.94)	10.43 (1.35)	9.73 (0.76)
5–6	37.80 (0.94)	37.30 (1.15)	31.89 (1.96)	31.49 (2.20)	24.90 (1.37)	24.11 (2.62)	27.24 (0.98)	26.17 (1.11)	10.26 (0.97)	10.20 (0.97)
6–7	39.11 (1.45)	37.87 (0.91)	32.05 (1.08)	32.08 (1.96)	25.75 (2.66)	24.91 (1.68)	28.86 (2.09)	27.75 (1.06)	11.35 (0.65)	10.46 (0.83)
7–9	41.45 (1.69)	38.50 (1.89)	34.90 (2.07)	32.35 (1.91)	27.64 (1.85)	26.38 (2.56)	30.81 (1.74)	27.84 (2.05)	12.84 (1.25)	10.41 (1.25)
9–11	42.24 (1.70)	38.52 (1.93)	35.49 (2.33)	32.92 (2.58)	30.28 (3.13)	26.10 (3.05)	33.36 (1.86)	28.85 (0.95)	13.34 (2.03)	10.94 (0.68)
11–13	44.90 (1.73)	38.32 (1.28)	38.83 (2.16)	32.48 (1.98)	32.38 (2.18)	28.00 (1.67)	34.38 (2.18)	28.36 (1.49)	15.13 (1.03)	10.34 (0.95)
13–15	44.40 (1.62)	38.46 (1.23)	37.84 (2.78)	31.53 (1.24)	31.62 (1.15)	26.93 (1.84)	34.15 (1.36)	27.75 (0.93)	14.60 (1.66)	10.41 (0.91)
15+	44.83 (2.03)	39.47 (1.75)	36.87 (3.01)	33.56 (2.02)	32.59 (3.19)	29.27 (2.41)	34.54 (2.04)	29.25 (1.53)	14.50 (1.42)	10.55 (1.09)

Age	Calf circumference M	F	Thigh circumference M	F	Ventral trunk length M	F	Shoulder width M	F	Arm length M	F
0–1	6.40 (0.67)	6.05 (0.74)	8.21 (0.97)	8.18 (1.40)	14.35 (1.68)	14.90 (1.68)	5.23 (0.60)	5.44 (0.73)	6.72 (0.80)	6.67 (0.80)
1–2	8.27 (0.87)	7.49 (0.52)	12.39 (1.47)	11.73 (1.69)	19.93 (1.47)	20.17 (1.12)	6.80 (0.71)	6.31 (0.74)	8.79 (0.24)	8.86 (0.75)
2–3	8.71 (0.63)	8.61 (0.71)	13.26 (1.40)	13.66 (0.99)	22.67 (1.42)	21.93 (1.14)	7.21 (0.80)	6.82 (0.70)	9.77 (0.44)	9.33 (0.62)
3–4	8.81 (0.67)	8.55 (0.65)	14.07 (1.01)	13.76 (1.62)	23.21 (3.94)	23.55 (1.03)	7.63 (0.79)	7.38 (0.81)	10.59 (0.82)	10.37 (0.64)
4–5	10.26 (1.13)	9.33 (0.93)	16.22 (1.53)	14.53 (0.80)	26.50 (2.07)	26.03 (1.70)	7.91 (1.01)	7.44 (1.02)	11.45 (0.90)	11.28 (0.35)
5–6	10.13 (1.42)	10.06 (0.49)	16.73 (1.81)	15.40 (0.64)	27.31 (0.97)	27.24 (1.27)	8.37 (0.80)	8.44 (1.14)	11.98 (0.50)	11.53 (0.71)
6–7	10.98 (1.03)	10.06 (0.72)	17.33 (1.53)	15.50 (1.00)	28.29 (1.78)	27.51 (1.18)	9.00 (0.62)	7.88 (0.91)	12.80 (0.62)	12.01 (0.18)
7–9	11.56 (1.01)	10.32 (1.08)	18.76 (1.85)	16.57 (2.09)	30.78 (2.35)	28.04 (1.82)	9.32 (0.83)	8.15 (0.48)	13.38 (0.69)	12.11 (0.70)
9–11	11.88 (1.03)	10.37 (0.55)	19.29 (1.26)	16.73 (1.37)	31.40 (2.29)	28.60 (1.62)	10.33 (0.61)	8.54 (0.61)	14.13 (1.18)	12.07 (0.49)
11–13	13.22 (0.76)	9.89 (0.36)	22.33 (2.26)	15.38 (1.11)	31.70 (1.13)	28.23 (1.08)	12.15 (0.44)	8.50 (0.81)	13.57 (0.88)	11.70 (0.73)
13–15	12.60 (0.42)	10.14 (0.98)	20.35 (1.00)	15.56 (2.55)	32.40 (1.46)	27.94 (1.11)	11.23 (1.32)	8.90 (0.71)	14.07 (0.33)	11.96 (0.54)
15+	12.58 (0.88)	9.92 (0.84)	19.68 (2.44)	15.73 (1.60)	32.42 (1.39)	28.92 (1.46)	10.91 (1.18)	8.75 (0.79)	14.24 (0.75)	12.30 (0.59)

Appendix 2 Continued.

Age	Forearm length M	Forearm length F	Hand length M	Hand length F	Hand width M	Hand width F	Hip width M	Hip width F	Thigh length M	Thigh length F
0–1	6·96 (0·71)	6·75 (0·76)	5·51 (0·60)	5·37 (0·47)	1·75 (0·14)	1·64 (0·16)	4·71 (0·44)	4·64 (0·53)	7·02 (0·77)	7·03 (0·87)
1–2	8·99 (0·54)	8·87 (0·61)	6·52 (0·32)	6·57 (0·58)	2·04 (0·07)	1·86 (0·15)	6·17 (0·41)	5·93 (0·37)	9·54 (0·82)	9·56 (0·49)
2–3	9·82 (0·50)	9·47 (0·50)	7·01 (0·40)	6·99 (0·50)	2·20 (0·18)	2·07 (0·18)	6·76 (0·57)	6·53 (0·33)	10·83 (0·74)	10·35 (0·70)
3–4	10·69 (0·69)	10·72 (0·68)	7·36 (0·43)	7·43 (0·39)	2·16 (0·12)	2·14 (0·18)	7·07 (0·85)	7·07 (0·31)	11·72 (0·96)	11·52 (0·42)
4–5	11·81 (0·47)	11·56 (0·55)	8·20 (0·32)	7·63 (0·33)	2·44 (0·14)	2·21 (0·17)	7·63 (1·00)	7·79 (0·60)	13·00 (0·76)	12·64 (0·47)
5–6	12·11 (0·86)	11·91 (0·26)	8·39 (0·36)	7·66 (0·44)	2·31 (0·16)	2·22 (0·16)	7·85 (0·71)	8·28 (0·33)	13·30 (1·29)	12·94 (0·67)
6–7	13·32 (0·87)	12·09 (0·34)	8·75 (0·52)	7·61 (0·37)	2·60 (0·05)	2·28 (0·24)	8·75 (0·77)	8·55 (0·42)	14·38 (0·85)	13·11 (0·51)
7–9	13·61 (0·66)	11·74 (0·59)	8·77 (0·54)	7·49 (0·48)	2·67 (0·18)	2·35 (0·18)	9·55 (0·64)	8·58 (0·56)	15·02 (0·50)	13·04 (0·35)
9–11	13·88 (1·08)	12·12 (0·66)	8·76 (0·57)	7·88 (0·45)	2·62 (0·17)	2·24 (0·19)	10·01 (0·74)	8·64 (0·33)	15·29 (0·75)	12·96 (0·77)
11–13	14·03 (0·76)	11·95 (0·53)	8·63 (0·15)	7·78 (0·36)	2·50 (0·16)	2·34 (0·21)	10·20 (0·26)	8·60 (0·27)	15·50 (1·02)	12·93 (0·73)
13–15	14·22 (0·97)	11·54 (0·46)	8·87 (0·55)	7·44 (0·45)	2·78 (0·19)	2·30 (0·20)	10·50 (0·45)	8·66 (0·18)	15·20 (1·08)	12·75 (0·72)
15+	13·81 (1·13)	12·15 (0·57)	9·13 (0·79)	7·82 (0·43)	2·74 (0·21)	2·38 (0·17)	10·66 (0·53)	9·04 (0·41)	15·73 (1·28)	13·17 (0·76)

Age	Leg length M	Leg length F	Foot length M	Foot length F	Foot width M	Foot width F	Tail length M	Tail length F	Head length M	Head length F
0–1	7·47 (0·94)	7·42 (1·26)	7·82 (0·52)	7·70 (0·61)	1·63 (0·11)	1·67 (0·17)	33·44 (2·81)	32·37 (4·32)	6·30 (0·40)	6·21 (0·59)
1–2	9·86 (0·66)	9·76 (0·72)	9·69 (0·47)	9·38 (0·53)	2·08 (0·10)	1·96 (0·16)	43·27 (3·24)	41·92 (2·24)	6·08 (0·27)	6·62 (0·46)
2–3	10·80 (0·50)	10·59 (0·76)	10·28 (0·37)	9·91 (0·37)	2·15 (0·13)	2·06 (0·17)	46·71 (4·01)	45·31 (6·08)	7·05 (0·19)	6·90 (0·25)
3–4	11·46 (0·66)	11·97 (1·04)	10·70 (0·61)	10·20 (0·58)	2·21 (0·19)	2·26 (0·28)	45·76 (4·27)	47·77 (2·31)	7·26 (0·27)	6·93 (0·33)
4–5	12·82 (0·69)	12·53 (0·76)	11·75 (0·54)	11·19 (0·58)	2·45 (0·13)	2·28 (0·09)	52·81 (3·06)	53·50 (4·76)	7·46 (0·21)	7·18 (0·28)
5–6	13·31 (0·89)	12·96 (0·61)	11·79 (0·49)	11·01 (0·21)	2·50 (0·21)	2·30 (0·15)	54·31 (2·51)	51·25 (1·51)	7·51 (0·28)	7·37 (0·24)
6–7	14·63 (1·34)	13·35 (0·73)	12·39 (0·56)	11·08 (0·43)	2·54 (0·15)	2·33 (0·16)	57·33 (5·58)	51·69 (4·18)	7·75 (0·36)	7·39 (0·22)
7–9	14·99 (0·62)	13·02 (0·95)	12·71 (0·48)	11·18 (0·50)	2·68 (0·15)	2·33 (0·18)	60·59 (2·90)	53·44 (3·28)	8·00 (0·38)	7·50 (0·33)
9–11	15·23 (1·03)	12·83 (0·69)	12·57 (0·41)	11·22 (0·57)	2·64 (0·20)	2·32 (0·22)	57·28 (3·54)	54·19 (3·42)	8·04 (0·44)	7·58 (0·24)
11–13	14·90 (0·57)	12·83 (0·74)	12·72 (0·21)	11·04 (0·57)	2·60 (0·27)	2·35 (0·19)	60·25 (3·28)	53·06 (5·12)	8·45 (0·17)	7·65 (0·22)
13–15	14·87 (0·86)	12·47 (0·61)	12·73 (0·53)	10·79 (0·32)	2·68 (0·20)	2·39 (0·16)	59·92 (3·51)	53·19 (1·64)	8·48 (0·42)	7·38 (0·09)
15+	14·98 (1·21)	13·40 (0·64)	12·84 (0·61)	11·06 (0·53)	2·71 (0·18)	2·44 (0·13)	60·35 (3·47)	54·32 (4·27)	8·42 (0·35)	7·58 (0·38)

Appendix 2 Continued.

Age	Head breadth M	Head breadth F	Biorbital width M	Biorbital width F	Bizygomatic width M	Bizygomatic width F	Jaw width M	Jaw width F	Upper facial height M	Upper facial height F
0–1	5·23 (0·20)	5·00 (0·37)	3·86 (0·22)	3·78 (0·23)	3·99 (0·54)	3·89 (0·52)	2·37 (0·59)	2·32 (0·35)	2·04 (0·15)	1·87 (0·21)
1–2	5·62 (0·26)	5·44 (0·17)	4·46 (0·12)	4·32 (0·20)	4·90 (0·53)	4·94 (0·31)	2·93 (0·52)	2·98 (0·36)	2·46 (0·20)	2·44 (0·15)
2–3	5·86 (0·20)	5·57 (0·22)	4·64 (0·17)	4·57 (0·16)	5·33 (0·28)	5·02 (0·47)	3·03 (0·49)	3·12 (0·52)	2·73 (0·20)	2·73 (0·17)
3–4	5·90 (0·25)	5·76 (0·25)	4·83 (0·18)	4·81 (0·28)	5·57 (0·34)	5·26 (0·42)	3·33 (0·47)	3·06 (0·71)	2·90 (0·42)	2·84 (0·20)
4–5	6·12 (0·28)	5·73 (0·42)	5·03 (0·26)	4·95 (0·20)	5·62 (0·53)	5·73 (0·37)	3·07 (0·44)	3·54 (0·23)	3·26 (0·17)	3·10 (0·12)
5–6	6·15 (0·22)	5·99 (0·17)	5·19 (0·19)	5·09 (0·19)	6·03 (0·60)	5·89 (0·42)	3·50 (0·38)	3·40 (0·78)	3·63 (0·17)	3·37 (0·13)
6–7	6·50 (0·23)	6·07 (0·13)	5·64 (0·34)	5·19 (0·15)	6·10 (0·33)	6·04 (0·28)	3·53 (0·55)	3·53 (0·50)	3·84 (0·34)	3·60 (0·13)
7–9	6·97 (0·44)	6·01 (0·41)	5·81 (0·32)	5·30 (0·29)	6·75 (0·66)	5·88 (0·50)	3·98 (0·74)	3·19 (0·44)	4·31 (0·28)	3·69 (0·24)
9–11	7·01 (0·31)	6·10 (0·23)	6·28 (0·24)	5·35 (0·19)	7·15 (0·62)	6·24 (0·33)	4·10 (0·69)	3·67 (0·43)	4·65 (0·44)	3·70 (0·31)
11–13	7·30 (0·10)	6·11 (0·16)	6·43 (0·59)	5·35 (0·21)	8·05 (0·24)	6·25 (0·42)	4·40 (0·20)	3·33 (0·67)	4·80 (0·12)	3·66 (0·18)
13–15	7·52 (0·20)	6·08 (0·18)	6·65 (0·24)	5·45 (0·26)	7·82 (0·42)	6·28 (0·30)	3·94 (0·20)	3·46 (0·72)	4·83 (0·28)	3·60 (0·22)
15+	7·27 (0·38)	5·94 (0·42)	6·51 (0·27)	5·53 (0·23)	7·91 (0·39)	6·42 (0·40)	4·37 (0·48)	3·50 (0·61)	5·04 (0·31)	3·89 (0·28)

Age	Lower facial height M	Lower facial height F	Nasal width M	Nasal width F	Weight M	Weight F	Sample size M	Sample size F
0–1	1·49 (0·23)	1·41 (0·23)	0·93 (0·12)	0·92 (0·12)	0·63 (0·13)	0·59 (0·19)	19	22
1–2	1·68 (0·35)	1·71 (0·20)	1·08 (0·14)	1·07 (0·10)	1·40 (0·15)	1·26 (0·16)	9	18
2–3	2·02 (0·18)	1·83 (0·30)	1·19 (0·14)	1·15 (0·14)	1·79 (0·20)	1·69 (0·22)	17	12
3–4	1·98 (0·45)	1·81 (0·14)	1·18 (0·14)	1·22 (0·10)	2·11 (0·28)	1·89 (0·28)	9	11
4–5	2·37 (0·11)	1·92 (0·31)	1·42 (0·25)	1·24 (0·10)	2·60 (0·38)	2·45 (0·20)	8	10
5–6	2·58 (0·20)	2·07 (0·31)	1·42 (0·13)	1·38 (0·14)	2·89 (0·28)	2·71 (0·27)	8	7
6–7	2·74 (0·29)	2·31 (0·21)	1·53 (0·11)	1·33 (0·15)	3·54 (0·33)	2·89 (0·18)	8	8
7–9	2·93 (0·41)	2·34 (0·14)	1·54 (0·21)	1·46 (0·14)	4·32 (0·55)	3·08 (0·46)	13	8
9–11	3·11 (0·19)	2·40 (0·20)	1·70 (0·24)	1·37 (0·12)	5·02 (0·72)	3·16 (0·32)	8	13
11–13	3·14 (0·15)	2·59 (0·30)	1·83 (0·20)	1·50 (0·22)	5·71 (0·52)	3·09 (0·26)	4	8
13–15	3·10 (0·28)	2·46 (0·20)	1·77 (0·07)	1·60 (0·12)	5·71 (0·52)	3·02 (0·24)	5	8
15+	3·54 (0·33)	2·61 (0·27)	1·91 (0·18)	1·53 (0·14)	5·66 (0·76)	3·30 (0·40)	14	26

Anthony B. Falsetti
Department of Ecology and Evolution, Division of Biological Sciences, State University of New York at Stony Brook, Stony Brook, New York 11794-5245, U.S.A.

Theodore M. Cole III
Doctoral Program in Anthropological Sciences, State University of New York at Stony Brook, Stony Brook, New York 11794, U.S.A.

Received 1 July 1991
Revision received 27 January 1992 and accepted 22 March 1992

Keywords: growth, postcranial skeleton, callitrichines, locomotion, positional behavior.

Relative growth of the postcranial skeleton in callitrichines

This study presents the results and suggested functional implications of a metric analysis of skeletal size and shape variation in ontogenetic samples of three callitrichine species: *Saguinus oedipus*, *Saguinus fuscicollis* and *Callithrix jacchus*. Adult interspecific shape differences distinguish *S. fuscicollis* from the other species. The most notable of these are the intermembral and brachial indices, both of which are significantly higher in *S. fuscicollis*. Ontogenetic growth patterns are highly conserved across species and the species-specific patterns of growth-related change are virtually identical. *Saguinus fuscicollis* exhibits a significantly higher intermembral index throughout the period of ontogeny observed in this study, and has a relatively longer radius at any given size or age. The elongation of the forelimb in *S. fuscicollis* is best seen as an adaptation for increasing the lateral positioning of the forelimbs when foraging on the trunks of large trees.

Journal of Human Evolution (1992) **23,** 79–92

Introduction

Previous morphometric studies of the postcranial skeleton of adult tamarins and marmosets (Glassman, 1983; Falsetti *et al.*, 1989) have demonstrated that consistent differences in interspecific shape reflect both phylogeny and the locomotor and postural behaviors related to differential substrate preference and utilization (Moynihan, 1970; Garber, 1980, in press). Significant interspecific diversity has been documented for the postcranial skeleton, primarily in limb proportions, hip and shoulder morphology, and the positioning of muscle attachments at the elbow (Glassman, 1983). These differences are all functionally consistent with morphological adaptations in *Saguinus oedipus* for the generation of greater propulsive forces in leaping than are generated by *Saguinus fuscicollis*. In contrast, phenetic analyses of the appendicular skeleton for *Saguinus* and *Callithrix* species suggest that differences in inter-specific scaling patterns of the forearm and shoulder girdle are best resolved as phylogenetic distinctions, rather than arising from differential locomotor behavioral patterns (Falsetti *et al.*, 1989).

Previous results and conclusions have been derived from "static" data, or data limited to fully adult individuals; thus, little is known about how species-specific differences arise developmentally. Analyses of ontogenetic data can provide a means to inspect the pathways along which growth-related changes in form occur and can afford insights into similarities and differences in locomotor adaptations (e.g., Alberch *et al.*, 1979; Aiello, 1981; Jungers & Fleagle, 1980; Shea, 1981; Jungers & Hartman, 1988). Furthermore, inspection of onto-genetic scaling trends may provide a means by which to reconstruct past selective forces (Gould, 1977; Alberch *et al.*, 1979; Jungers & Fleagle, 1980; Shea, 1983, 1984). Therefore, in order to better comprehend these adult interspecific differences and perhaps shed light on how they arise, a comparative investigation of postcranial ontogenetic growth was undertaken for two *Saguinus* species, *S. oedipus* (cotton-top tamarin) and *S. fuscicollis* (saddle-back tamarin) and for *Callithrix jacchus* (common marmoset).

At this point, some explicit research questions may be posed:

(1) Do the proportions of the adult postcranial skeleton of these three species differ significantly?

(2) If adult interspecific postcranial differences exist, what are their ontogenetic histories?

(3) Do proportional differences among taxa reflect functional variations in locomotor and postural behaviors?

Species-specific growth patterns will be compared to evaluate several different hypotheses regarding the origin of adult shape variation. The following models of postnatal growth (Figure 1) are descriptive of the different conditions under which shape differences among the adults of different species can arise:

(1) Model I or divergent growth trajectories: descriptive of a case in which there are no significant shape differences present at birth, with adult shape differences attained through divergent patterns of growth-related shape change.

(2) Model II or parallel allometry: interspecific shape differences present at birth are maintained via similar allometric growth trajectories.

(3) Model III or parallel isometry: adult shape differences are maintained throughout growth via common isometric growth patterns.

(4) Model IV or ontogenetic scaling: adult shape differences arise through change in the rates or duration (timing) of growth along a similar allometric growth trajectory.

Materials and methods

The adult sample consists of 36 total specimens, 12 each of *S. fuscicollis*, *S. oedipus* and *C. jacchus*. The non-adult sample consists of 69 individuals (29 *S. fuscicollis*, 26 *S. oedipus* and 14 *C. jacchus*). Figure 2 provides a graphic display of the age distributions for the immature samples. The criterion for differentiating adults from non-adults is closure of all long bone epiphyses. The sample is a mixture of wild-caught and captive-born animals obtained from the Oak Ridge Associated Universities (ORAU) *Saguinus* collection and from the U.S. National Museum of Natural History (USNM). Previous morphometric analyses of postcranial variation for these species (Glassman, 1983; Falsetti, 1986; Cole *et al.*, 1988; Falsetti *et al.*, 1989) report no significant difference between wild- and captive-born animals; thus, specimens have been pooled within species. Furthermore, since these species exhibit little or no cranial, dental or postcranial sexual dimorphism (Swindler, 1976; Hershkovitz, 1977; Falsetti, 1986; Cole *et al.*, 1988), sexes are pooled for all analyses.

The lengths of the scapula, humerus, radius, ulna, ilium, femur and tibia (Martin, 1928; Trotter & Gleser, 1952; Bass, 1987) are used in the calculation of all functional indices, variance–covariance matrices, and size and shape variables (Table 1). To make adult and non-adult data comparable, all linear measurements of the long limb bones for the adult sample are diaphyseal lengths. These measurements were recorded to the nearest 0·01 mm with sliding calipers.

Comparative analyses of species-specific growth trajectories can aid in determining which of these models (either alone or in combination) best describes the ontogenetic origins of adult shape differences. Species-specific patterns of postcranial growth are examined via a multivariate generalization of the bivariate allometry equation $(y=bx^k)$, following Jolicoeur (1963*a,b*). This method summarizes multivariate proportionality through use of a principal components analysis of the variance–covariance matrix of log-transformed data. If the first principal axis accounts for an overwhelming majority of the total variance, and if all loadings are of the same sign, this axis may be interpreted as a "vector of relative growth" (Sprent, 1972; Shea, 1985). For p variables, multivariate isometry exists when all p variables increase at the same rate as the internally defined size variable (explicitly defined as log-

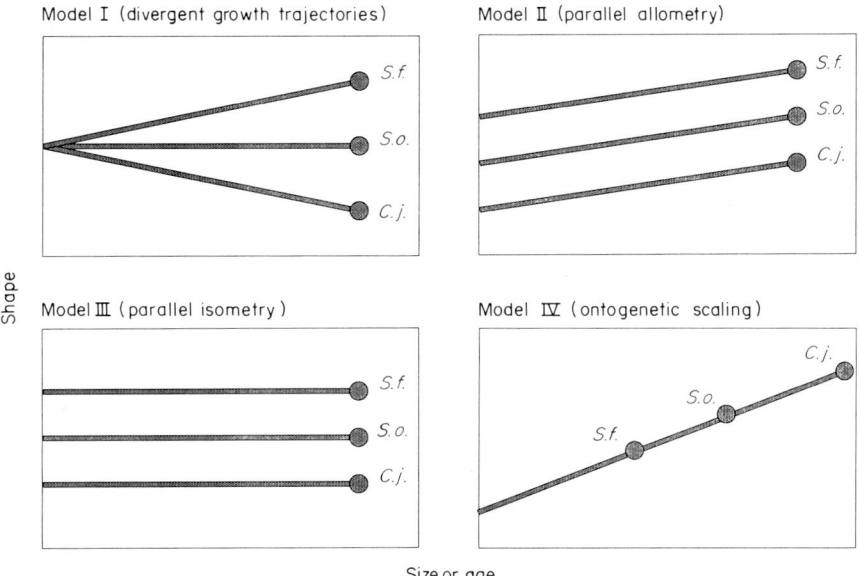

Figure 1. Models of ontogenetic interspecific growth.

transformation of the geometric mean of all variables), such that all p loadings are equal to $p^{-1/2}$. For the seven linear dimensions used in this analysis $p^{-1/2} = 0.377$; all loadings can be divided by this value so that standardized coefficients greater than 1·0 indicate positive allometry and those less than 1·0 indicate negative allometry. Within each species, the null hypothesis is that all seven juvenile postcranial measurements grow at the same rate, thus preserving proportions throughout growth (i.e., a condition of global isometry). Failure to reject the null hypothesis of global isometry would lend support to Model III (parallel isometry); rejection of the null hypothesis would support one of the remaining models, where growth-related shape change is expected. Intraspecific departures from proportional similarity can be examined by calculating an angle θ between the hypothetical (isometric) vector and the observed species-specific vectors of relative growth; the larger the angle, the greater the divergence from overall isometry. A small-sample direction test approximating an F-distribution is used for testing the statistical significance of noted departures from isometry (Jolicoeur, 1984).

Direction cosines and vector angles can also be used to quantify differences among species-specific growth trajectories. More divergent growth patterns are indicated by larger direction cosines and larger values of θ. If growth patterns are highly dissimilar, then Model I (divergent growth trajectories) is suggested as an ontogenetic source of adult shape differences. Similar growth trajectories would suggest one of the remaining models (parallel allometry, parallel isometry or ontogenetic scaling).

Results

Species-specific adult shape variation
Initially, adult interspecific differences in limb proportions were examined via the calculation of the intermembral (humerus + radius)/(femur + tibia), brachial (radius/humerus)

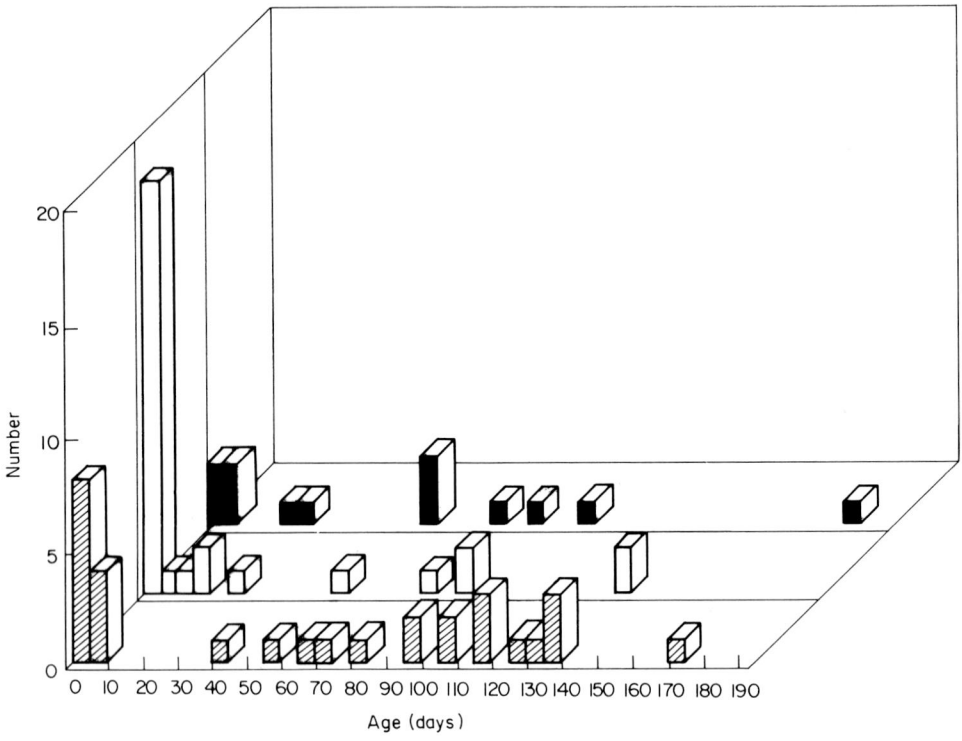

Figure 2. Distribution of immature specimens. 29 *S. fuscicollis* (▨), 26 *S. oedipus* (□) and 14 *C. jacchus* (■).

Table 1 **Forelimb measurements[1] and abbreviations**

Measurement	Abbreviation
Maximum length of the scapula	SML
Humerus maximum length	HML
Ulna maximum length	UML
Radius maximum length	RML
Femur maximum length	FML
Tibia maximum length	TML
Maximum height of the ilium[2]	ILH

[1]See Martin (1928) and Bass (1987) for definitions and illustrations of those measurements.

[2]Often referred to as "length"; this measurement is made from the ischial tuberosity to the iliac crest (Bass, 1987:191).

and crural (tibia/femur) indices. The ratio of ulna to radius lengths was also derived to provide information regarding the relative lengths of the distal elements of the forelimb. Tests for interspecific differences (Table 2) were made using a Bonferroni *t*-test (Sokal & Rohlf, 1981). *Saguinus fuscicollis* was found to have a significantly *higher* intermembral index than

Table 2 Comparisons of adult skeletal proportions in *Saguinus fuscicollis*, *Saguinus oedipus* and *Callithrix jacchus*

Variable	n	S. fuscicollis (S.f.) x	S.D.	S. oedipus (S.o.) x	S.D.	C. jacchus (C.j.) x	S.D.	Bonferroni (Dunn) t-test[1] α<0·05
Intermembral index	12	0·943*	0·003	0·932	0·003	0·933	0·003	S.f. > C.j. > S.o.
Brachial index	12	0·976	0·006	0·972	0·009	0·962*	0·005	S.f. > S.o. > C.j.
Crural index	12	0·996	0·006	0·994	0·002	0·992	0·002	N.S.
Ulna/radius index	12	1·038*	0·005	1·044	0·004	1·043	0·003	S.o. > C.j. > S.f.

[1]Underlined means are *not* significantly different.
*Significantly different.
N.S. = not significant.

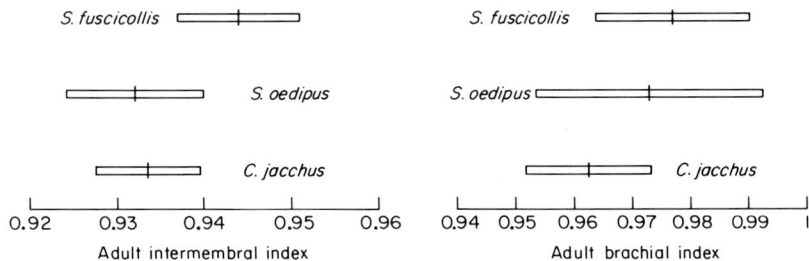

Figure 3. Metric comparison of adult limb proportions in (A) *S. fuscicollis*, (B) *S. oedipus* and (C) *C. jacchus*. Error bars are two standard deviations.

either *S. oedipus* or *Callithrix* (Figure 3). In other words, adult *S. fuscicollis* have longer forelimbs relative to their hindlimbs. *Callithrix* presents a significantly *lower* brachial index, and thus has a shorter radius, relative to the humerus, than either species of *Saguinus* (Figure 3). The crural index showed no heterogeneity among species. Finally, adult *S. fuscicollis* have a significantly longer radius relative to ulna, and therefore a *lower* ulna/radius index (i.e., the ulna is relatively shorter in *S. fuscicollis*). Overall, our results are in general agreement with those of previous studies, suggesting that adult shape differences are present and may be used to distinguish members of these groups. We next examined an ontogenetic series of tamarins and marmosets to determine the mechanisms of growth by which these adult differences are produced.

Species-specific ontogenetic growth

For species-specific multivariate growth allometry, the first principal axis for each species accounts for over 99% of its respective sample variance and all loadings are positive in sign (Table 3). The angle of divergence between isometry and the observed vectors of relative growth is 7·3° for *S. fuscicollis*, 6·3° for *S. oedipus* and 5·3° for *C. jacchus*. For each species, the null hypothesis of overall isometric growth is easily rejected, particularly for *S. fuscicollis*. Therefore, with increasing size and age, the long bones of the **lower** limb and pelvis (femur,

Table 3   **Multivariate growth allometry of the postcranial skeleton in tamarins and marmosets**

Variable[1]	S. fuscicollis (n=29)	S. oedipus (n=26)	C. jacchus (n=14)
SML	0·835	0·844	0·877
HML	0·928	0·920	0·960
UML	0·918	0·953	0·918
RML	0·861	0·899	0·925
FML	1·139	1·125	1·103
TML	1·135	1·116	1·120
ILH	1·126	1·095	1·065
% of variance	0·994	0·997	0·997
Angle of divergence (θ)	7·32°	6·33°	5·37°
Test of isometric null hypothesis[2]	$F=45·62$ $p<0·001$ Reject	$F=101·28$ $p<0·001$ Reject	$F=78·85$ $p<0·001$ Reject

[1]All measurements recorded in millimeters and transformed to natural logarithms.
[2]Test statistic defined in Jolicoeur (1984).

Table 4   **Vector cosines and angles of divergence (θ) between first principal axes**

	S. fuscicollis	S. oedipus
S. oedipus	0·9997 1·40°	
C. jacchus	0·9991 2·43°	0·9995 1·81°

tibia and ilium) grow at a relatively faster rate than the elements of the upper limb and shoulder girdle (scapula, humerus, radius and ulna).

Within the upper extremity itself, the scapula appears to change faster throughout growth than do the brachial elements. Within the lower limbs of the *Saguinus* species, the femur changes faster than the tibia throughout ontogeny. However, growth-related change is particularly strong. In addition, ilium length changes more slowly with respect to overall growth than do the other lower limb segments. The finding of allometric growth patterns in all three taxa is evidence against the hypothesis that isometric growth preserves shape differences that are present from birth (rejecting Model III or parallel isometry).

Table 4 presents a comparison of the intraspecific growth vectors. All three species diverge significantly from isometry; however, they appear to grow along virtually identical allometric trajectories. Greater similarity in growth patterns between *Saguinus* species is indicated by more congruent values of θ. These results suggest that ontogenetic growth trajectories are highly conserved across species. Therefore, the hypothesis of divergent growth trajectories (Model I) can be rejected, leaving parallel allometry (Model II) and ontogenetic scaling (Model IV) as the remaining viable alternatives.

Table 5 Comparisons of immature[1] skeletal proportions in *S. fuscicollis*, *S. oedipus* and *C. jacchus*

Variable	n	*S. fuscicollis (S.f.)* x	S.D.	*S. oedipus (S.o.)* x	S.D.	*C. jacchus (C.j.)* x	S.D.	Bonferroni (Dunn) t-test[2] $\alpha < 0.05$
Intermembral index	10	0.998*	0.009	0.976*	0.006	0.956*	0.001	S.f. > S.o. > C.j.
Brachial index	19	0.985*	0.015	0.955	0.007	0.958	0.004	S.i. > C.j. > S.o.
Crural index	6	1.018	0.015	1.004	0.005	1.012	0.002	S.f. > C.j. > S.o.

[1] Age ≤ 7 days.
[2] Underlined means are significantly different.
*Significantly different.

Models II and IV are distinguished by the presence of significant shape differences at the onset of postnatal growth. In Model II (parallel allometry), shape differences are apparent from birth and are maintained throughout the remainder of growth because of the similar patterns of growth-related shape change. In Model IV (ontogenetic scaling), the taxa would grow along the same allometric trajectory but would differ in shape due to different adult size. Because the starting point for growth and the rate of shape change would be the same, the taxa then could be thought of as "the same animals at different sizes" (*sensu* Pilbeam & Gould, 1974:892).

To test for proportional differences at the onset of postnatal growth, species means of the log-transformed indices were compared for animals aging from zero to 7 days (Table 5). For the intermembral index, all three species express significantly different means from birth. *Saguinus fuscicollis* has the highest intermembral index and maintains this proportional distinction throughout ontogeny with the ratio of upper to lower limb lengths being relatively greater at any given size or age (Figure 4). *Callithrix jacchus* possesses the lowest intermembral index throughout ontogeny, while *S. oedipus* occupies a consistently intermediate position. As previously indicated by the multivariate analysis of relative growth, the intermembral index decreases with increasing size for all three taxa. It is also very interesting to note that these callitrichines depart from the usual non-human primate pattern of growth-related increases in the intermembral index (Jungers, 1977, 1979, 1984, 1985).

For the brachial index, *S. fuscicollis* expresses the highest mean value and is significantly distinct from the other taxa. This suggests that the relative length of the radius is a primary contributor to the overall upper limb *vs.* lower limb variation seen in these species. *Saguinus oedipus* and *C. jacchus* do not, however, exhibit significantly different mean values from each other for this ratio. Crural indices for these groups show relatively little variation, nonetheless, *S. fuscicollis* possesses a significantly greater mean value than either *S. oedipus* or *C. jacchus*, and thus has relatively longer tibia early in ontogeny. Again, *S. oedipus* and *C. jacchus* do not exhibit significant differences, and are thus more similar in shape at the onset of growth.

Because the brachial index differed significantly among these species in the adult sample, the relative growth of its various components were examined in greater detail for the non-adult sample. Reduced major axis regression (RMA; Sokal & Rohlf, 1981) of log radius diaphysis length on the log geometric mean of all seven measurements indicates that the relationship between radius length and overall postcranial size is negatively allometric for all three species (Figure 5 and Table 6). Examination of the differences between slopes suggest

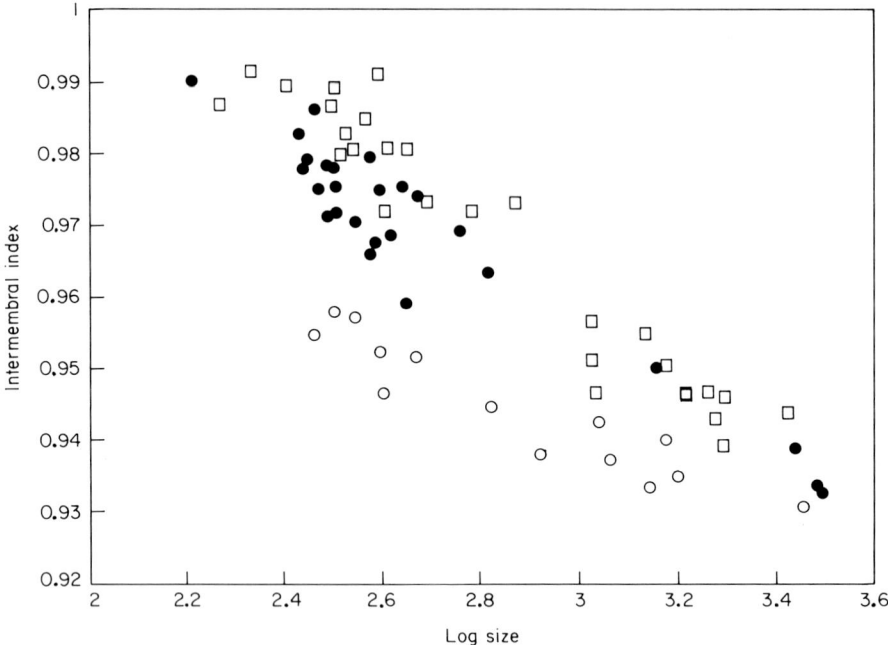

Figure 4. Bivariate log plot of immature intermembral index on size. (□) *S. fuscicollis*; (●) *S. oedipus*; (○) *C. jacchus*.

Table 6 Reduced major axis regression statistics

Variable	Species	n	Slope	95% confidence limits[1]		Intercept	r
Radius length vs. log size	(1) *S. fuscicollis*	28	0·882	0·861	−0·903	0·4503	0·998
	(2) *S. oedipus*	26	0·905	0·888	−0·923	0·3242	0·998
	(3) *C. jacchus*	14	0·930	0·904	−0·956	0·2443	0·997

Clarke t-test slopes[2]		Non-parametric rank comparison intercepts[3,4]			
Group			Group	Q	
$t_{1-2}=1·746$	$P=0·091$		1 vs. 2	8·78	Reject
$t_{1-3}=1·113$	$P=0·007$*		1 vs. 3	15·67	Reject
$t_{2-3}=1·131$	$P=0·116$		2 vs. 3	6·89	Reject

[1] Jolicoeur & Mosimann (1968).
[2] Clarke (1980) RMA slope test.
[3] Non-parametric multiple sample comparison of ranked bootstrap intercept values (Zar, 1984:200; Dunn, 1964).
[4] Q = Mean ranks of group A minus group B divided by S.E. (Dunn, 1964).
*Significant at $P=0·05$.

that *S. fuscicollis* is most strongly negative and that the *Callithrix* slope is the greatest, with *S. oedipus* again taking an intermediate position. In Figure 4, the ontogenetic trajectory for *S. fuscicollis* appears to be transposed above those of *S. oedipus* and *C. jacchus*, indicating that

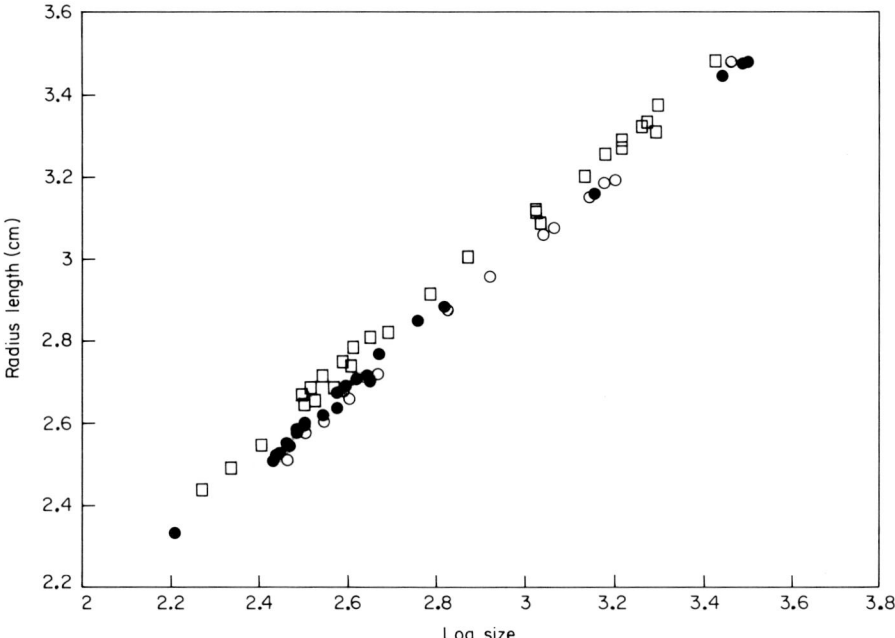

Figure 5. Bivariate log plot of immature radius diaphyseal length on size. Symbols as in Figure 4.

S. fuscicollis radial elongation is present early in postnatal ontogeny. Bootstrap estimates of RMA intercepts were generated following Plotnick (1989), producing a sample of 100 y-intercept estimates for each sample. Dunn's (1964; Zar, 1984) non-parametric multiple comparisons test was then applied to test for differences in the mean ranks of the species estimates. Table 6 shows the results of the Dunn's test indicating that the relative length of the radius is significantly greater in *S. fuscicollis* throughout postnatal ontogeny. These results correspond to the multivariate generalization of allometry, and suggests that the length of the radius, while decreasing with increasing size or age, is nevertheless longer for *S. fuscicollis* at any given age or size.

The ontogenetic model that best describes how adult differences in shape arise is Model II, which describes shape differences present at birth being maintained through similar, but in this case significantly transposed, growth trajectories (Figure 6). It therefore appears as if a significant component of the shape differences measured in this study are present prenatally and that postnatal growth serves to preserve them. This pattern is consistent with previous results reported for orang-utans, siamangs and gibbons (Jungers & Hartman, 1988; Jungers & Cole, 1992).

Discussion

Field studies of free-ranging members of these species suggest that adult proportional differences are associated with variations in their locomotor and positional behaviors. *Saguinus oedipus* is essentially a quadrupedal walker and runner, frequently performing series of long and acrobatic leaps [Erikson, 1963; Napier & Napier, 1967; Hershkovitz, 1977; Garber,

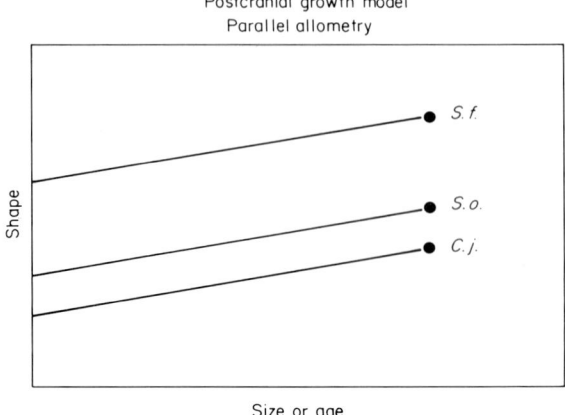

Figure 6. Proposed model of postcranial ontogenetic interspecific growth (parallel allometry).

1980, 1988; see also Fleagle & Mittermeier (1980) for accounts of similar behavior in *S. midas*]. These behaviors are all functionally consistent with their relatively *low* intermembral index. A low intermembral index indicates greater hindlimb length relative to that of the forelimb, facilitating the animal's ability to generate greater propulsive forces in leaping and "springing" (Erikson, 1963; Napier & Napier, 1967; Fleagle, 1977), because increasing the overall length of the hindlimb increases the distance over which accelerating forces can be expressed (Rodman, 1979).

Current field studies conducted by Garber suggest that *S. fuscicollis* devotes proportionally more time locating and consuming insects than its congeners *S. geoffroyi* and *S. mystax*. In these pursuits, a greater amount of time is spent clinging to vertical supports in the undercanopy, while feeding on gums and foraging for insects (Garber, 1988, 1991; see also Terborgh, 1983). *Saguinus fuscicollis* performs these behaviors regardless of the season, while other tamarins have moved on to other more readily available foodstuffs. Furthermore, Garber (1991) noted that the support preference for *S. fuscicollis* is most often greater than 10 cm, compared to more frequent use of smaller support by its congeners, *S. mystax* and *S. geoffroyi*. The significantly *higher* intermembral index exhibited by *S. fuscicollis* would facilitate this behavioral repertoire by lengthening the upper limb relative to the hindlimb. This would allow for positioning the body's center of gravity relatively closer to a horizontal substrate and increase the span of the upper limbs, increasing clinging ability on large vertical trunks (Jungers, 1979).

A similar condition has been reported for the American pygmy squirrels *Microsciurus* and *Sciurillus* (Thorington & Thorington, 1989). These species possess relatively longer limbs than their larger-sized relative *Sciurus* (the gray squirrel; Thorington, 1972). The longer forelimb provides the smaller animals a greater lateral reach around large supports (Figure 7). A greater reach around a trunk is beneficial for increasing the normal component of adductive force, and thus increases the efficiency by which a smaller animal can gain a wider purchase (Cartmill, 1974, 1985; Thorington & Thorington, 1989). The observed forelimb elongation in *S. fuscicollis* could thus be part of an overall adaptation to vertical climbing and clinging. Elongation of the forearm is an efficient method by which to lengthen the entire upper limb and could be viewed as an adaptation similar to that of small squirrels that climb

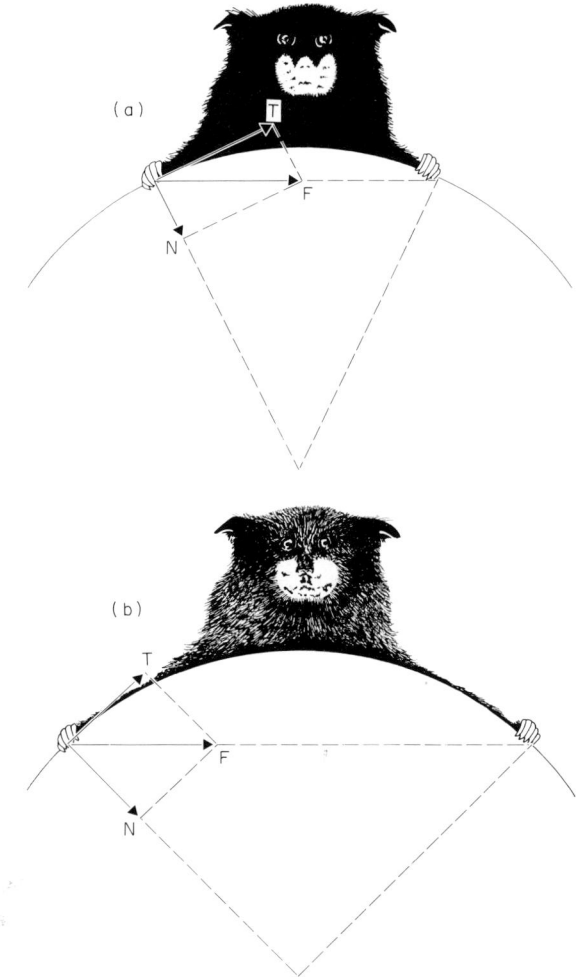

Figure 7. Schematic of forces generated by a climbing tamarin. F is a given adductive force which may be resolved into two components; T is tangential to the support surface and N is normal or perpendicular to the surface. In (a), the normal component is relatively smaller than the tangential component. When the length of the forearm is increased (b), the proportional of the normal component is increased. Vector diagrams after Thorington & Thorington (1989).

large vertical trunks. In concert with the presence of claw-like nails, this small-bodied animal could increase its clinging ability with an elongated forelimb, and at the same time bring its lower limb closer to the center of gravity on a vertical substrate (Cartmill, 1985). Thus, this small-bodied animal, by increasing its lateral reach through forelimb elongation, could be equally competent on wide substrates.

We recognize the tentative nature of this hypothesis with regard to the functional importance of forelimb elongation as a means of increasing clinging ability. An alternative hypothesis suggests that the possession of relatively longer forelimbs is related to the complex suite of locomotor behaviors necessary for trunk-to-trunk leaping. Generally, small-bodied primates who utilize trunk-to-trunk modes of locomotion require long hindlimbs to generate the propulsive forces necessary to accomplish the movement (Peters & Preuschoft, 1984; Demes & Gunther, 1989). Concomitantly, the forelimb may be used as a means of initiating body rotation and as a decelerating mechanism (Peters & Preuschoft, 1984). In a comparison with *S. geoffroyi* and *S. mystax*, Garber (1991) argued that the relatively long forelimbs in *S. fuscicollis* are possibly related to increasing the braking distance during observed trunk-to-trunk leaps. He reported that in the undercanopy, trunk-to-trunk leaping accounts for 19·2% of all leaps for *S. fuscicollis*. We concur that this is a viable hypothesis and that "filmed sequences and biomechanical analyses" are required in order to elucidate the functional significance of forelimb elongation in *S. fuscicollis* relative to the other callitrichines (Garber, 1991:227).

Understanding the adaptational significance of forelimb elongation in *S. fuscicollis* is a complex process which is most certainly related to its behavioral repertoire, which is in turn affected by substrate usage (including support size), ecological niche and socialization. *Callithrix jacchus*, one of the smallest callitrichines (measured by body weight), spends a considerable amount of time clinging to large vertical supports in search of exudates and insects, but does not exhibit the extreme postcranial morphology of *S. fuscicollis*. Conceivably, it is the *size* of the substrate that is the most crucial element. More generally, it appears the unique shape and elongation of the forelimb in *S. fuscicollis*, which lacks the specialized incisor morphology of the primary exudate feeders (including *C. jacchus*), is part of the suite of characteristics which enable it to be regarded as one of the most behaviorally flexible of the tamarin species in terms of geographical distribution, socialization and niche exploitation (Kinzey, 1982; Garber, 1989, in press).

Summary

The overall pattern of growth that ultimately leads to significant interspecific differences in adult proportions in the study taxa can best be described as "parallel allometry". Patterns of growth-related change in shape tend to be conserved, varying little among taxa. These patterns tend to preserve shape differences that are apparently present from birth. Adult differences in postcranial size and shape primarily distinguish *S. fuscicollis* from both *S. oedipus* and *C. jacchus*. The most important of these distinctions are the intermembral and brachial indices, which may be related to varying position and locomotor behaviors. Perhaps, the most significant trend is that the radius of *S. fuscicollis* is longer, relative to body size, at any given age. This morphological pattern is best seen as an overall adaptation for increasing the ability to exploit varied ecological niches which are characterized by varied substrates and foodstuffs and which require a wide range of locomotor skills to successfully navigate.

Acknowledgements

We would like to extend thanks to Matthew J. Ravosa and Anne M. Gomez for inviting us to contribute to their symposium on *Ontogenetic Perspectives on Primate Evolutionary Biology*. Richard W. Thorington, Jr (U.S. National Museum of Natural History) and Suzette Tardif (Marmoset Research Center of the Oak Ridge Associated Universities and University of Tennessee-Knoxville) are thanked for access to their collections. Paul A. Garber is especially thanked for his correspondence and critique of this manuscript. We also thank William L. Jungers, Richard W. Thorington Jr, Daniel O. Schmitt, Susan Mitnick Falsetti and

Maria S. Cole for their helpful comments on earlier drafts of this manuscript. Catherine Sexton adapted Figure 7 with permission from C. M. Thorington.

This is contribution No. 812 in Ecology and Evolution from the State University of New York at Stony Brook. Partial funding for this research was provided by NSF Grant BNS 9020562.

References

Aiello, L. C. (1981). The allometry of primate body proportions. *Symp. Zool. Soc. (Lond.)* **48,** 331–358.

Alberch, P., Gould, S. J., Oster, G. & Wake, D. (1979). Size and shape in ontogeny and phylogeny. *Paleobiology* **5,** 296–317.

Bass, W. M. (1987). *Human Osteology: A Laboratory and Field Manual*, 3rd edn. Special Publication No. 2. Columbia: Missouri Archaeological Society.

Cartmill, M. (1974). Pads and claws in arboreal locomotion. In (F. A. Jenkins, Ed.) *Primate Locomotion*, pp. 45–83. New York: Academic Press.

Cartmill, M. (1985). Climbing. In (M. Hildebrand, D. M. Bramble, K. E. Liem & D. B. Wake, Eds) *Functional Vertebrate Anatomy*, pp. 73–88. New York: Harvard University Press.

Clarke, M. R. B. (1980). The reduced major axis of a bivariate sample. *Biometrika* **67,** 441–446.

Cole, T. M., Falsetti, A. B. & Cole, M. S. (1988). Relationships between body size and dental, cranial, and postcranial variables in saddle-backed tamarins. *Am. J. phys. Anthrop.* **75,** 197.

Demes, B. & Gunther, M. M. (1989). Biomechanics and allometric scaling in primate locomotion and morphology. *Folia primat.* **53,** 125–141.

Dunn, O. J. (1964). Multiple contrasts using ranked sums. *Technometrics* **6,** 241–252.

Erickson, G. E. (1963). Brachiation in New World monkeys and in anthropoid apes. *Symp. Zool. Soc. (Lond.)* **10,** 135–164.

Falsetti, A. B. (1986). Allometric variation of the postcranial skeleton in two South American tamarins *Saguinus oedipus oedipus* and *Saguinus fuscicollis illigeri* (Callitrichidae, Primates). M.A. Dissertation, Department of Anthropology, The University of Tennessee.

Falsetti, A. B., Cole, T. M. & Jungers, W. L. (1989). Size and shape of the forelimb of tamarins. *Am. J. phys. Anthrop.* **78,** 219.

Fleagle, J. G. (1977). Locomotor behavior and skeletal anatomy of sympatric Malaysian leaf-monkeys (*Presbytis obscura* and *Presbytis melalophos*). *Yearb. phys. Anthrop.* **20,** 440–453.

Fleagle, J. G. & Mittimeier R. A. (1980). Locomotor behavior, body size, and comparative ecology of seven Surinam monkeys. *Am. J. phys. Anthrop.* **52,** 301–314.

Garber, P. A. (1980). Locomotor behavior and feeding ecology of the Panamanian tamarin (*Saguinus oedipus geoffroyi*, Callitrichidae, Primates). *Int. J. Primatol.* **1,** 185–201.

Garber, P. A. (1988). Diet, foraging patterns, and resource defense in a mixed species troop of *Saguinus mystax* and *Saguinus fuscicollis* in Amazonian Peru. *Behavior* **105,** 18–34.

Garber, P. A. (1989). Role of spatial memory in primate foraging patterns: *Saguinus mystax* and *Saguinus fuscicollis*. *Am. J. Primat.* **19,** 203–216.

Garber, P. A. (1991) A comparative study of positional behavior in three species of tamarin monkeys. *Primates* **32,** 219–230.

Garber, P. A. (in press). Feeding ecology and behaviour of the genus *Saguinus*. In (A. B. Rylands, Ed.) *Marmosets and Tamarins: Systematics, Ecology, and Behaviour*. Oxford: Oxford University Press.

Glassman, D. M. (1983). Functional implications of skeletal diversity in two South American tamarins. *Am. J. phys. Anthrop.* **61,** 291–298.

Gould, S. J. (1977). *Ontogeny and Phylogeny*. Cambridge: Harvard University Press.

Hershkovitz, P. (1977). *Living New World Monkeys* (Platyrrhini), Vol. I. Chicago: University of Chicago Press.

Jolicoeur, P. (1963a). The degree of generality of robustness in *Martes americana*. *Growth* **27,** 1–27.

Jolicoeur, P. (1963b). The multivariate generalization of the allometry equation. *Biometrics* **19,** 497–499.

Jolicoeur, P. (1984). Principal components, factor analysis, and multivariate allometry: a small sample direction test. *Biometrics* **40,** 685–690.

Jolicoeur, P. & Mosimann, J. E. (1968). Intervalles de confiance pour la pente de l'axe majeur d'une distribution normale bidimensionnelle. *Biometrie-praximetrie* **9,** 121–140.

Jungers, W. L. (1977). Hindlimb and pelvic adaptations to vertical climbing and clinging in *Megaladapis*, a giant subfossil prosimian from Madagascar. *Yearb. phys. Anthrop.* **20,** 508–524.

Jungers, W. L. (1979). Locomotion limb proportions and skeletal allometry in lemurs and lorises. *Folia primat.* **32,** 8–28.

Jungers, W. L. (1984). Scaling of the hominoid locomotor skeleton with special reference to the lesser apes. In (H. Preuschoft, D. Chivers, W. Brockelman & N. Creel, Eds) *The Lesser Apes: Evolutionary and Behavioral Biology*, pp. 146–169. Edinburgh: Edinburgh University Press.

Jungers, W. L. (1985). Body size and scaling of limb proportions in primates. In (W. L. Jungers, Ed.) *Size and Scaling in Primate Biology*, pp. 345–381. New York: Plenum Press.

Jungers, W. L. & Cole, M. S. (1992). Relative growth and shape of the locomotor skeleton in lesser apes. *J. hum. Evol.* **23,** 93–105.

Jungers, W. L. & Fleagle, J. G. (1980). Postnatal growth allometry of the extremities in *Cebus albifrons* and *Cebus apella*: a longitudinal and comparative study. *Am. J. phys. Anthrop.* **53,** 471–478.

Jungers, W. L. & Hartman, S. E. (1988). Relative growth of the locomotor skeleton in orang-utans and other large-bodied hominoids. In (J. H. Schwartz, Ed.) *Orang-utan Biology*, pp. 347–359. Oxford: Oxford University Press.

Kinzey, W. G. (1982). Distribution of primates and forest refuges. In (G. T. Prance, Ed.) *Biological Diversification in the Tropics*, pp. 455–482. New York, Columbia University Press.

Martin, R. (1928). *Lehrbuch der Anthropologie, Zeiter Band: Kraniologie, Osteologie*. Jena, Germany: Verlag Gustav Fischer.

Moynihan, M. (1970). Some behavioral patterns of platyrrhine monkeys. II. *Saguinus geoffroyi* and some other tamarins. *Smithsonian Contributions to Zoology*, No. 28. Smithsonian Institution Press, Washington D.C.

Napier, J. R. & Napier, P. H. (1967). *A Handbook of Living Primates*. London: Academic Press.

Peters, A. & Preuschoft, H. (1984). External biomechanics of leaping in *Tarsius* and its morphological and kinematic consequences. In (C. Niemitz, Ed.) *Biology of Tarsiers*, pp. 227–255. New York: Gustav Fischer.

Pilbeam, D. & Gould, S. J. (1974). Size and scaling in human evolution. *Science* **186,** 892–901.

Plotnick, R. E. (1989). Application of bootstrap methods to reduced major axis line fitting. *Syst. Zool.* **38,** 144–153.

Rodman, P. S. (1979). Skeletal differentiation of *Macaca fasicularis* and *Macaca nemestrina* in relation to arboreal and terrestrial quadrupedalism. *Am. J. phys. Anthrop.* **51,** 51–62.

Shea, B. T. (1981). Relative growth of the limbs and trunk of the African apes. *Am. J. phys. Anthrop.* **56,** 179–202.

Shea, B. T. (1983). Paedomorphosis and neoteny in the pygmy chimpanzee. *Science* **222,** 521–522.

Shea, B. T. (1984). An allometric perspective on the morphological and evolutionary relationships between pygmy (*Pan paniscus*) and common (*Pan troglodytes*) chimpanzees. In (R. L. Susman, Ed.) *The Pygmy Chimpanzee: Evolutionary Morphology and Behavior*, pp. 89–130. New York: Plenum Press.

Shea, B. T. (1985). Bivariate and multivariate growth allometry: statistical and biological considerations. *J. Zool. (Lond.)* **206,** 367–390.

Sokal, R. R. & Rohlf, F. J. (1981). *Biometry*, 2nd edn. San Francisco: W. H. Freeman and Co.

Sprent, P. (1972). The mathematics of size and shape. *Biometrics* **28,** 23–27.

Swindler, D. R. (1976). *Dentition of Living Primates*. New York: Academic Press.

Terborgh, J. (1983). *Five New World Primates*. Princeton: Princeton University Press.

Thorington, R. W. (1972). Proportions and allometry in the gray squirrel, *Sciurus carolinensis*. *Nemouria* **8,** 1–17.

Thorington, R. W. & Thorington, E. M. (1989). Postcranial proportions of *Microsciurus* and *Sciurillus*, the American pygmy squirrels. *Adv. Neotrop. Mamm.* **xx,** 125–136.

Trotter, M. & Gleser, G. C. (1952). Estimation of stature from long bones of American whites and Negroes. *Am. J. phys. Anthrop.* **19,** 213–227.

Zar, J. H. (1984). *Biostatistical Analysis*, 2nd edn. Englewood Cliffs, N.J.: Prentice-Hall, Inc.

William L. Jungers
& Maria S. Cole

Department of Anatomical Sciences, School of Medicine, State University of New York at Stony Brook, Stony Brook, NY 11794-8081, U.S.A.

Received 1 July 1991
Revision received 22 January 1992 and accepted 24 January 1992

Keywords: lesser apes, relative growth, evolution, postcranial skeleton.

Relative growth and shape of the locomotor skeleton in lesser apes

This study addresses the question of whether the siamang can be accurately described as an overgrown or peramorphic version of the gibbon in terms of its locomotor skeleton. Lesser ape relative growth and skeletal design are also placed in a broader comparative context of all extant hominoids. With respect to the *lar*-gibbon and the siamang, the answer to this question is clearly negative; *lar*-gibbon ontogenetic trajectories do a poor job of predicting siamang body shape. Although there are some similarities in postnatal relative growth between the two, transpositional differences are most common. Siamang and *lar*-gibbon never overlap in overall shape regardless of size/age. This finding may be related to the highly derived position of the *lar*-gibbons relative to the evolutionary affinities of other lesser apes. The possibility cannot be ruled out that other gibbons such as *Hylobates hoolock*, *H. concolor* and possibly *H. klossi* are ontogenetically scaled-down (truncated) versions of the siamang. When all hominoids are considered together, patterns of relative growth in the postcranium appear to be much more closely related to locomotor function than to phylogenetic affinities.

Journal of Human Evolution (1992) **23**, 93–105

Introduction

The fundamental relationship between development and evolution has long been appreciated (Gould, 1977, 1988). Recent work has served to reaffirm the great significance of this pervasive link (Alberch *et al.*, 1979; Bonner, 1982; McKinney, 1988; Raff & Wray, 1989), including major contributions in the areas of biological anthropology and primatology (e.g., Gould, 1977; Shea, 1981, 1983, 1988; Cheverud, 1982; Cheverud & Richtsmeier, 1986; Jungers & Hartman, 1988; Ravosa, 1991). Most analyses to date have focused on the development and evolution of cranial form, but the application of a heterochronic framework to the study of the locomotor skeleton has also proved fruitful (Lumer, 1939; Shea, 1981; Buschang, 1982; Jungers & Susman, 1984; Jungers & Hartman, 1988).

In the present study, we extend this type of inquiry to a consideration of relative growth and skeletal design in the postcranium of lesser apes (Hylobatidae). Our primary focus will be on the smallest (*Hylobates lar*) and largest (*H. syndactylus*) representatives of this group. The latter is roughly twice the body mass of the former (approximately 11 kg for the siamang compared to under 6 kg for gibbons; Jungers, 1984) and proportional differences between the two are well known (Schultz, 1933; Jungers, 1984). It also seems probable that the siamang takes several years longer to reach maturity (Geissmann, 1991). In very general terms, therefore, the question we address is whether the siamang can be considered an overgrown or ontogenetically scaled-up version of its smaller congener. This issue could be framed in the converse way; perhaps gibbons are scaled-down versions of the siamang. In a real sense, how the question is phrased is rather arbitrary because we must assume that one of the two represents the ancestral condition (Fink, 1982) if we are to derive either through allometric progenesis or hypermorphosis (*sensu* McKinney, 1988). According to the definitive study on hylobatid systematics by Creel & Preuschoft (1984), the siamang is characterized by a higher frequency of primitive character states in comparison to members of the *lar*-gibbon group (*contra* Schultz, 1933); we will return to this last point in the discussion section. Because *lar*-gibbons are perhaps the most derived group of lesser apes, we also cannot rule out the possibility that both the *lar*-gibbon and siamang are derived from some other gibbon morph for which we have no ontogenetic data.

In terms of skull size and shape, Shea's work (1988:240) on hylobatids would answer our question largely in the affirmative: "most of the shape differences between the gibbons and the siamangs appear to result from ontogenetic scaling, and siamangs may be described as peramorphic gibbons". Lumer's (1939) analysis of relative growth of the limb bones in non-human hominoids had also hinted at this possibility for hylobatid extremities; he concluded that relative growth within the limbs of lesser apes is essentially the same for all species, whereas interlimb scaling deviated markedly between siamang and the gibbons. On balance, he asserted that "the evolution of the Hylobatidae has involved chiefly mutations affecting adult body size but not relative growth of the limb segments" (Lumer, 1939:391). It should be noted that Lumer's sample was limited to a rather small number of non-adult cadavers (he used only one juvenile siamang point in his interlimb bivariate plot!).

We propose to re-examine this question with larger samples of non-adult skeletal material, by addition of information on the pectoral and pelvic girdles and by application of alternative methods including multivariate analyses of size and shape. In addition to addressing the question of whether or not the siamang is an overgrown or "peramorphic" (McNamara, 1986) version of the gibbon in its locomotor skeleton, we hope to better document the extent to which differences in postnatal growth contribute to the observed differences in adult shape that distinguish these two species. In addition, we will place relative growth in lesser apes into a broader comparative context of growth-related shape changes in larger bodied hominoids.

Materials and methods

We have chosen Carpenter's white-handed gibbon (*H. lar carpenteri*; Groves, 1968) as the representative of the *lar*-group for this study, because it is characterized by the lowest average adult body mass of any natural gibbon group (5·5 kg; Schultz, 1944) and an extensive ontogenetic series of wild-shot skeletal material exists from the collections of the Asian Primate Expedition, housed primarily in the Museum of Comparative Zoology (Harvard University). Our non-adult sample size is 40, 34 of which possessed records of body mass. Specimens of non-adult siamang with "complete" postcranial skeletons are surprisingly rare in European and U.S. museums, but we have done somewhat better than Lumer (1939) in locating 14 individuals, most of which were wild-collected. By "complete", we mean that the specimens were sufficiently complete to permit the full suite of nine linear measurements discussed below. None of the non-adult siamang had body mass data; an additional six adult siamang skeletons with wild-collected records of body mass were also measured using non-adult landmarks (see below).

The nine linear variables sample broadly from the appendicular skeleton and include lengths of four long bones (humerus, radius, femur and tibia), two measures on the pectoral girdle (clavicle length and spinal axis "breadth" of the scapula) and three from the bony pelvis (ilium length, pubis length and ischial shank length). These measurements were designed to permit the inclusion of bones lacking epiphyses (figured in Jungers & Susman, 1984), thereby extending the sampled size range down to small infants. Long bone lengths are therefore diaphyseal lengths, iliac height excluding the epiphyseal crest and ischial length excluding the tuberosity.

A variety of statistical methods were enlisted to describe and compare patterns of relative growth. Initially, we extrapolated the gibbon's ontogenetic bivariate relationship of each variable relative to body mass into the adult siamang size range. Such extrapolations were examined to determine the degree to which the siamang condition was predicted accurately by an extension of gibbon ontogeny. The reduced major axis (RMA; Ricker, 1984; Rayner, 1985) line-fitting technique was employed for each extrapolation. Selected ontogenetic RMA trends were then compared pair-wise (e.g., humerus on femur) for the nine log-transformed linear variables in order to test for differences in slopes and elevations; Clarke's (1980) t-test was used to compare slopes and Tsutakawa & Hewett's (1977) two-sample "quick test" was employed to test for differences in elevations (i.e., for transpositions). The latter test gave identical results to Plotnick's (1989) bootstrap method for RMA analyses of covariance.

We next examined the overall pattern of relative growth in each ontogenetic sample by application of Jolicoeur's (1963) multivariate generalization of the allometry equation. This principal components approach is based on the relative loadings (directional cosines) of the first eigenvector of the logged variance–covariance matrix of each species. Departures from overall or global isometry (i.e., no shape change) can be assessed by a small-sample direction test (Jolicoeur, 1984). This method implicitly selects the geometric mean of all nine variables as "size" (Mosimann & James, 1979).

Interspecific patterns of size and shape were subsequently examined using Darroch & Mosimann's (1985) technique based on the principal components of "log size-and-shape" (i.e., logged raw data) *vs.* "log shape" only. Size is defined here explicitly as the geometric mean and dimensionless shape variables are constructed as the ratio of each measurement to the geometric mean. This method is closely related to that of Jolicoeur (Jungers *et al.*, 1988). In this system, shape may or may not be correlated with size, and this can be determined empirically. We wish to point out that this approach yields identical results to a version (Rohlf & Bookstein, 1987) of Burnaby's (1966) size-invariant method, where an isometric vector is swept from the logged data. This procedure adjusts for differences in isometric size, but avoids the undesirable effect (seen in residuals from empirical regression lines) of discarding important shape information simply because it is correlated with size (cf. Jungers, 1988; Bookstein, 1989).

Finally, we compare the within-group patterns of multivariate relative growth in the two species of lesser apes to those trends already documented for larger-bodied hominoids (Jungers & Hartman, 1988). This was accomplished by first computing correlations between the species-specific eigenvectors of relative growth and then by summarizing these correlations via clustering algorithms (Rohlf, 1990).

Results

The results of extrapolating bivariate ontogenetic scaling trends of logged variables on log mass in *lar*-gibbons into the adult siamang size range are summarized in Table 1. In some cases (clavicle breadth, ilium length and pubis length), the adult siamang condition was accurately predicted by gibbon lines of extrapolation (Figure 1). Most of the time, however, the gibbon ontogenetic base lines overpredicted values for siamang (Figure 2). The hypothetical animal that would result from the combined effects of these extrapolations, therefore, would not look terribly much like an adult siamang (or any other known lesser ape for that matter). Six of the nine variables scale significantly allometrically in *lar*-gibbon, indicating that, relative to body mass, growth introduces predictable proportional changes in the

Table 1 **Extrapolation of *Hylobates lar* ontogenetic scaling (reduced major axis) into the adult *Hylobates syndactylus* size range**

Log variable on log body mass	Equation slope/intercept/r	Extrapolation
Humeral diaphysis	0·368*/4·7967/0·983	overpredicts
Radial diaphysis	0·410*/4·8277/0·987	overpredicts
Femoral diaphysis	0·353/4·6556/0·982	overpredicts
Tibial diaphysis	0·372*/4·4770/0·983	overpredicts
Scapula breadth	0·361*/3·5953/0·989	overpredicts
Clavicle breadth	0·358/3·8389/0·978	good fit
Ilium length	0·392*/3·8642/0·988	good fit
Pubis length	0·392*/2·8467/0·977	good fit
Ischium length	0·360/2·8676/0·974	overpredicts

$n=34$ for *H. lar* and $n=6$ for adult *H. syndactylus*.
*95% Confidence limits do not include isometry (0·333).

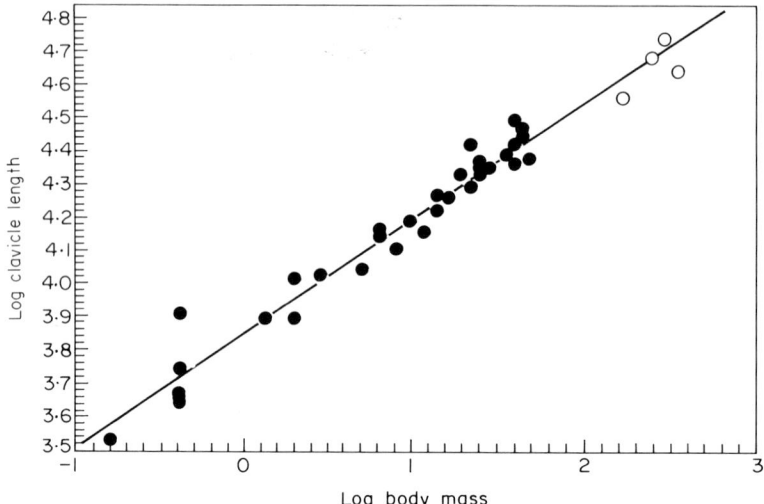

Figure 1. Bivariate scatter of log clavicle length on log body mass. (●) Gibbons; (○) adult siamang. The reduced major axis line for the gibbon sample accurately predicts clavicle length in siamang. Gibbon ontogeny: $\log y = 0.358 \log x + 3.83895$, $r = 0.978$.

locomotor skeleton, but that these changes do not culminate in the overall body shape seen in siamang.

Of the 36 possible pair-wise ontogenetic scaling relationships among the nine variables, we present the results of eight such trends in lesser apes (Table 2). We discovered only one relationship that could be described best as ontogenetic extrapolation of gibbons into siamang: tibia diaphyseal length on femoral diaphyseal length (Figure 3). Neither slope nor elevation differed between these two congeners in this case. In both species, the crural index increases slightly and in similar fashion (the slope is significantly positively allometric). In all

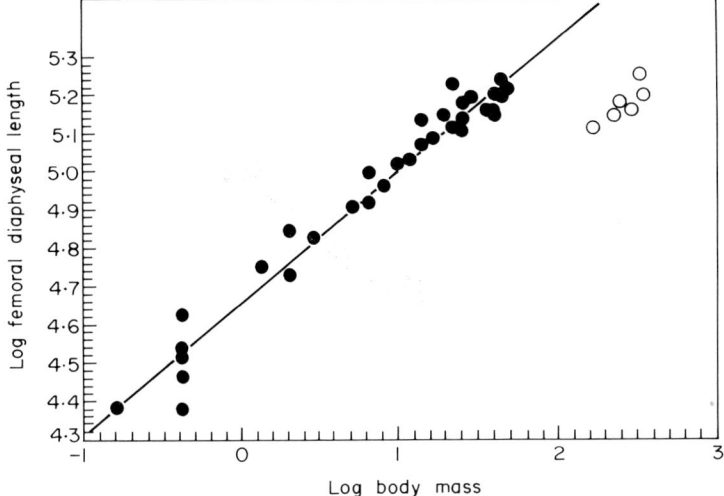

Figure 2. Bivariate scatter of log femoral diaphyseal length on log body mass. Symbols as in Figure 1. The gibbon trajectory greatly overestimates femoral length in the adult siamang. Gibbon ontogeny: $\log y = 0.353 \log x + 4.65565$, $r = 0.978$.

Table 2 Pair-wise variable scaling (reduced major axis) in ontogenetic series of *Hylobates lar* (LAR) and *Hylobates syndactylus* (SYN)

Variable pair ($\log Y$ on $\log X$)	*H. lar* ($n=40$) slope/intercept/r	*H. syndactylus* ($n=14$) slope/intercept/r	Slope differences[1]	Elevation differences[2]
Radial diaphysis on humeral diaphysis	1·114*/−0·5105/0·995	1·136*/−0·5962/0·993	NS	SYN > LAR
Tibial diaphysis on femoral diaphysis	1·050*/−0·4024/0·995	1·112*/−0·7224/0·990	NS	NS
Humeral diaphysis on femoral diaphysis	1·039*/−0·0431/0·996	0·994/0·2837/0·987	NS	SYN > LAR
Clavicle length on humeral diaphysis	0·948/−0·6966/0·983	1·003/−0·8571/0·992	NS	SYN > LAR
Scapula breadth on clavicle length	1·071/−0·5454/0·975	1·004/−0·4147/0·972	NS	LAR > SYN
Ischium length on ilium length	0·927*/−0·7190/0·980	0·917/−0·7761/0·962	NS	LAR > SYN
Pubis length on ischium length	1·091*/−0·2921/0·978	1·260*/−0·6913/0·964	NS	SYN > LAR
Pubis length on femoral diaphysis	1·112*/−2·3452/0·979	1·239*/−2·7393/0·964	NS	SYN > LAR

[1] t-test for reduced major axis slopes (Clark, 1980).
[2] "Quick test" for independence in 2×2 matrix (Tsutakawa & Hewett, 1977).
*95% Confidence limits do not include isometry (1·0).
NS = not significant ($P > 0.05$). > = significantly transposed above ($P < 0.05$).

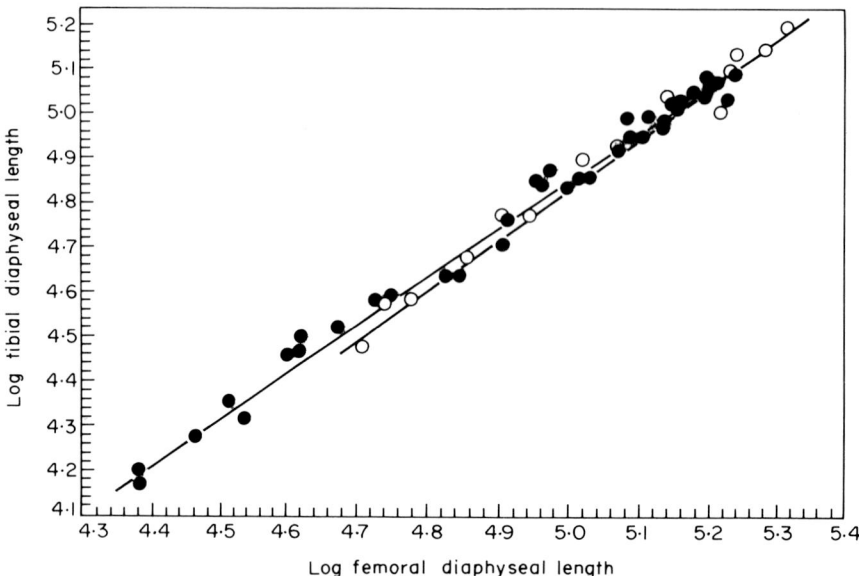

Figure 3. Ontogenetic scaling of log tibial diaphyseal length on log femoral diaphyseal length in the gibbon (●) and siamang (○). Neither the slopes nor the elevations of the RMA lines are significantly different. Gibbon: $\log y = 1\cdot050 \log x - 0\cdot40241$, $r = 0\cdot995$; siamang: $\log y = 1\cdot112 \log x - 0\cdot72236$, $r = 0\cdot990$.

other instances examined, siamang do not appear to be overgrown gibbons. Although no significant slope differences were disclosed, significant transpositions appeared to be the rule. For example, at any given femur length the humerus is predictably longer in the siamang (Figure 4). Bivariate scaling patterns were predominantly allometric and were similar in both species, but postnatal starting points were usually quite different.

The results of the multivariate investigation into patterns of relative growth are presented in Table 3. For each species the first eigenvector of the principal components analyses accounted for the overwhelming majority of the variance. Coefficients are scaled so that isometry is indicated by a value of $1\cdot0$. The angle between an isometric vector and the observed growth vector in gibbons in only 3°, but this departure from global isometry is still significant. The angular departure is slightly greater in the siamang sample and approaches significance ($P < 0\cdot07$) despite the relatively small number of individuals. In other words, subtle but predictable growth-related changes characterize both species. As might be expected from the bivariate scaling results summarized above, there is considerable similarity, but not identity, between the growth vectors; the correlation coefficient between them is $0\cdot8$. In both species, pubis and radius lengths are positively allometric ($> 1\cdot0$), whereas femoral, clavicular, scapular and ischial dimensions are all negatively allometric ($< 1\cdot0$).

The first two principal components of log size-and-shape in the pooled interspecific analysis account for almost all of the variance, with axis 1 explaining $97\cdot0\%$ (Table 4). All loadings are positive on axis 1, and scores along this axis obviously reflect overall size to a great degree (Figure 5). Axis 2 appears to summarize various bivariate transpositions (Shea, 1985) and serves to separate the two species. Gibbons are removed from siamang by virtue of relatively longer femora and tibiae and relatively shorter pubis, ilia and clavicles.

The total variance in the log shape analysis is only $3\cdot7\%$ of that seen in its log size-and-shape counterpart. However, removal of isometric size differences does leave sufficient

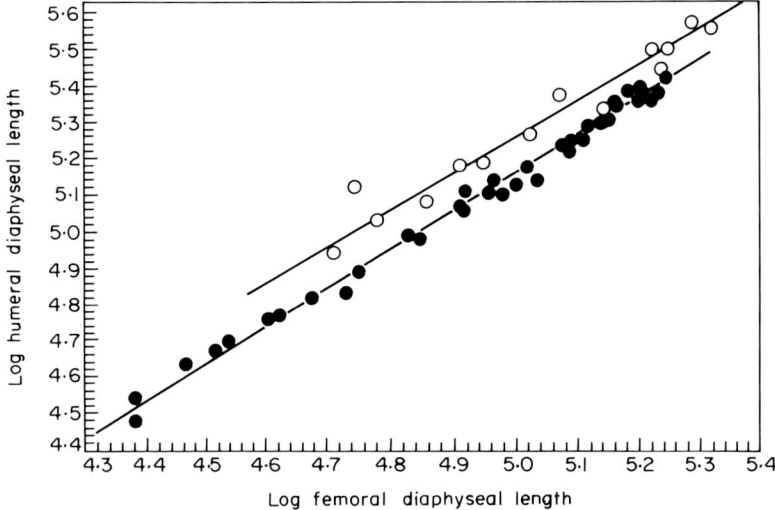

Figure 4. Ontogenetic scaling of log humeral diaphyseal length on log femoral diaphyseal length in the gibbon (●) and siamang (○). The slopes are not significantly different, but the siamang trajectory is significantly transposed above that of the gibbon. Siamang: $\log y = 0.994 \log x + 0.28373$, $r = 0.987$; gibbon: $\log y = 1.039 \log x - 0.04309$, $r = 0.996$.

Table 3 **Multivariate growth allometry of hylobatid postcrania**

Variable	*Hylobates lar* ($n=40$)	*Hylobates syndactylus* ($n=14$)
Humeral diaphysis	0·982	0·944
Radial diaphysis	1·097	1·066
Femoral diaphysis	0·945	0·942
Tibial diaphysis	0·988	1·042
Scapula breadth	0·971	0·947
Clavicle length	0·939	0·941
Ilium length	1·050	1·015
Pubis length	1·052	1·170
Ischium length	0·963	0·906
% Variance (PC1)	98·5%	97·2%
Theta	3·0°	4·6°
Small-sample direction test	$F = 18·923$ $P < 0·001$ Reject	$F = 4·122$ $P = 0·067$

variation in shape such that gibbons and siamang never overlap in overall form regardless of animal size or age (Figure 6). Axis 1 of log shape is very similar to axis 2 of log size-and-shape, with a correlation of 0·98 between their respective loadings. Axis 1 of the log size-and-shape analysis is therefore very close to the isometric vector that was explicitly removed in the

Table 4 **Principal components of log size-and-shape and log shape only in pooled ontogenetic series**

Variable	Log size-and-shape		Log shape	
	Axis I	Axis II	Axis I	Axis II
Humeral diaphysis	0·3215	0·1024	−0·0945	−0·2070
Radial diaphysis	0·3656	0·0104	0·1185	−0·3520
Femoral diaphysis	0·2939	0·4330	−0·4362	−0·1034
Tibial diaphysis	0·3087	0·5365	−0·4694	−0·2189
Scapula breadth	0·3130	0·1687	−0·1755	0·1901
Clavicle length	0·3344	−0·4002	0·3316	−0·2624
Ilium length	0·3613	−0·2513	0·3065	−0·0492
Pubis length	0·3801	−0·5012	0·5587	0·2147
Ischium length	0·3106	0·1053	−0·1397	0·7881
% Total variance	97·0	1·9	63·2	13·5
Total variance	0·7073956		0·02632299	

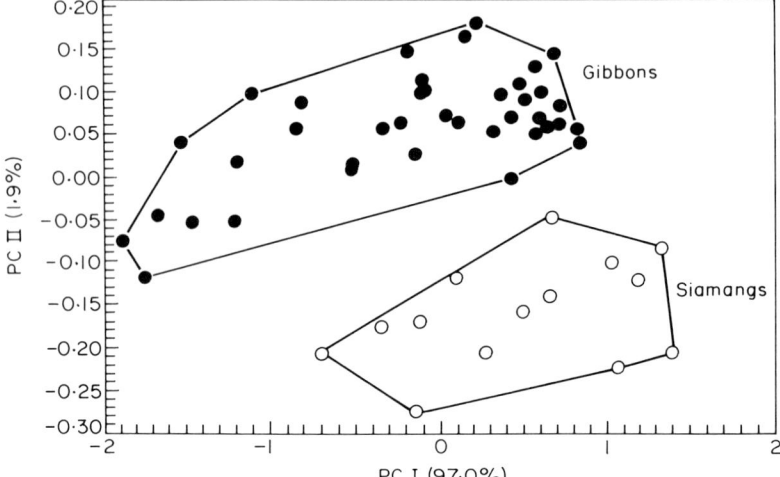

Figure 5. The first two principal components (PC) of log size-and-shape (logged raw data).

consideration of log shape only. Axis 2 of shape orders individuals *within* each species more or less by age/size.

Similarities and differences in patterns of relative growth among all living hominoids are depicted by the phenogram in Figure 7. The phenogram that is illustrated is the UPGMA attempt to summarize the correlation matrix among hominoid vectors of multivariate allometry (non-hylobatid data from Jungers & Hartman, 1988). The cophenetic correlation of almost 0·92 suggests that UPGMA did a good job of summarizing the information about similarity contained in the original correlation matrix. Alternative algorithms yielded the same clusters. The lesser ape cluster includes the orang-utan (with the siamang slightly more

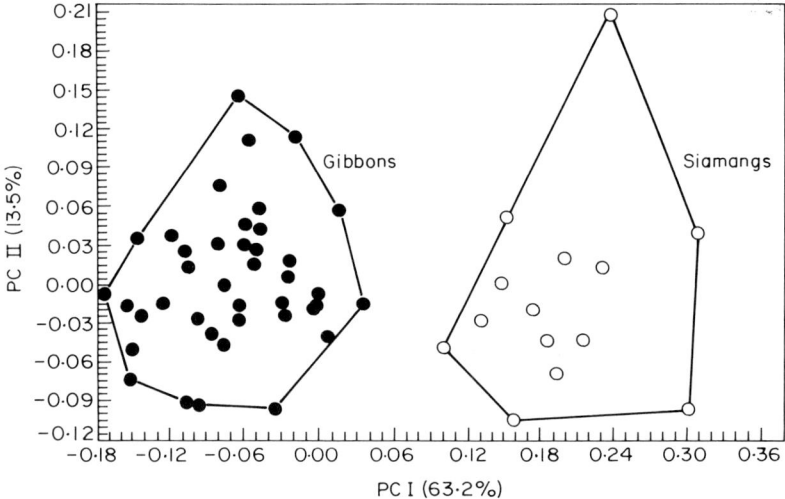

Figure 6. The first two principal components of log shape only.

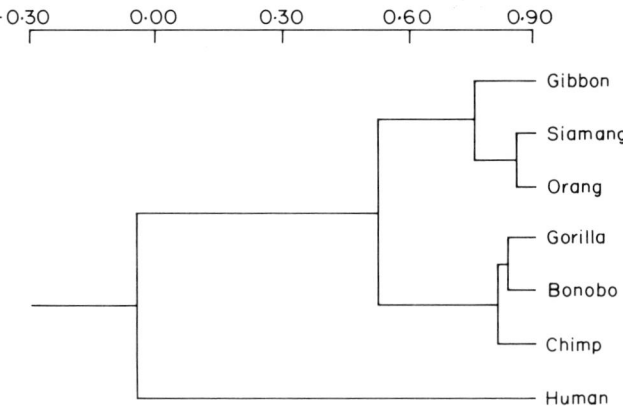

Figure 7. A UPGMA phenogram that illustrates hominoid clusters based on similarity in patterns of multivariate relative growth. Cophenetic $r = 0.916$.

similar overall to *Pongo*); this agglomeration is connected to an African ape cluster (within which bonobos connect first to gorillas) and humans join the overall non-human hominoid cluster quite distantly.

Discussion

Extrapolation of *lar*-gibbon scaling trends, whether with respect to body mass or in pair-wise variable combinations, into the siamang size range did a poor job overall of predicting siamang postcranial proportions. Relative to body mass, gibbon ontogenetic trajectories most frequently overestimated the target dimension in siamang. Although there was a great

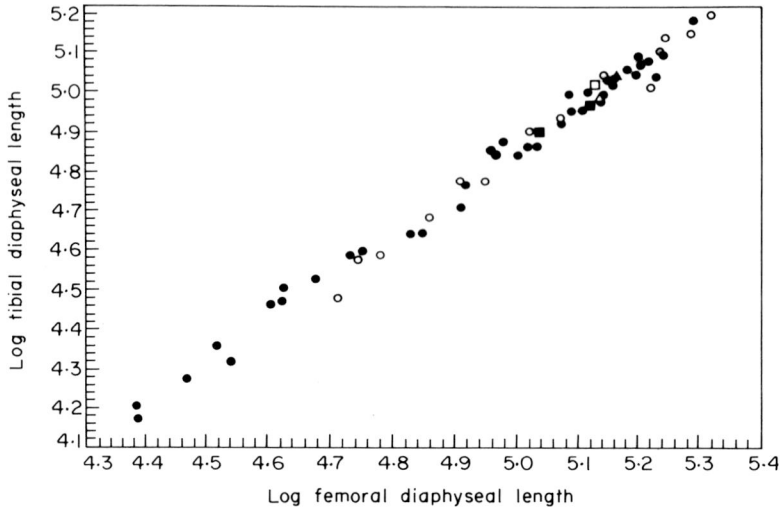

Figure 8. Bivariate scatter that is identical to Figure 3 except for the addition of specimens of *H. hoolock* (□), *H. klossi* (■) and *H. concolor* (△).

similarity in bivariate slopes such that shape change was similar postnatally in both species, transpositions in elevations were the rule. These displacements (Alberch *et al.*, 1979) or differences in process initiation (Raff & Wray, 1989) varied from dimension to dimension, and they point out that much of the developmental action relevant to observed shape differences in lesser apes takes place prenatally. Although there are noteworthy similarities in the multivariate patterns of relative growth in the *lar*-gibbon and siamang, their overall shapes never overlap at any age/size. Moreover, those aspects of shape that serve to distinguish adults of the two species (e.g., relatively longer elements of the hindlimb in the gibbon *vs.* relatively longer ilia and pubes in the siamang; Schultz, 1933) are already more or less present early in postnatal development. Because the siamang is larger than the gibbon, many of these persistent differences in shape are by definition correlated with size; however, it needs to be emphasized that this does not mean that those aspects of shape that are significantly size-related are simply the same thing as size (Oxnard, 1978). Furthermore, ontogenetic shape changes are necessarily correlated with size. As a consequence, if we were to focus exclusively on those developmental aspects of shape that are statistically uncorrelated with size, we would have little more than noise with which to work.

Taken together, and in contrast to the conclusions of Lumer's (1939) more limited study, these findings lead us to reject the hypothesis that the locomotor skeleton of the siamang is simply an overgrown or peramorphic version of the gibbon postcranium. We need to qualify this conclusion in several important ways. A more precise inference is that the siamang postcranial skeleton is not an ontogenetically scaled-up version of the *lar*-gibbon. This does not exclude the possibility that other species of *Hylobates* might be scaled-down models of siamang, especially in view of the evidence suggesting that the *lar*-group "has the largest number of supposedly progressive character states and is presumably most distant from the root" (Creel & Preuschoft, 1984:606). Perhaps the less derived *H. hoolock*, *H. concolor* and/or *H. klossi* would make better candidates from which to derive the siamang body shape via ontogenetic extrapolation.

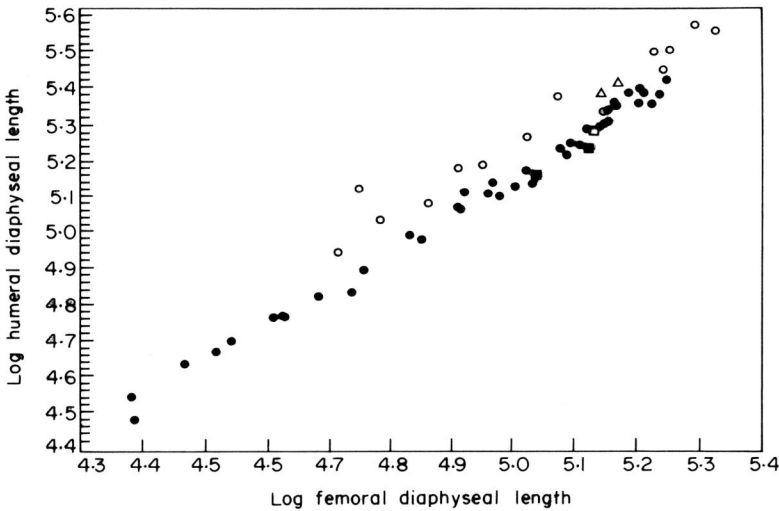

Figure 9. Bivariate scatter that is identical to Figure 4 except for the addition of *H. hoolock* (□), *H. klossi* (■) and *H. concolor* (△).

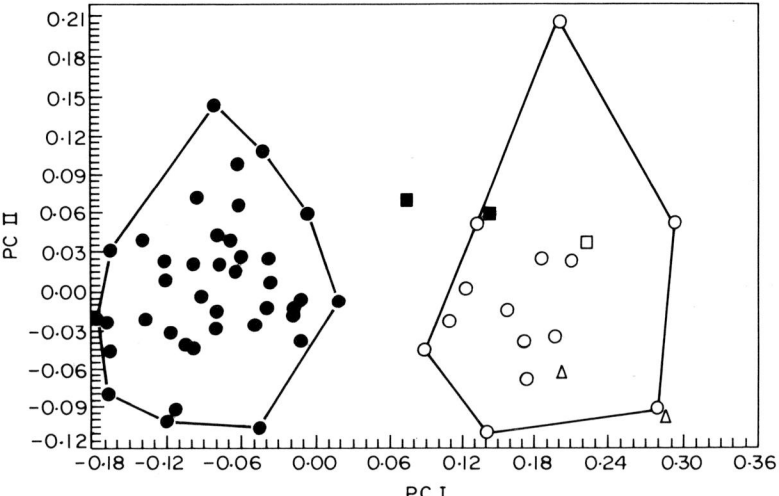

Figure 10. The first two principal components of log shape only (as in Figure 6) with the addition of *H. hoolock* (□), *H. klossi* (■) and *H. concolor* (△).

Regrettably, there are no non-adult skeletal samples of which we are aware for any of these species that could be used to test this alternative possibility in a rigorous fashion. In order to provide a few tentative insights, we have collected comparable data on several adult specimens (two *H. klossi*, two *H. concolor* and one *H. hoolock*); again, the non-adult measurement set was used. With respect to the scaling of tibial diaphyseal length on femoral diaphyseal length (Figure 8), all five specimens fall on the common hylobatid trajectory that is also shared by siamang and *lar*-gibbons. With respect to other pair-wise contrasts, sometimes *H. concolor* alone resembled the siamang (e.g., humerus on femur, Figure 9), and in others all three species fell much closer to the siamang than to the *lar*-gibbon. A new principal components analysis of log shape that included these five specimens suggests that adult *H. hoolock* and *H. concolor* (and perhaps even some *H. klossi*) share an overall body shape that one might find in non-adult siamang (Figure 10). The siamang may not be an overgrown *lar*-gibbon in terms of its skeletal shape, but the intriguing possibility remains that it is indeed an overgrown gibbon of another variety.

When relative growth patterns in all hominoids are considered together via correlation and clustering, the picture that emerges appears to be related weakly to currently recognized phylogenetic affinities among the taxa. The orang-utan pattern is more similar to that of the lesser apes than to the African apes (and bonobos are connected to gorillas rather than to common chimpanzees). Humans, with their strongly positive long bone allometries (especially in the hindlimb), share little similarity in relative growth to any other group. Ruling out some inexplicable geographic factor, ontogenetic similarities appear to more closely reflect *functional affinities*; suspensory forms are united (orang-utan, siamang and gibbon), knuckle-walking quadrupeds cluster together (gorillas, chimpanzees and bonobos) and human bipeds are isolated from all other hominoids. We recognize that this is simply a description of shared *vs.* disparate or unique trends rather than a functional explication of relative growth rates. The human pattern of relative growth with an emphasis on elongation of the hindlimbs is clearly linked mechanically to the attainment of efficient bipedal gait throughout postnatal ontogeny (Jungers *et al.*, 1988). Positive allometry of forearm length is one of the trends clearly shared by lesser apes and orang-utans, related perhaps to enlarging the feeding sphere (Grand, 1972) of growing individuals in each species. These Asian apes also all grow in such a way that their intermembral indices increase with increasing body size, a finding that has been related in interspecific analyses of primate limb proportions to frictional constraints on climbing in mammals lacking claws (e.g., Jungers, 1984). As was noted earlier, those proportional differences that distinguish between adult *lar*-gibbons and adult siamang are already in place to a considerable degree at birth. As such, there are probably corresponding differences in the mechanics of brachiation and other suspensory behaviors even at comparable body sizes.

Acknowledgements

We wish to thank Matt Ravosa and Anne Gomez for their invitation to participate in their AAPA symposium. Special thanks are also due to those museum curators and collection managers who kindly made material in their charge available to us. We also thank Tim Cole for his assistance with SASGRAPH and for writing the program that tests for significant differences in RMA slopes. We also appreciate the care and thought that went into evaluating our manuscript by two anonymous reviewers. This research was supported by NSF grants BNS 8606781 and BNS 8819621.

References

Alberch, P., Gould, S. J., Oster, G. F. & Wake, D. B. (1979). Size and shape in ontogeny and phylogeny. *Paleobiology* **5,** 296–317.
Bookstein, F. L. (1989). "Size and shape": a comment on semantics. *Syst. Zool.* **38,** 173–180.
Bonner, J. T. (Ed.) (1982). *Evolution and Development*. New York: Springer-Verlag.

Burnaby, T. P. (1966). Growth-invariant discriminant functions and generalized distances. *Biometrics* **22**, 96–110.
Buschang, P. H. (1982). The relative growth of the limb bones for *Homo sapiens*—as compared to anthropoid apes. *Primates* **23**, 465–468.
Cheverud, J. M. (1982). Relationships among ontogenetic, static, and evolutionary allometry. *Am. J. phys. Anthrop.* **59**, 139–149.
Cheverud, J. M. & Richtsmeier, J. T. (1986). Finite-element scaling applied to sexual dimorphism in rhesus macaque (*Macaca mulatta*) facial growth. *Syst. Zool.* **35**, 381–399.
Clarke, M. R. B. (1980). The reduced major axis of a bivariate sample. *Biometrika* **67**, 441–446.
Creel, N. & Preuschoft, H. (1984). Systematics of the lesser apes: a quantitative taxonomic analysis of craniometric and other variables. In (H. Preuschoft, D. L. Chivers, W. Y. Brockelman & N. Creel, Eds) *The Lesser Apes. Evolutionary and Behavioural Biology*, pp. 562–613. Edinburgh: Edinburgh University Press.
Darroch, J. N. & Mosimann, J. E. (1985). Canonical and principal components of shape. *Biometrika* **72**, 241–252.
Fink, W. L. (1982). The conceptual relationship between ontogeny and phylogeny. *Paleobiol.* **8**, 254–264.
Geissmann, T. (1991). Reassessment of age of sexual maturity in gibbons (*Hylobates* spp.). *Am. J. Primatol.* **23**, 11–22.
Gould, S. J. (1977). *Ontogeny and Phylogeny*. Cambridge: Harvard University Press.
Gould, S. J. (1988). The uses of heterochrony. In (M. L. McKinney, Ed.) *Heterochrony in Evolution. A Multidisciplinary Approach*, pp. 1–13. New York: Plenum Press.
Grand, T. I. (1972). A mechanical interpretation of terminal branch feeding. *J. Mammal.* **53**, 198–201.
Groves, C. P. (1968). A new subspecies of white-handed gibbon from northern Thailand, *Hylobates lar carpenteri* new subspecies. *Proc. Biol. Soc. Wash.* **81**, 625–627.
Jolicoeur, P. (1963). The multivariate generalization of the allometry equation. *Biomechanics* **19**, 497–499.
Jolicoeur, P. (1984). Principal components, factor analysis, and multivariate allometry: a small sample direction test. *Biometrics* **40**, 685–690.
Jungers, W. L. (1984). Scaling of the hominoid locomotor skeleton with special reference to lesser apes. In (H. Preuschoft, D. J. Chivers, W. Y. Brockelman & N. Creel, Eds) *The Lesser Apes. Evolutionary and Behavioural Biology*, pp. 146–169. Edinburgh: Edinburgh University Press.
Jungers, W. L. (1988). Relative joint size and hominoid locomotor adaptations with implications for the evolution of hominid bipedalism. *J. hum. Evol.* **17**, 247–265.
Jungers, W. L. & Hartman, S. E. (1988). Relative growth of the locomotor skeleton in orang-utans and other large-bodied hominoids. In (J. H. Schwartz, Ed.) *Orang-utan Biology*, pp. 347–359. Oxford: Oxford University Press.
Jungers, W. L. & Susman, R. L. (1984). Body size and skeletal allometry in African apes. In (R. L. Susman, Ed.) *The Pygmy Chimpanzee: Evolutionary Biology and Behavior*, pp. 131–178. New York: Plenum Press.
Jungers, W. L., Cole, T. M. III & Owsley, D. W. (1988). Multivariate analysis of relative growth in the limb bones of Arikara Indians. *Growth, Develop. & Aging* **52**, 103–107.
Lumer, H. (1939). Relative growth of the limb bones in the anthropoid apes. *Hum. Biol.* **11**, 379–392.
McKinney, M. L. (Ed.) (1988). *Heterochrony in Evolution. A Multidisciplinary Approach*. New York: Plenum Press.
McNamara, K. J. (1986). A guide to the nomenclature of heterochrony. *J. Paleontology* **60**, 4–13.
Mosimann, J. E. & James, F. C. (1979). New statistical methods for allometry with aplications to Florida red-winged blackbirds. *Evolution* **23**, 444–459.
Oxnard, C. E. (1978). One biologist's view of morphometrics. *Ann. Rev. Evol. Syst. 1978* **9**, 219–241.
Plotnick, R. E. (1989). Application of bootstrap methods to reduced major axis line fitting. *Syst. Zool.* **38**, 144–153.
Raff, R. A. & Wray, G. A. (1989). Heterochrony: developmental mechanisms and evolutionary results. *J. Evol. Biol.* **2**, 409–434.
Ravosa, M. J. (1991). Ontogenetic perspective on mechanical and nonmechanical models of primate circumorbital morphology. *Am. J. phys. Anthrop.* **85**, 95–112.
Rayner, J. M. V. (1985). Linear relations in biomechanics: the statistics of scaling functions. *J. Zool. Lond.* **206**, 415–439.
Ricker, W. E. (1984). Computation and uses of central trend lines. *Can. J. Zool.* **62**, 1897–1905.
Rohlf, F. J. (1990). *NTSYS-PC. Numerical Taxonomy and Multivariate Analysis System*. Setauket: Exeter Software.
Rohlf, F. J. & Bookstein, F. L. (1987). A comment on shearing as a method for "size correction". *Syst. Zool.* **36**, 356–367.
Schultz, A. H. (1933). Observations on the growth, classification, and evolutionary specializations of gibbons and siamangs. *Hum. Biol.* **5**, 212–255, 385–428.
Schultz, A. H. (1944). Age changes and variability in gibbons. A morphological study on a population sample of man-like ape. *Am. J. phys. Anthrop.* **2**, 1–129.
Shea, B. T. (1981). Relative growth of the limb and trunk in the African apes. *Am. J. phys. Anthrop.* **56**, 179–202.
Shea, B. T. (1983). Allometry and heterochrony in the African apes. *Am. J. phys. Anthrop.* **62**, 275–289.
Shea, B. T. (1985). Bivariate and multivariate growth allometry. *J. Zool. Lond.* **206**, 367–390.
Shea, B. T. (1988). Heterochrony in primates. In (M. L. McKinney, Ed.) *Heterochrony in Evolution. A Multidisciplinary Approach*, pp. 237–266. New York: Plenum Press.
Tsutakawa, R. K. & Hewett, J. E. (1977). Quick test for comparing two populations with bivariate data. *Biometrics* **33**, 215–219.

Sandra E. Inouye
*Department of Anthropology,
Northwestern University, Evanston,
IL 60208, U.S.A.*

Received 21 June 1991
Revision received 5 December
1991 and accepted 3 January
1992

Keywords: ontogeny, allometry, size, African apes, metacarpals, phalanges.

Ontogeny and allometry of African ape manual rays

Comparative morphology of the African ape hand is receiving renewed interest in the wake of recent discoveries of fossil hominid hands and the increasing number of biomolecular studies that question the traditional chimpanzee/gorilla clade. This paper compares ontogenetic sequences of *Pan paniscus*, *Pan troglodytes troglodytes*, *Gorilla gorilla gorilla*, *Gorilla gorilla beringei* and *Pongo pygmaeus* in order to assess whether or not previously established interspecific differences between adult chimpanzee and gorilla metacarpals and phalanges are due to a differential extension/truncation of a common growth allometry. Furthermore, inter-ray comparisons are made within the African apes to evaluate the functional role of individual rays. Results indicate that most dimensions of metacarpals and phalanges are ontogenetically scaled between *Pan* and *Gorilla* and also among pongids. Metacarpal and phalangeal lengths, however, depart from the general pattern of ontogenetic scaling. In interspecific comparisons of metacarpal and phalangeal lengths regressed on pubic lengths, the slopes are not significantly different between chimpanzees and gorillas, but there are significant transpositions in the growth trajectories such that *Gorilla* has shorter rays than *Pan* at common sizes. Inter-ray comparisons between genera reveal that gorillas have less variation between rays in almost all dimensions compared to chimpanzees. On the basis of these results, it is concluded that most differences in metacarpal, phalangeal, and inter-ray morphology between chimpanzees and gorillas that are not attributed to ontogenetic scaling may be related to the following: (1) kinematically distinct types of knuckle-walking, (2) a greater proportion of body weight borne on the hands of gorillas than chimpanzees at common sizes, or (3) a higher frequency of arboreal behavior of chimpanzees than gorillas at common sizes.

Journal of Human Evolution (1992) **23**, 107–138

Introduction

The comparative morphology of the hominoid hand has been a subject of considerable anthropological interest in discussions of hominid phylogeny (e.g., Gregory, 1928; Huxley, 1893; Straus, 1940, 1949). Due to discoveries of fossil hominid hand bones during the past few decades (Brain *et al.*, 1988; Bush *et al.*, 1982; Napier, 1959; Ricklan, 1986; Susman, 1988*a,b*, 1989; Susman & Creel, 1979) and the sudden growth in the number and popularity of biomolecular studies on hominoid phylogeny (e.g., Goldman *et al.*, 1987; Goodman *et al.*, 1990; Hasegawa *et al.*, 1989; Kishino & Hasegawa, 1989; Koop *et al.*, 1989; Miyamoto *et al.*, 1987; Williams & Goodman, 1989), researchers have renewed an interest in the functional morphology of hominoid hands (e.g., Inouye, 1989, 1990, 1991*a,b*; Lewis, 1969, 1972*a,b*, 1973, 1977; Napier, 1959, 1960*a,b*, 1962; Sarmiento, 1988; Smith, 1990; Susman, 1978, 1979, 1988*a,b*, 1989; Susman & Creel, 1979; Susman & Stern, 1980; Susman *et al.*, 1982; Tuttle, 1967, 1969*a,b,c*).

Most differences in the adult morphology of hands among African apes, orang-utans and early hominids are attributed to differences in function related to the degree of arboreal and terrestrial locomotion (e.g., Preuschoft, 1973; Sarmiento, 1988; Schultz, 1927, 1936, 1956; Stern & Susman, 1983; Straus, 1940; Susman, 1979; Tuttle, 1967, 1969*a,b*; Tuttle & Watts, 1985). Clinical and comparative research on human and non-human primate hands have also determined morphological differences among primates related to manipulative capabilities in extant and extinct primates and humans (e.g., Marzke, 1983; Marzke & Markze,

1987; Marzke & Shackley, 1986; Musgrave, 1969, 1970; Napier, 1956, 1959, 1960a,b, 1962, 1980; Smith, 1990; Susman 1988a,b, 1989; Susman et al., 1982; Susman & Stern, 1980). Most studies on primate hands highlight the comparative morphology among African apes as a group, orang-utans and humans, and don't directly address functional arguments associated with variation within these groups. Documented differences in the morphology between chimpanzees and gorillas exist (Erikson, 1963; Schultz, 1927, 1936, 1956; Susman, 1979; Tuttle, 1967, 1969a,b,c); however, functional discussion of the variation between chimpanzee and gorilla hands is often underemphasized, perhaps because both African apes utilize their hands in a unique form of quadrupedalism called knuckle-walking.

Early studies on hominoid hands noted that gorillas have the relatively shortest and broadest hands (Erikson, 1963; Midlo, 1934; Schultz, 1927, 1936, 1956) of all great apes. In contrast, chimpanzees were noted as intermediate between gorillas and orang-utans in relative hand length and breadth. These differences in gross proportions of the great ape hand led to subsequent studies on the relative lengths of the individual segments of the manual rays (Napier, 1959; Schultz, 1956; Susman, 1979). Some of these studies noted that phalangeal length relative to metacarpal length is shortest in gorillas, longest in orang-utans and intermediate in chimpanzees.

Susman (1979) extended these studies by documenting among adult chimpanzees and gorillas, as well as all hominoids, the variation in the shape of individual segments of each manual ray and among segments of different rays. Susman noted that in gorillas, the length of the fifth metacarpal more closely approximates the length of the fourth metacarpal than other pongids. Susman also described gorilla metacarpals and phalanges as "short and stout", in contrast to the "long and thin" metacarpals of chimpanzees (Susman, 1979). In a discriminant analysis using the combined metacarpal and proximal phalanx III lengths as the covariate, Susman (1979) indicated that gorillas, and to a lesser extent chimpanzees, are distinguished by greater relative metacarpal head depth, metacarpal dorsopalmar midshaft width, proximal phalangeal trochlear width and proximal phalangeal base widths from orang-utans. Furthermore, Susman (1979) described the features used in the discriminant analysis with a variety of indices. These indices reflect an increasing trend in the relative length of the proximal phalanx and depth of the metacarpal heads from the African apes through the orang-utan (Susman, 1979). Gorillas have the relatively and absolutely deepest metacarpal heads and relatively shortest proximal phalanges of the great apes; whereas chimpanzees differ from gorillas in the direction of orang-utans, which have the relatively smallest metacarpal heads and absolutely and relatively longest proximal phalanges (Susman, 1979).

Functional differences in hominoid manual rays

The documented differences in the hand between adult chimpanzees and gorillas are most commonly described as adaptations for differing degrees of arboreal and terrestrial behaviors (Erikson, 1963; Preuschoft, 1973; Straus, 1940; Schultz, 1927, 1936, 1956; Susman, 1979; Tuttle, 1967, 1969a,b; Tuttle & Cortright, 1988; Tuttle & Watts, 1985). From analyses of the proportions of the great ape hand (Erikson, 1963; Schultz, 1927, 1936, 1956; Susman, 1978, 1979) many claim that most manual features exhibit a continuum of shape change related to the extreme terrestriality of the gorillas at one pole, to the strict arboreality of the orang-utans at the other. Compared to other primates, the relatively long hands (compared to hand breadth) of great apes are described as elongated, short-thumbed, grasping hooks adapted for arboreal, suspensory postures (Erikson, 1963; Napier, 1967; Preuschoft, 1973; Straus, 1940; Susman, 1979). Within the great apes, the expression of these arboreal trends is weakest in the gorilla hand and culminates in the orang-utan hand. Chimpanzees are described as morphologically and behaviorally intermediate between gorillas and orang-utans; chimpanzee hands show a compromise between knuckle-walking and arboreal, grasping behaviors (Erikson, 1963; Schults, 1927, 1936, 1956; Susman, 1978, 1979).

A few biomechanical models exist which explain the relationship of adult hand proportions to different locomotor behaviors among the great apes (Preuschoft, 1973; Susman, 1979). One biomechanical model proposes that the lengthening of the ray segments of the orang-utan increases the compass of the hand and thereby enhances the effectiveness of the hook grip when grasping branches of relatively large diameter during climbing (Susman, 1979; Napier, 1967).

Assuming equal frequencies of arboreal behavior at common sizes and an increase in terrestrial behavior with increasing size, this model predicts that gorillas and chimpanzees at common sizes have ray segments of equal length, but adult gorillas have relatively shorter rays than adult chimpanzees. If chimpanzees are more arboreal than gorillas at all comparable sizes, then we expect chimpanzees to have absolutely longer rays at common sizes with gorillas and relatively longer rays as adults compared to gorilla adults.

Another biomechanical explanation of the proportions of the great ape hand argues that the shorter proximal phalanges of gorillas, and to a lesser extent chimpanzees, helps to preserve the integrity of the metacarpophalangeal joint in knuckle-walking postures (Susman, 1979). The joint reaction force which must be resisted by passive support of the long flexor muscles and bony and ligamentous structures (Tuttle, 1967; Tuttle et al., 1972; Tuttle & Basmajian, 1974; Susman, 1979) is increased in knuckle-walking postures as compared to grasping postures. Compared to orang-utans, the shorter proximal phalanges of gorillas and chimpanzees will reduce the moment arm of the torque (Figure 1) produced about the metacarpophalangeal joints by the ground reaction forces during knuckle-walking (Susman, 1979).

Extending this model to differences between chimpanzees and gorillas, the model predicts that given equal joint reaction forces between gorillas and chimpanzees at common sizes, proximal phalangeal length of adult gorillas should be relatively shorter than that of adult chimpanzees due to the larger body size (and increasing joint reaction force) of the former, but equal in length at common sizes. If there is a difference between chimpanzees and gorillas in proximal phalangeal lengths at common sizes, the model indicates a difference between chimpanzees and gorillas in joint reaction forces. This difference may be linked to using kinematically distinct types of knuckle-walking, using different frequencies of knuckle-walking, or bearing differential proportions of body weight on their hands.

This biomechanical model is similar to a model used to describe the condition in the terrestrial foot. In a digitigrade primate, the ground reaction force is concentrated at the metatarsal heads and the joint reaction force which must be resisted by the extrinsic digital flexors is increased (Strasser, 1992). Like hands, the pedal phalanges will produce a torque at the metatarsophalangeal joint during the latter part of the stance phase as the foot is about to leave the ground. The extrinsic flexor muscles must resist the joint reaction force by either increasing the power-arm of the flexors (metatarsal head depth) or decrease the load arm (Strasser, 1992). More terrestrial primates have relatively short pedal phalanges (Schultz, 1963) in order to decrease the load arm, thus increasing the ratio between the power arm of the foot and the load arm (Schultz, 1963).

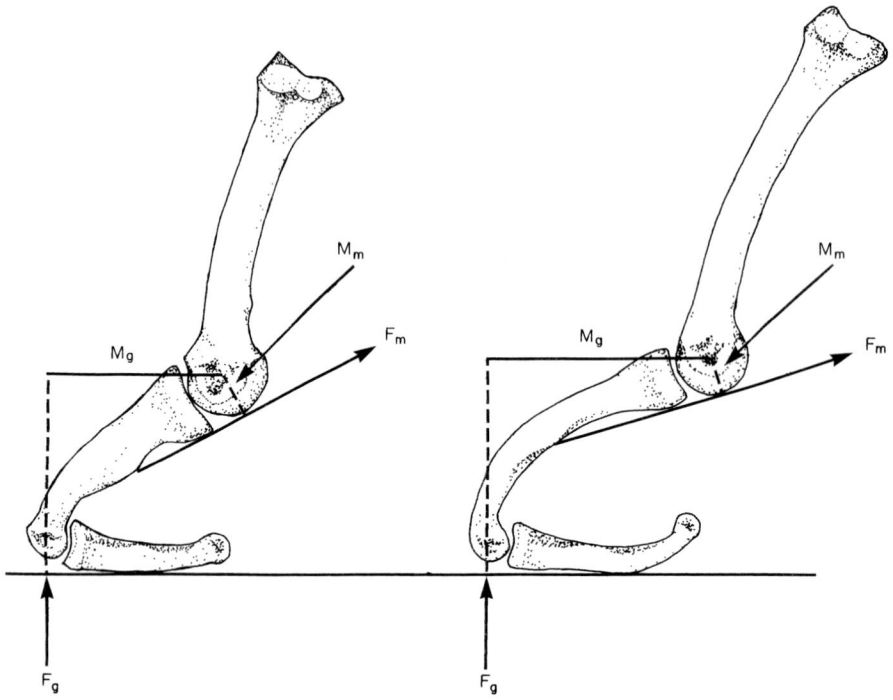

Figure 1. Right third rays of *Gorilla* and *Pongo* in knuckle-walking and hypothetical knuckle-walking postures. F_g = ground reaction force; F_m = passive or active force produced by the long flexor muscles; M_g = moment arm F_g; M_m = moment arm of F_m. To maintain the integrity of the metacarpophalangeal joint in knuckle-walking postures, the torque produced by the ground reaction force ($F_g \times M_g$) must be opposed by the passive resistance of the long flexor muscles ($F_m \times M_m$) or by joint ligaments. After Susman (1979).

A separate but related biomechanical model suggests that extension at the metacarpophalangeal joint during knuckle-walking would place considerable bending stresses on the proximal phalanges (Preuschoft, 1973). The shortening of the proximal phalanges as well as a widening of midshaft widths of gorillas and chimpanzees, relative to the orang-utan, helps to reduce the bending moments on the proximal phalanges during knuckle-walking. A comparable biomechanical argument for the shortened pedal phalanges in terrestrial cercopithecines has been suggested by Meldrum (1991) and Strasser (1992).

The metacarpals of pongids are purported to always be exposed to some compressive stresses in knuckle-walking and suspensory postures (Preuschoft, 1973). Apart from the compression, the metacarpals are exposed to dorsally convex bending stresses during knuckle-walking (Preuschoft, 1973). In knuckle-walking postures, the metacarpals are more likely to fail by bending and by Euler buckling, given comparable metacarpal shaft dimensions, as they increase in length (see Currey, 1984). Therefore, we might expect knuckle-walkers to have shorter metacarpals and proximal phalanges as well as wider metacarpal and midshaft dimensions than orang-utans at common body sizes.

Extension of this model to African apes would predict that given equal amounts of weight on the hands of gorillas and chimpanzees in knuckle-walking postures, there should be no differences in metacarpal and phalangeal lengths and midshaft widths at common sizes. If gorillas bear a greater proportion of their body weight on their hands than chimpanzees at common sizes, we might expect gorillas to have absolutely and relatively shorter metacarpals and phalanges and/or absolutely and relatively broader metacarpals and phalanges than comparable sized chimpanzees in order to resist increased bending stresses in the former.

Allometry and ontogeny
Much of what has been previously described and functionally interpreted about the proportions of hominoid hands has been based on the adult morphologies. However, differences in overall body size may have a significant influence on hand proportions. For a variety of functional and developmental reasons, many interspecific differences in shape are often causally related to changes in body size (Gould, 1966; Shea, 1988); this differential growth and shape change is called allometry.

Among similarly sized animals there is less of a concern about the effects of size. However, in the case of hominoid hands, there is an obvious disparity in size between adult gorillas and chimpanzees and among the great ape adults. In absolute terms, adult gorillas and orangutans are commonly greater in dimensions of the manual rays than chimpanzees, simply because of their sheer size; a full-grown, adult male gorilla can easily weigh three or four times as much as an adult male chimpanzee. In previous studies on African ape hands others have attempted to minimize the confounding effects of body size by expressing shape characters as indices (Erikson, 1963; Schultz, 1927, 1936, 1956; Susman, 1979). For example, whole hand lengths, breadths, metacarpal and phalangeal lengths, widths, etc., were expressed as ratios of trunk length, limb length, ray length, metacarpal length, phalangeal length, etc. Many of the functional interpretations describing variation between gorillas and chimpanzees and among great apes are based on these indices; however, indices are problematic for a number of reasons (Strasser, 1992; Weil, 1962).

The relationship of the indices to body size was undetermined in these studies. In the case of hominoid hands, we do not know if the denominator, metacarpal length, tracks size differently in gorillas, chimpanzees and orang-utans. If metacarpal length scales differently with increasing weight among the great apes, then the indices become misleading because it does not follow that apes with the same metacarpal length have the same body size. Furthermore, the relationship of the indices numerators and denominators with body size have to be examined to determine how the indices are correlated with body size (Strasser, 1992). For example, if the numerator, dorsopalmar metacarpal midshaft width, is more positively allometric than the denominator, metacarpal length, then the index will have a positive correlation with body size. Conversely, if the dorsopalmar metacarpal midshaft width is more negatively allometric than metacarpal length then the index will have a negative correlation with body size. If dorsopalmar metacarpal midshaft width and metacarpal length have the same allometric relationship with body size then the index will have no correlation with body size.

Also, the relationship of the numerator and the denominator to body size was undetermined in previous studies. From the indices, we cannot tell if the values reflect changes in the numerator, changes in the denominator, or both. For example, according to Susman's (1979) indices, gorillas have the stoutest metacarpals, orang-utans the most slender metacarpals and chimpanzees are intermediate in shape. However, we do not know how metacarpal length and metacarpal dorsopalmar midshaft widths scale among the great apes with increasing size. Without allometric clarification, the stoutness of adult gorilla metacarpals, compared to chimpanzees, may be due to any of the following: (1) chimpanzees and gorillas are ontogenetically scaled for metacarpal midshaft diameters but are positively allometric with body size, producing relatively larger shaft dimensions in gorilla adults simply because of bigger body size; (2) chimpanzees and gorillas are ontogenetically scaled for metacarpal length but are negatively allometric with body size, producing relatively smaller shaft dimensions in gorilla adults simply because of their bigger body size; (3) gorillas have larger midshaft diameters at common sizes with chimpanzes; (4) gorillas have shorter midshaft diameters at common sizes with chimpanzees; (5) any combinations of the above; or (6) metacarpal length and shaft dimensions are not influenced by increasing size.

Indices, without proper consideration, are clearly not the most appropriate method to standardize for body size. How then do we know whether or not chimpanzees and gorillas have similar hand proportions at comparable sizes and how can we determine if the variation in the adult morphologies is produced by differential growth patterns?

Ontogenetic scaling provides the most appropriate solution for the problem of size-correction in closely related species. Ontogenetic scaling of shape between species results from the sharing and differential extension of a common growth allometry; divergence of ontogenetic allometries suggests interspecific shape difference which require explanation apart from differences in body size (Shea, 1983a). Ontogenetic scaling as a criterion of subtraction can be used to identify features which depart from a common size/shape pattern, likely signifying adaptive shape changes to new functional requirements (Gould, 1975; Jungers & Susman, 1984; Shea, 1983a, 1988). Interspecific differences that result from the differing endpoints along a common growth trajectory may be the correlated result for selection for overall size, and not the selection for altered shape (Shea, 1988). This does not imply that interspecific differences attributed to ontogenetic scaling are non-adaptive or not without functional significance, since allometric growth can be related to the changing functional requirements during growth (cf. Clutton-Brock & Harvey, 1979; Shea, 1985). Nonetheless, a dissociation of growth trends between closely related species to produce new size/shape forms likely indicates selection for novel proportions (Gould, 1977; Shea, 1988).

Most features of adult chimpanzee hands are interpreted as adaptive morphological and functional intermediates between adult gorillas and orang-utans. By comparing ontogenetic allometries of manual features between chimpanzees and gorillas, we can elucidate whether documented differences in the adult morphologies are due to differing endpoints along a common growth pattern or due to divergent, adaptive growth patterns. For example, Susman (1979) noted that the height of the ridges on the dorsum of metacarpals, purported knuckle-walking adaptations, have greater expression in gorilla adults compared to chimpanzee adults and are not present in a moderate number of adult and subadult chimpanzees and bonobos. However, Inouye (1990) demonstrated that a common growth trajectory underlies this differential expression; gorillas simply extend the common growth trend into larger size ranges. Thus, gorilla adults have a greater degree of expression of the dorsal metacarpal ridges than chimpanzees simply because of their greater body size. Ontogenetic allometric correction of manual characters is important, especially since Shea and others (Shea, 1981, 1983a,b,c, 1985; Jungers, 1984; Jungers & Susman, 1984) have demonstrated that many differences in the proportions of the skull, torso and limbs within adult African apes are attributed to pervasive ontogenetic scaling.

This study is a component of a larger project which attempts to quantify and interrelate the kinematic patterns of knuckle-walking and its associated ray morphology. This paper documents the ontogenetic variation in standard linear measurements of the manual rays within and between chimpanzees and gorillas, and evaluates whether or not the interspecific variation in manual morphology is a result of ontogenetic scaling. Furthermore, as Susman appropriately noted (1979:233), "morphological patterns of one ray cannot be generalized to the others as both the pattern of loading and the manipulatory role of each ray differ within a single hand." Therefore, the scaling patterns are compared among different rays of the same hand within and between chimpanzees and gorillas to determine whether or not differential growth patterns between fingers exist, and if these patterns vary between genera. *Pongo* is included in the study as a logical outgroup; as the only non-knuckle-walking pongid genus, *Pongo* serves to identify and elucidate phylogenetic and gross locomotor effects.

This study extends earlier quantitative work by Susman (1978, 1979) on the morphology of hominoid fingers, but differs significantly by its ontogenetic and allometric approach. Unlike Susman's earlier work, this study focuses on whether or not the quantitative differences among the African apes in the adult finger morphology are due to ontogenetic scaling or represent size/shape dissociations adaptive to a particular type of locomotion or habitat.

The null hypotheses tested are:
(A) There are no differences in linear dimensions of the manual rays between *Pan* and *Gorilla* once size factors have been properly considered.
(B) There are no differences in linear dimensions between different metacarpals of the same hand.

These hypotheses will be tested by examining the variation in the growth of bony dimensions of metacarpals and phalanges such as maximum lengths, midshaft diameters and articular surface diameters, in interspecific and intraspecific, ontogenetic comparisons of great apes. The previous biomechanical interpretations are evaluated in light of the results from this study.

Materials and methods

Samples
This study is based on measurements taken from a total of 300 pongid hands from wild-shot populations of differing age and sex. Measurements on cross-sectional samples of common chimpanzees (*Pan troglodytes troglodytes*, $n=107$) and western lowland gorillas (*Gorilla gorilla gorilla*, $n=103$) are taken from the collection at the Powell Cotton Museum and Quex House. Measurements of bonobo (*Pan paniscus*, $n=29$) and mountain gorilla (*Gorilla gorilla beringei*, $n=23$) hands are taken from the collections at the Musee Royale de l'Afrique Centrale and the National Museum of Natural History. Measurements of orang-utan (*Pongo pygmaeus pygmaeus/abelli*, $n=38$) hands are taken solely from the National Museum of Natural History.

Measurements
A set of variables were selected that reflect the basic shape of the manual rays. Features such as maximum length, midshaft diameters and articular surface diameters were measured on each individual metacarpal, proximal phalanx and middle phalanx of rays I–V, with 150 and 300 mm digital calipers.

The points that define the measurements are illustrated in Figure 2. Measurements taken from the metacarpals included maximum length (D:a–b), dorsopalmar midshaft depth (D:e–f), mediolateral midshaft width (A:a–b), dorsopalmar head depth (D:c–d), radioulnar head width (A:c–d) and biepicondylar width (A:e–f). Measurements taken from the proximal and middle phalanges included maximum length (E,F:a–b), dorsopalmar base depth

114 S. E. INOUYE

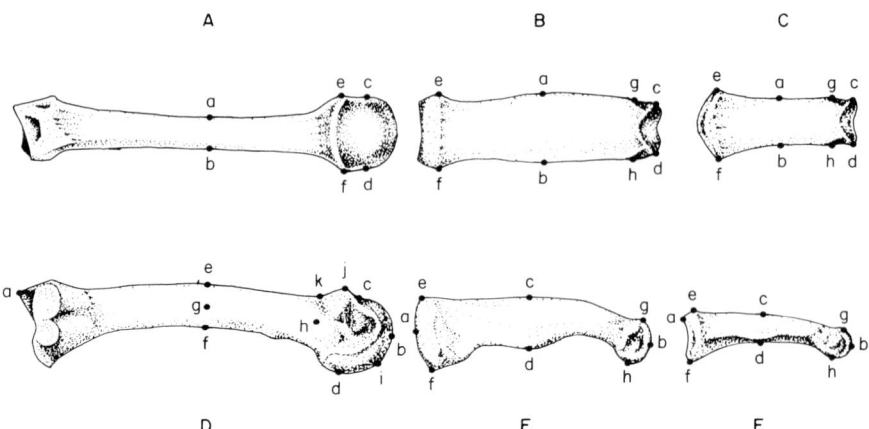

Figure 2. Right third ray of *Gorilla gorilla beringei*. Dorsal views of the (A) metacarpal, (B) proximal phalanx and (C) middle phalanx. Medial views of the (D) metacarpal, (E) proximal phalanx and (F) middle phalanx.

(E,F:e–f), mediolateral base width (B,C:e–f), dorsopalmar midshaft depth (E,F:c–d), mediolateral midshaft width (B,C:a–b), dorsopalmar trochlear depth (E,F:g–h), mediolateral trochlear width (B,C:c–d) and biepicondylar width (B,C:g–h).

Statistics

All analyses were performed on log transformed data. Bivariate ordinary least squares (OLS) and reduced major axis (RMA) regression were used to describe scaling trajectories and analysis of covariance (ANOCOVA) of OLS regressions were used to detect differences in the patterns of relative growth among the groups. Analysis of differences in RMA slope was based on a method described by Clarke (1980) and positional differences in the RMA regression lines were examined using Tsutakawa & Hewitt's (1977) "quick test". Visual inspection of the bivariate plots was also used to assess the degree of difference in patterns of relative growth between the groups.

The primary goal of this study was to establish whether or not any differences among group trajectories existed; functional interpretations of the empirical value of the slope are beyond the scope of this study. OLS regression was the preferred line-fitting technique to quantify growth trajectories, because analysis of covariance can be used with OLS regression to provide a powerful statistical test for slope and intercept differences between different allometric trajectories. However, since there is much debate over the appropriate line-fitting technique in allometric regressions, that has been discussed elsewhere (Jolicoeur, 1990; Kuhry & Marcus, 1977; McArdle, 1988; Plotnick, 1989; Rayner, 1985), RMA regression slopes are reported. Furthermore, results from RMA and OLS regressions on a subset of variables were compared to insure that the x variable was not influencing OLS results and to see if the two line-fitting techniques yielded similar scaling patterns. Since multiple comparisons of variables were made, Bonferroni criterion (Wilkinson, 1990), which divides the critical level by the number of comparisons being made, was used to adjust the probabilities to insure that the Type I error rate was no greater than the chosen significance level ($P<0.01$).

Since actual body weights were unavailable for most specimens, other skeletal variables were used as measurements of size in all regression analyses. For all growth comparisons, humeral diaphyseal length was used as a size surrogate, because it gave reasonably high correlations with the variables tested and it scales isometrically with body size in interspecific bivariate analyses of African apes (Jungers & Susman, 1984; Steudal, 1985). To insure that variation in x variable did not determine the scaling patterns, the relative growth statistics and results from ANOCOVAs using humeral diaphyseal length were compared with relative growth statistics and ANOCOVAs using pubic length, humeral mediolateral midshaft diameter and a composite size variable for regressions of all y variables, at all levels of comparison. The composite size variable was the mean of humeral diaphyseal length, humeral mediolateral midshaft diameter, humeral anteroposterior midshaft diameter, pubic length and femoral diaphyseal length. Protecting the null hypothesis, only results from ANOCOVAs that were significant and overlapping for all size surrogates were believed to accurately portray significant differences in the y variables; these results are reported. Relative growth statistics (OLS and RMA) for humeral diaphyseal length at the species/subspecies level only are reported for simplicity (Appendix).

Data is depicted using Gaussian bivariate ellipses (P-value = 0·95) instead of regression lines because of the large number of species and the sheer number of data points in the comparisons. The bivariate ellipse is centered on the sample means of X and Y variables. The unbiased sample standard deviations of X and Y establish the major axes of the ellipse and the sample covariance between X and Y determines its orientation (Wilkinson, 1990). The 95% confidence ellipses clearly portray the growth trend as well as the variation in the data sample without an overwhelming spray of overlapping data points. Ontogenetic comparisons were made between:

1. Sexes within each pongid group.
2. *P. troglodytes* and *P. paniscus*; *G. g. gorilla* and *G. g. beringei*.
3. All *Pan* and all *Gorilla*.
4. The African great apes and *Pongo*.

Results

To avoid a five-fold repetition of results for all rays, Ray III was selected to describe the general scaling patterns in ontogenetic and interspecific comparisons. Moreover, a single metacarpal is probably as representative of scaling patterns as using a larger number of metacarpals (Garn *et al.*, 1991). For clarity, inter-ray differences are addressed separately in a subsequent section.

Regression analyses for most proximal and all middle phalangeal measurements are omitted for *Pongo*, due to low correlations and an inadequate number of specimens.

Comparison of line-fitting techniques
For all variables, RMA regression yields consistently higher slopes and lower y-intercepts than does OLS regression (Appendix). In cases where the correlation coefficient (r) is less than 0·92, most RMA slopes do not fall within the 95% confidence intervals of the OLS slopes and are significantly higher than the OLS slopes. Clarke's test (1980) for slope differences and Tsutakawa & Hewett's "quick test" (1977) for y-intercept differences was used on RMA

Table 1 **A comparison between OLS and RMA regression of tests for slope and y-intercept differences**

	ANOCOVAs between taxa for regressions of metacarpal III dimensions against humeral diaphyseal length $(P<0.01)$[1]		
	Pp/Pt[2]	Ggg/Ggb	Pan/Gor
Metacarpal length	NS	*	*
Mediolateral midshaft width	*	NS	NS
Dorsopalmar midshaft width	NS	NS	NS
Mediolateral head width	*	NS	NS
Dorsopalmar head width	NS	NS	*
Biepiconylar width	NS	*	*

	Differences between taxa in RMA slopes from regressions of metacarpal III dimensions against humeral diaphyseal lengths[3]		
Metacarpal length	NS	NS	NS
Mediolateral midshaft width	NS	NS	NS
Dorsopalmar midshaft width	NS	NS	NS
Mediolateral head width	NS	NS	NS
Dorsopalmar head width	NS	NS	NS
Biepicondylar width	NS	§	NS

	Differences between taxa in RMA y-intercepts from regressions of metacarpal III dimensions against humeral diaphyseal lengths[4]		
Metacarpal length	NS	†	†
Mediolateral midshaft width	†	NS	NS
Dorsopalmar midshaft width	NS	NS	NS
Mediolateral head width	†	NS	NS
Dorsopalmar head width	NS	NS	†
Biepicondylar width	NS	†	†

[1] * = Significant difference in y-intercept at $P<0.01$; NS = no significant difference in y-intercept. All comparisons yielded NS in slopes.
[2] Pp = *Pan paniscus*; Pt = *Pan troglodytes troglodytes*; Ggg = *Gorilla gorilla gorilla*; Ggb = *Gorilla gorilla beringei*; Pan = Pp & Pt; Gor = Ggg & Ggb.
[3] Results are calculated using the technique described by Clarke (1980); § = significant T statistic, $P=0.01$, NS = no significant difference in slopes.
[4] Results are calculated using Tsutakawa & Hewett's "Quick Test" (1977); † = significant X^2 statistic, $P=0.01$, NS = no significant difference in y-intercept.

regression statistics for a subset of variables. Using humeral diaphyseal length as the size surrogate, RMA statistics for metacarpal III dimensions were compared to OLS statistics at all taxonomic levels (Table 1). ANOCOVAs among taxa for OLS slope differences produced patterns similar to those produced by comparisons among taxa for RMA slope differences using the method described by Clarke (1980). In all but one comparison, lowland versus mountain gorilla for biepicondylar width, the results of ANOCOVAs and Clarke's method agree. For those taxonomic comparisons that showed no differences in slopes, significant differences in y-intercepts among taxa from both OLS and RMA regressions were tested using ANOCOVA and Tsutakawa & Hewett's "quick test", respectively. Comparison of the intertaxonal patterns produced by ANOCOVA and the "quick test" yields matching results

Table 2 A comparison between relative growth statistics using different size surrogates relative growth statistics for *Pan paniscus*

	n	r	OLS[1] slope (95% CI)	RMA[1] slope (95% CI)	Y-intercept[2]
MCL3*HUML[3]	26	0.97	0.995 (0.906–1.084)	1.026 (0.919–1.134)	−1.014
MCL3*HMLD	26	0.96	0.760 (0.680–0.840)	0.794 (0.696–0.892)	1.623
MCL3*PUBICL	26	0.96	0.693 (0.621–0.765)	0.723 (0.636–0.811)	0.853
MCL3*SIZE	26	0.98	0.927 (0.857–0.997)	0.949 (0.863–1.034)	−0.135

[1] OLS = Ordinary least squares; RMA = reduced major axis regression.
[2] Y-intercept = y-intercept for least squares regression.
[3] MCL3 = metacarpal III length; HUML = humeral diaphyseal length; HMLD = humeral M–L midshaft diameter; PUBICL = pubic length; SIZE = composite size surrogate.

(Table 1). Since the two line-fitting techniques yielded similar scaling patterns among groups, only results from OLS regressions and ANOCOVAs are discussed.

Comparison of size surrogates

Depending on the choice of size surrogate as the x variable, relative growth statistics may vary widely for any y variable. Even though correlations (r) were all equally high, a single metacarpal dimension yields significantly different slope and y-intercept values when regressed against different size surrogates, regardless of the line-fitting technique used (Table 2). For example, metacarpal III length of bonobos regressed against humeral diaphyseal length and a composite size surrogate yield slopes close to isometry, while regressions using humeral midshaft diameter and pubic length yield significantly lower slopes (Table 2). Similarly, y-intercepts from regressions of metacarpal III length against humeral diaphyseal length and a composite size surrogate yield significantly lower y-intercepts than regressions against humeral midshaft diameter and pubic length.

In addition to differences in relative growth statistics, different size surrogates produce different intertaxonal patterns of relative growth. With some size surrogates, differential patterns of growth between species are evident for certain ray dimensions; however, other size surrogates produce patterns of ontogenetic scaling for the exact same interspecific comparisons. For example, significant positional shifts in the growth allometries of mountain and lowland gorillas for metacarpal mediolateral head width regressed against pubic lengths exist, whereas regression of this variable against humeral diaphyseal lengths yields ontogenetic scaling between mountain and lowland gorillas.

Regression slopes and y-intercepts vary too much among size surrogates, as well as between line-fitting techniques, to confidently and accurately interpret the meaning of the absolute value of the slope. However, intertaxonal patterns of relative growth can still be determined, in part, because OLS and RMA regressions yield similar relationships between different groups. Even though disparity in patterns using different size surrogates exists, the intertaxonal patterns of relative growth can be reliably determined by comparing ANOCOVAs for each size surrogate against one another for all ray dimensions. Intertaxonal relationships are reported and discussed for each variable only when the results for all size surrogates were in agreement. In these cases, the patterns are believed to reflect the relationship of the y variable among groups without being influenced by the size surrogate. Differences in patterns that occur between groups using all size surrogates are corroborated by visual inspection of the plots.

Sex differences

Within *P. paniscus*, *P. troglodytes*, *G. g. gorilla*, *G. g. beringei* and *Pongo* there are no significant differences ($P<0.001$) between males and females during ontogeny in both slopes and y-intercepts of regressions of each metacarpal measurement on size surrogates. For each group, scaling trajectories of males and females overlap extensively, with males extending the female's growth trajectories into larger size ranges. For all metacarpal variables, the sexes were pooled within each group for all subsequent analyses.

Manual ray III

Within Pan *and within* Gorilla. ANOCOVAs for most metacarpal measurements regressed against size surrogates reveal pervasive ontogenetic scaling between bonobos and chimpanzees and between lowland and mountain gorillas (Table 3).

Compared to common chimpanzees, there is a single downward shift in the bonobo growth trajectory for metacarpal mediolateral midshaft width against size surrogates. Within *Gorilla*, there is a significant downward shift of the lowland gorilla growth trajectory from the mountain gorilla trajectory for metacarpal length against all size surrogates.

Tests for differences between *P. paniscus* and *P. troglodytes* in bivariate regressions of phalangeal dimensions on size surrogates show a couple of differences in the y-intercepts of the regression lines (Table 3). As with metacarpals, ANOCOVAs indicate pervasive ontogenetic scaling for all variables except proximal phalangeal mediolateral midshaft width. For this regression against size surrogates, there is a downward transposition of the bonobo growth trajectories below those of the common chimpanzee trajectories. ANOCOVAs testing for differences between mountain and western gorillas, in comparison, reveal the pervasive pattern of ontogenetic scaling for all middle phalangeal and proximal phalangeal measurements. Slope values of all African apes for all proximal and middle phalangeal dimensions regressed on pubic lengths show significant negative allometry, while regressions against humeral diaphyseal length yield slopes near or including isometry (Appendix).

Pan vs. Gorilla. For bivariate regressions of all metacarpal and phalangeal III dimensions on size surrogates, ANOCOVAs reveal predominantly ontogenetic scaling between chimpanzees and gorillas. With the exception of metacarpal length, proximal and middle phalangeal lengths, proximal and middle phalangeal mediolateral midshaft widths and middle phalangeal dorsopalmar basal width, ANOCOVAs of regressions against humeral diaphyseal and pubic lengths reveal the pattern of ontogenetic scaling of *Pan* and *Gorilla* (Figure 3).

Metacarpal and phalangeal lengths are the most dramatic departures from the pattern of ontogenetic scaling (Figures 4 and 5). Slopes for regressions of metacarpal and phalangeal lengths on size surrogates are not significantly different ($P<0.001$) between *Pan* and *Gorilla* (Table 3); however, the growth trajectories of *Pan* and *Gorilla* clearly differ in y-intercepts. The ontogenetic trajectories of gorillas are shifted below those of chimpanzees. Similarly, gorilla proximal and middle phalangeal mediolateral midshaft widths and middle phalangeal dorsopalmar basal widths are similar in slope to chimpanzees, but their growth trajectories are shifted above those of chimpanzees.

Pan *and* Gorilla vs. Pongo. ANOCOVAs reveal that *Pongo* is ontogenetically scaled with *Pan* for all metacarpal dimensions, except mediolateral midshaft width, regressed on size surrogates (Table 3). For this variable, the growth trajectory of orang-utans is shifted below the

Table 3 Ontogenetic differences in slope and y-intercepts among taxa for manual ray dimensions regressed against size surrogates

ANOCOVAs between taxa for metacarpal III dimensions against humeral diaphyseal length, humeral A–P midshaft diameter and pubic length[1]

	Pp/Pt[2]	Ggg/Ggb	Pan/Gor	Pan/Or	Gor/Or
Metacarpal length		*	*		*
Mediolateral midshaft width	*			*	*
Dorsopalmar midshaft width					*
Mediolateral head width					*
Dorsopalmar head width					
Biepiconylar width					

ANOCOVAs between taxa for proximal phalanx III dimensions against humeral diaphyseal length, humeral A–P midshaft diameter and pubic length

	Pp/Pt	Ggg/Ggb	Pan/Gor	Pan/Or	Gor/Or
Phalangeal length			*	*	*
Mediolateral midshaft width	*		*		*
Dorsopalmar midshaft width					*
Mediolateral basal width					
Dorsopalmar basal width					
Mediolateral trochlear width					
Dorsopalmar trochlear width					
Biepicondylar width	*				

ANOCOVAs between taxa for middle phalanx III dimensions against humeral diaphyseal length, humeral A–P midshaft diameter and pubic length

	Pp/Pt	Ggg/Ggb	Pan/Gor	Pan/Or	Gor/Or
Phalangeal length			*		
Mediolateral midshaft width			*		
Dorsopalmar midshaft width			*		
Mediolateral basal width					
Dorsopalmar basal width			*		
Mediolateral trochlear width					
Dorsopalmar trochlear width					
Biepicondylar width					

[1] * = significant difference in y-intercept at $P<0.001$.
[2] Pp = *Pan paniscus*; Pt = *Pan troglodytes troglodytes*; Ggg = *Gorilla gorilla gorilla*; Ggb = *Gorilla gorilla beringei*; Or = *Pongo pygmaeus*; Pan = Pp & Pt, Gor = Ggg & Ggb.

growth trajectory of chimpanzees. Compared to *Gorilla*, *Pongo* exhibits several differences in ontogenetic allometries. For midshaft diameters and mediolateral head width, there are significant transpositions between *Gorilla* and *Pongo* growth trajectories such that the growth trajectories of *Pongo* are shifted below those of *Gorilla* (Figure 3). Furthermore, the growth trajectories of metacarpal III length for *Gorilla* are shifted below those of *Pongo* and *Pan*, which are ontogenetically scaled (Figure 4).

For proximal phalangeal lengths, *Pongo* is markedly different than any African ape. The slopes of proximal phalangeal lengths regressed on size surrogates are similar between orang-utans and the African apes (Table 3); however, the growth trajectories of *Pongo* are transposed above all the African apes' growth trajectories. In other words, for a given size,

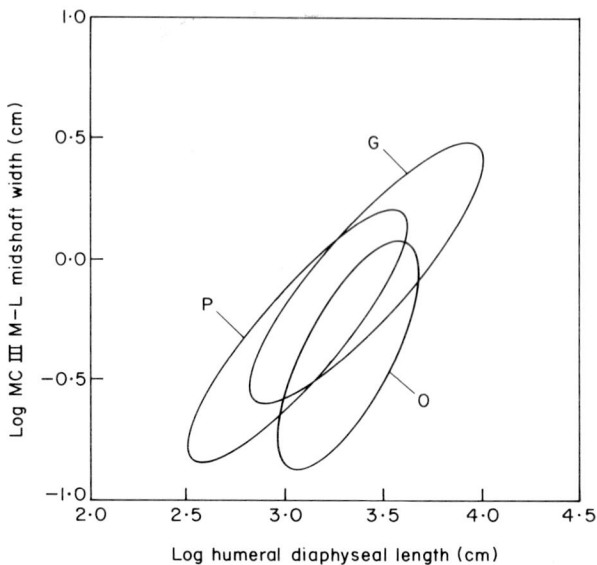

Figure 3. An intergeneric plot of the bivariate ontogenetic allometry of metacarpal III mediolateral midshaft width *vs.* humeral diaphyseal length. Ellipses contain 95% of the data points. P = *Pan*, G = *Gorilla*, O = *Pongo*.

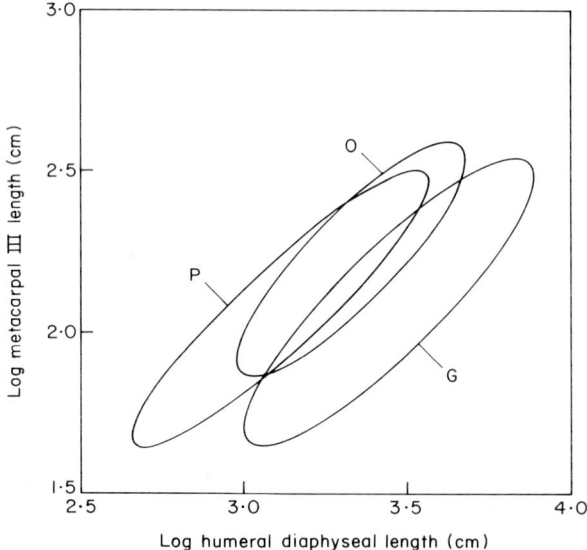

Figure 4. An intergeneric plot of the bivariate ontogenetic allometry of metacarpal III length *vs.* humeral diaphyseal length. Ellipses contain 95% of the data points. For labels, see Fig. 3.

Pongo has the longest proximal phalanges, *Gorilla* has the shortest and *Pan* has phalanges intermediate in length (Figure 5).

In contrast to proximal phalangeal length, orang-utan growth trajectories are shifted below gorilla growth trajectories for proximal phalangeal midshaft diameters; however, orang-utans are ontogenetically scaled with chimpanzees for these variables (Table 3).

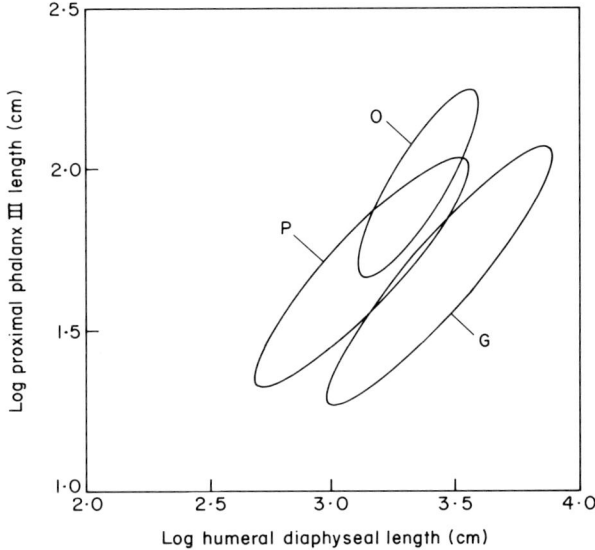

Figure 5. An intergeneric plot of the bivariate ontogenetic allometry of proximal phalanx III length *vs.* humeral diaphyseal length. Ellipses contain 95% of the data points. For labels, see Fig. 3.

Among manual rays
Within Pan *and within* Gorilla. *Post hoc* tests for homogeneity of slopes and *y*-intercepts between different metacarpals yield similar patterns for *P. paniscus* and *P. troglodytes* (Table 4). Intermetacarpal analyses for chimpanzees and bonobos reveal that the ontogenetic allometries of metacarpals IV and V are similar in slope to metacarpals II and III for length and dorsopalmar midshaft widths, but are shifted beneath them (Figure 6).

Post hoc analyses for homogeneity of slopes and intercepts among different metacarpals for *G. g. gorilla* and *G. g. beringei* yield patterns different from both chimpanzees; analyses reveal no significant differences in slopes and *y*-intercepts among metacarpals II–V for most measurements for *G. g. gorilla* and *G. g. beringei* (Figure 7).

Comparisons among rays for slope differences in phalangeal variables regressed on humeral diaphyseal or pubic lengths reveal some differences between chimpanzees and bonobos (Table 4). There is no transpositioning of growth trends among rays for *P. paniscus* in virtually all proximal phalangeal dimensions regressed on size surrogates. In other words, bonobos are ontogenetically scaled among different phalanges for almost any dimension on size surrogates. In contrast, *P. troglodytes* shows great interphalangeal variation for most measurements. For many of these phalangeal dimensions, phalanges III and IV were usually significantly larger than phalanx V.

Post hoc analyses for slope and intercept differences in all proximal phalangeal measurements for the mountain gorilla reveal ontogenetic scaling among rays (Table 4). The western lowland gorilla, in contrast, exhibits a common pattern of variation among the phalanges for most regressions of phalangeal dimensions on humeral diaphyseal or pubic lengths. In virtually all contrasts of phalangeal variables against size surrogates, phalanges II, III and IV are ontogenetically scaled. However, the growth allometry of phalanx V, which has a slope similar to those of phalanges II, III and IV, is significantly ($P<0.001$) transposed beneath the growth allometries of phalanges II, III and IV. In other words, there is a shift of

Table 4 **Ontogenetic differences in slope and *y*-intercepts among manual rays regressed against size surrogates**

	Differences between metacarpals				
	Pp^2	Pt	Ggg	Ggb	Or
Metacarpal length					
II/III					
II/IV	*				
II/V	*	*			*
III/IV					
III/V	*	*			
IV/V					*
Mediolateral midshaft width					
II/III					
II/IV					
II/V					
III/IV					
III/V	*				
IV/V					
Dorsopalmar midshaft width					
II/III		*			
II/IV					
II/V	*				
III/IV					
III/V	*	*			
IV/V		*			
Mediolateral head width					
II/III					
II/IV					
II/V					
III/IV					
III/V					
IV/V					
Dorsopalmar head width					
II/III					
II/IV					
II/V					
III/IV					
III/V	*				
IV/V	†				*
Biepicondylar width					
II/III		†			
II/IV					
II/V				*	
III/IV	*				
III/V	*	†		*	
IV/V				*	

Differences between proximal phalanges					
Phalangeal length					
II/III		*			
II/IV					
II/V				*	
III/IV	*				
III/V				*	
IV/V				*	

Table 4 *(Continued)*

Mediolateral midshaft width				
II/III				
II/IV				
II/V				*
III/IV				
III/V			*	*
IV/V			*	*
Dorsopalmar midshaft width				
II/III			†	
II/IV				
II/V				
III/IV				
III/V			†	
IV/V			†	
Mediolateral basal width				
II/III			*	
II/IV			*	
II/V		*		
III/IV				
III/V		*	*	*
IV/V			*	
Dorsopalmar basal width				
II/III				
II/IV				
II/V				
III/IV				
III/V				*
IV/V			*	
Mediolateral trochlear width				
II/III				
II/IV				
II/V				†
III/IV				†
III/V			*	†
IV/V			*	*
Dorsopalmar trochlear width				
II/III			*	
II/IV				
II/V				
III/IV				
III/V				
IV/V				
Biepicondylar width				
II/III				
II/IV				
II/V				
III/IV			*	
III/V			*	
IV/V				

Differences between middle phalanges

	Pt	Ggg
Phalangeal length		
II/III		
II/IV		

Table 4 (*Continued*)

	Pt	Ggg
II/V		
III/IV		
III/V		
IV/V		*
Mediolateral midshaft width		
II/III		
II/IV		
II/V		
III/IV		
III/V		
IV/V		
Dorsopalmar midshaft width		
II/III		
II/IV		
II/V		
III/IV		
III/V		
IV/V		
Mediolateral basal width		
II/III	*	
II/IV		
II/V	*	
III/IV	*	
III/V	†	*
IV/V	*	*
Dorsopalmar basal width		
II/III		
II/IV		
II/V		
III/IV		
III/V	*	
IV/V		
Mediolateral trochlear width		
II/III		*
II/IV		
II/V		*
III/IV	*	
III/V	*	
IV/V		*
Dorsopalmar trochlear width		
II/III		
II/IV		
II/V		
III/IV		
III/V	*	*
IV/V		
Biepicondylar width		
II/III		
II/IV		
II/V		
III/IV		
III/V	*	*
IV/V		

[1]† = Significant difference in slope at $P < 0.001$; * = significant difference in y-intercepts at $P < 0.001$.
[2]Pp = *Pan paniscus*; Pt = *Pan troglodytes troglodytes*; Ggg = *Gorilla gorilla gorilla*; Ggb = *Gorilla gorilla beringei*; Or = *Pongo pygmaeus*.

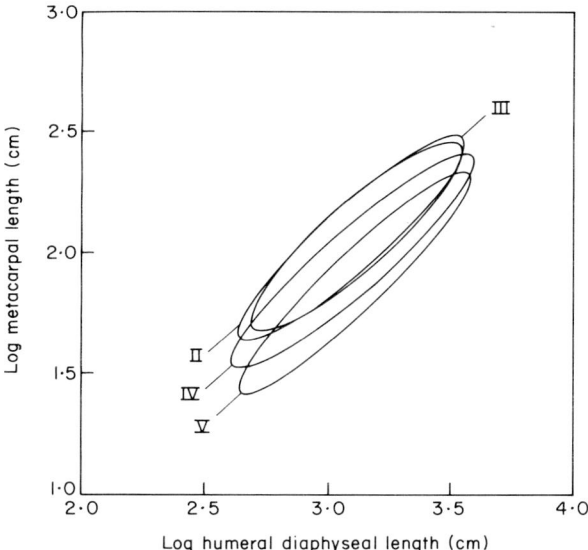

Figure 6. Bivariate ontogenetic allometry of metacarpal II, III, IV and V lengths *vs.* humeral diaphyseal length for *Pan troglodytes*. Ellipses contain 95% of the data points.

the growth trajectory of proximal phalanx V below the combined growth trajectory of phalanges II, III and IV for almost all bivariate regressions of phalangeal measurements on size surrogates.

Pan vs. Gorilla. Generally, gorilla metacarpals tend to be less variable among rays. In contrast, common chimpanzees and bonobos exhibit more prominent changes among rays in metacarpal dimensions. For most metacarpal dimension regressed on humeral diaphyseal or pubic lengths, both gorilla subspecies exhibit ontogenetic scaling between the different rays, whereas chimpanzees and bonobos both exhibit similar regression slopes among the rays, but show significant shifts in the metacarpal growth trajectories (Table 4). Inter-metacarpal analyses reveal that generally more lateral metacarpals (II, III) of chimpanzees have relatively larger metacarpal dimensions than the medial metacarpals (IV, V) at similar sizes.

Mountain gorillas and bonobos exhibit less interphalangeal variation than common chimpanzees and lowland gorillas. Common chimpanzees and lowland gorillas show significant transpositioning between rays for various proximal phalangeal dimensions regressed on humeral diaphyseal and pubic lengths (Table 4). The most common pattern of shifts for common chimpanzees among different phalangeal growth trajectories may be depicted as series of two downward shifts going from proximal phalanx III, to phalanges II/IV and finally phalanx V. For lowland gorillas, the growth trajectories are not significantly different among phalanges II, III and IV but are generally shifted above the growth trajectory of phalanx V.

Pan *and* Gorilla vs. Pongo. The scaling patterns among orang-utan metacarpals are similar to both chimpanzee and gorilla inter-metacarpal patterns. For comparisons among metacarpals, orang-utans are similar to gorillas, but are similar to chimpanzees for metacarpal lengths (Table 4). As with gorillas, different metacarpals of orang-utans are virtually indistinguishable for midshaft and head diameters at a given body size. However, like

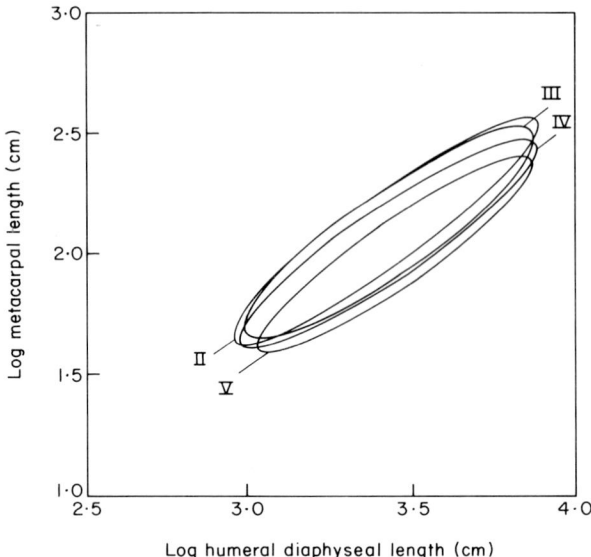

Figure 7. Bivariate ontogenetic allometry of metacarpal II, III, IV and V lengths *vs.* humeral diaphyseal length for *Gorilla gorilla gorilla*. Ellipses contain 95% of the data points.

chimpanzees, orang-utans exhibit intermetacarpal differences in scaling patterns for metacarpal length, with the length of the fifth metacarpal shorter relative to the second through fourth metacarpals.

Summary of manual ray III comparisons:
(1) there are no significant differences between male and female chimpanzees and gorillas for bivariate comparisons of metacarpal and phalangeal variables on size surrogates;
(2) for a given size surrogate, the African ape groups are generally ontogenetically scaled between species/subspecies and between genera for most metacarpal and phalangeal variables;
(3) for a given size surrogate, African apes generally have relatively larger manual ray midshaft dimensions than *Pongo*, although *Pongo* is ontogenetically scaled with the African apes for most articular surface diameters;
(4) at similar sizes, *Pongo* and both species of *Pan* have the longer metacarpals and phalanges than *Gorilla*, and *Pongo* has the longest phalanges.

Summary of inter-ray differences:
(1) the slopes and *y*-intercepts of metacarpals II–V are statistically indistinguishable among digits for almost all measurements regressed on size surrogate within both gorillas and sometimes orang-utans;
(2) at similar size surrogates, phalanges II, III and IV of lowland gorillas are generally relatively larger than phalanx V for most external dimensions;
(3) at similar size surrogates, metacarpals and phalanges II, III and IV of chimpanzees are generally relatively larger than digit V for all metacarpal and phalangeal dimensions;
(4) within orang-utans, metacarpals II, III and IV are longer than metacarpal V when compared to any size surrogate.

Discussion

Chimpanzees vs. *gorillas*

Previous studies on African ape hands have noted differences between adult gorillas and chimpanzees in proportions of the manual rays (Schultz, 1936, 1956; Susman, 1979). Susman, in particular, has documented the differences in the adult morphologies between chimpanzees and gorillas for each separate segment of the manual rays, as well as between rays (1979). Interesting differences between chimpanzees and gorillas have been noted using a variety of indices (Schultz, 1936, 1956; Susman, 1979). What is unclear from these studies is what is driving the differences in proportions between chimpanzees and gorillas—how do the individual characters used as numerators and denominators relate to increasing body size and how do characters relate to one another with increasing body size? Are differences in ray proportions between chimpanzees and gorillas simply due to differences in size and do gorillas exhibit proportions we would expect given their larger body size? Or, are the differences in ray proportions between chimpanzees and gorillas produced by dissociations of their growth patterns?

This study reveals that the majority of established variation between adult chimpanzees and gorillas in metacarpal and phalangeal morphology (Midlo, 1934; Susman, 1978, 1979; Tuttle, 1967, 1969a,b,c, 1970; Schultz, 1936) may be attributed to ontogenetic scaling. Metacarpal head depth, dorsopalmar metacarpal midshaft width, proximal phalangeal trochlear width and proximal phalangeal base widths, features that distinguish gorilla adults from chimpanzee adults (Susman, 1979), have equal expression in both genera at common sizes. In other words, even though there are relative differences in metacarpal and phalangeal characters between adult chimpanzees and gorillas, chimpanzees and gorillas share a common growth pattern and the differences in proportions are produced simply by the extension of gorillas into larger size ranges. Gorillas, compared to chimpanzees, have the proportions we would expect for most metacarpal and phalangeal characters given their greater body size.

Metacarpal and phalangeal lengths, however, show a significant departure from the pattern of ontogenetic scaling. Although gorillas have absolutely longer metacarpals and phalanges than chimpanzees, gorillas have shorter metacarpals and phalanges than chimpanzees at common sizes. The relatively shorter metacarpals and phalanges of adult gorillas, as well as gorillas at all comparable sizes with chimpanzees, are produced by the downward transposition of the metacarpal and phalangeal length growth trajectories below those of chimpanzees. This confirms previous findings that gorillas have the relatively shortest fingers of the great apes (Schultz, 1936; Susman, 1979), but clarifies that gorilla metacarpals and phalanges are not only relatively shorter for their body weight compared to chimpanzees, but are also absolutely shorter than chimpanzees at common body sizes.

Many of the previous studies use adjectives and indices that refer to both size and shape to describe proportions in hominoid hands (Midlo, 1934; Schultz, 1936, 1956; Susman, 1979). Descriptives such as "shorter", "longer" and "wider" imply differences in size for specific dimensions; whereas words such as "robust" and "stout" describe the relationship of one dimension to another, i.e., shape. Adult gorilla metacarpals have been described as "short and stout" and phalanges as "stout" and "heavily constructed" compared to the metacarpals and phalanges of adult chimpanzees and orang-utans (Midlo, 1934; Schultz, 1936; Susman, 1979); however, it is unclear from the descriptions whether or not one or both of these dimensions are changing with size, and what is the relationship of each dimension to size. It is unknown whether or not gorilla metacarpals and phalanges are more "stout" than the other pongids because (1) chimpanzees and gorillas are ontogenetically scaled but are negatively allometric for length producing a relatively smaller metacarpal length in gorillas, (2) chimpanzees and gorillas are ontogenetically scaled but are positively allometric for width, (3) gorillas have shorter lengths at common sizes with chimpanzees, or (4) gorillas have larger widths at common sizes with chimpanzees.

Results indicate that the previous metacarpal index values that separate chimpanzees and gorillas (Susman, 1979) are driven solely by changes in the denominator, length. For example, at common sizes there is no significant difference between chimpanzees and gorillas in metacarpal midshaft widths and midshaft widths scale isometrically with increasing body size. Metacarpal lengths, however, are shorter in gorillas compared to chimpanzees at common sizes. Similarly, gorillas have shorter phalanges at common sizes with chimpanzees. In addition, gorillas have larger proximal and middle mediolateral phalangeal midshaft widths and larger middle dorsopalmar phalangeal midshaft widths than chimpanzees at common sizes. Therefore, the "stout" appearance of gorilla metacarpals compared to adult chimpanzees is due to relatively shorter metacarpals in gorilla than chimpanzee adults and also to shorter metacarpals with no change in midshaft diameters at common sizes. On the other hand, the more "stout" appearance of adult gorilla proximal and middle phalanges compared to adult chimpanzees is due not only to relatively shorter phalanges at common sizes, but also greater phalangeal widths at comparable body sizes with chimpanzees.

Relative proximal phalangeal length, however, is produced by a different growth pattern. Susman (1979) has indicated that adult gorillas have the relatively shortest phalanges compared to metacarpals of the great apes. Results of this study indicate that there are significant transpositions among the great apes for both phalangeal and metacarpal lengths. However, when phalangeal length is regressed on metacarpal length, chimpanzees and gorillas are ontogenetically scaled with negatively allometric slopes. This confirms Susman's statements that in absolute terms, adult gorillas have relatively shorter phalanges compared to metacarpals than adult chimpanzees. However, it clarifies that these adult proportions are simply produced by negative allometry; at common metacarpal sizes, gorillas and chimpanzees have the same relative phalangeal length, but as gorillas surpass the size of chimpanzees, their phalanges progressively become relatively shorter.

African apes vs. *orang-utans*

It has been noted that the manual morphology of adult orang-utans differs from adult African apes in a number of ways (Midlo, 1934; Schultz, 1927, 1936; Susman, 1979). Some features thought to distinguish African apes from orang-utans are the stronger development of the metacarpal heads and dorsopalmar midshaft width in the former (Susman, 1979). However, contrary to previous notions, interspecific comparisons of articular surface measurements (e.g., metacarpal head diameters) and metacarpal dorsopalmar midshaft widths against size surrogates, reveal that orang-utans share a common growth pattern with the chimpanzees and the observed differences in the adult morphologies are due to differing endpoints along this common growth pattern. In contrast, orang-utans, compared to gorillas, have significantly smaller metacarpal and proximal phalangeal shaft dimensions and a smaller metacarpal mediolateral head width at common sizes.

It is well established that adult orang-utans have longer fingers than adult chimpanzees and gorillas (Midlo, 1934; Susman, 1979; Schultz, 1936). Differences in finger lengths between adult orang-utans and gorillas are obviously not thought to be related to simple differences in body size, since the adults clearly overlap in body size. However, it was unknown whether or not differences in finger lengths between African ape adults or between chimpanzee and orang-utan adults were attributed to differences in adult body size. Results from this study demonstrate that at common sizes, orang-utans have longer proximal phalanges compared to any African ape. However, contrary to previous thoughts, the metacarpal length of orang-utans is not significantly different from chimpanzees when size is properly considered (Figures 4 and 5). Therefore, the "long" and "slender" appearance of the orang-utan metacarpals and proximal phalanges, in contrast to gorillas, is due to greater metacarpal and proximal phalangeal lengths and smaller metacarpal and proximal phalangeal shaft diameters at common sizes. In contrast, the "long" and "slender" appearance of orang-utan metacarpals compared to chimpanzees is due only to smaller orang-utan metacarpal shaft diameters at common sizes; whereas the appearance of proximal phalanges is attributed to longer proximal phalanges at common sizes with chimpanzees.

In most metacarpal and phalangeal dimensions, chimpanzees were previously thought to occupy an intermediate position between gorillas and orang-utans (Midlo, 1934; Schultz, 1927, 1936; Susman, 1979). These results clarify that most metacarpal and phalangeal dimensions, with the exception of length, that purportedly differ between chimpanzee and gorilla adults are not significantly different after proper size correction. Therefore, the established relative differences between adult chimpanzee and gorilla metacarpals and proximal phalanges are predominantly the result of differing body size and functional association with body size. However, other characters that distinguish between gorillas and chimpanzees, as well as among gorillas, chimpanzees and orang-utans, even after size adjustment, are proximal phalangeal mediolateral midshaft width, middle phalangeal midshaft widths and middle phalangeal dorsopalmar base width. The differences among pongids in the ontogenetic allometries of metacarpal and phalangeal lengths and the indices that were constructed using these lengths, are perhaps what misled Susman (1979) and others to identify manual features that are allometrically based in gorilla, chimpanzee and orang-utan adults as taxonomically distinct features.

Re-evaluation of biomechanical models
All of the biomechanical models mentioned previously dealt with differences among pongid adults in metacarpal and phalangeal lengths and midshaft widths. Metacarpal and phalangeal lengths and widths may be affected by a variety of locomotor and non-locomotor requirements (Susman, 1979). The biomechanical explanations proposed at the outset of this paper to describe differences among adult chimpanzees, gorillas and orang-utans are not plausible when considering differences between African apes, as a group, and orang-utans. When comparing the relative proportions of orang-utan to gorilla hands, the expected differences are found, according to previously discussed biomechanical models, between a knuckle-walker and a more arboreal ape. However, comparison of orang-utan and chimpanzee hands yields fewer differences in their relative proportions than comparison to gorilla hands.

Gorilla metacarpals and phalanges, compared to those of comparably sized orang-utans, are shorter in lengths and greater in widths, as predicted by bending and lever models. In contrast, chimpanzee metacarpals and phalanges are different from those of orang-utans in ways that are predicted by models, but only in one dimension each. Compared to comparably sized orang-utans, chimpanzees have larger metacarpal mediolateral midshaft widths and shorter proximal phalanges. According to proposed models, chimpanzees appear to be biomechanically intermediate between gorillas and orang-utans, and thus should be treated as a distinct group from gorillas.

Since gorillas and chimpanzees are not ontogenetically scaled for all hand proportions, there must be an explanation for differing finger lengths and midshaft widths in chimpanzees and gorillas at similar sizes. Since gorillas and chimpanzees are both knuckle-walkers, the question we must then entertain is what would produce differences in metacarpal and phalangeal proportions between gorillas and chimpanzees at comparable sizes? One can posit that gorillas carry a greater proportion of body weight on their hands than chimpanzees, thus producing greater torques at the gorilla metacarpophalangeal joint. However, we must then explain why we find the metacarpal midshaft and articular surfaces, features that are related to weight bearing, to be similar at comparable sizes.

Another explanation for the differences in finger lengths within African apes is that chimpanzees and gorillas use kinematically (and biomechanically) distinct forms of knuckle-walking. Chimpanzees and gorillas appear to use distinct forms of knuckle-walking as adults (Inouye, 1989; Tuttle, 1967, 1969a,b); however, it is unknown whether chimpanzees and gorillas maintain these distinctions at similar sizes. Tuttle has noted that during knuckle-walking, gorillas tend to hyperextend their elbows and use more pronated hand positions than chimpanzees, which may flex their elbows and use a greater variety of hand positions (Inouye, 1989; Tuttle, 1967, 1969a,b). Tuttle also noted that juvenile gorillas use knuckle-walking postures stereotypical of gorilla adults (Tuttle, 1969a); this is of considerable importance, because it implies that gorillas and chimpanzees at common sizes use kinematically distinct types of knuckle-walking. Although differences in knuckle-walking postures between chimpanzees and gorillas exist, the variation has not yet been rigorously documented and thus makes difficult the evaluation of any biomechanical models.

One may also argue that differences in African ape finger lengths are related simply to the frequency of knuckle-walking. Shortened finger segments may be adaptive for repeated loading of the gorilla fingers, compared to chimpanzees, in knuckled-hand postures by lessening the stress at the metacarpophalangeal joints. Doran (1992) has demonstrated that the locomotor repertoires of pygmy and common chimpanzees change during ontogeny; however, the ontogeny of gorilla locomotion is less known. If the locomotor ontogenies of chimpanzees and gorillas demonstrate that they use different frequencies of locomotor behaviors at similar sizes, i.e., a greater frequency of knuckle-walking behavior in gorillas at common size with chimpanzees, then it could support the hypothesis that the observed ontogenetic differences in finger morphology are related to differences in locomotor repertoires throughout ontogeny.

An alternative functional explanation of the differences in phalangeal lengths between the African apes and *Pongo*, which implies opposite evolutionary directionality of this character, attributes the longer phalanges of *Pongo* as an adaptation for increasing the effectiveness of the hook grip when grasping large diameter branches during climbing (Susman, 1979). The longer orang-utan metacarpals and phalanges increase the compass of the hand during grasping behaviors (Preuschoft, 1973; Susman, 1979). If we agree that orang-utan fingers are lengthened as adaptations for arboreal, grasping behaviors, then the length of the fingers of chimpanzees, compared to gorillas, must similarly represent an evolutionary adaptation for arboreality. The differences among the great apes in metacarpal and phalangeal lengths may then be viewed as a part of a terrestrial–arboreal continuum, with gorillas representing the terrestrial pole, orang-utans the arboreal pole and chimpanzees the intermediate. This explanation is plausible, especially since Fleagle (1976) and Tuttle (1977, 1986) have summarized that behaviorally gorillas are the most terrestrial of the great apes, orang-utans clearly the most arboreal and chimpanzees are intermediate—spending a great deal of time engaged in both arboreal and terrestrial locomotor activitites. Further evidence that is needed to support this hypothesis are comparisons of ontogenetic locomotor data between chimpanzees and gorillas which show that chimpanzees are more frequently arboreal than gorillas at common sizes.

The extensions of these models to explain differences between chimpanzees and gorillas are largely compatible with the results. However, there are a few interspecific differences that deserve attention. According to the lever and bending models previously proposed, gorillas and chimpanzees have the expected differences in metacarpal and phalangeal lengths given some difference in knuckle-walking. Furthermore, the proximal phalangeal midshaft width is accordingly wider in gorillas compared to comparably sized chimpanzees; however, metacarpal midshaft width does not exhibit the same dissociation between genera. Even though metacarpals of gorillas are not wider than chimpanzee metacarpals at common sizes, they are still better adapted to resist bending stresses by reduction in their length. Why differential widening of the proximal phalanges compared to the metacarpals occurs in gorillas may be mechanically related to a variety of behaviors which is unclear at present.

There are several features of *P. paniscus* and *P. troglodytes* that depart from ontogenetic scaling and require functional interpretation in light of the proposed biomechanical models. Most of the characters that differ between bonobos and common chimpanzees are related to changes in shaft dimensions of rays. For metacarpal and proximal phalangeal mediolateral midshaft widths regressed against size and proximal phalangeal biepicondylar width, the growth trajectories of bonobos are shifted below those of common chimpanzees. Mountain and lowland gorillas, in contrast, show a departure from ontogenetic scaling in metacarpal lengths. At similar sizes, mountain gorillas have shorter metacarpals than lowland gorillas. These differences may be expected in light of the bending model for metacarpals and phalanges, especially since common chimpanzees are purportedly more terrestrial than pygmy chimpanzees (Doran, 1992) and mountain gorillas more than lowland gorillas. In other words, some of the differences that occur between chimpanzees and gorillas may also occur between bonobos and common chimpanzees and between lowland and mountain gorillas for similar reasons. However, the departure of ontogenetic allometries of chimpanzees and gorillas is visually and statistically more remarkable than departures between bonobos and common chimpanzees and between lowland and mountain gorillas.

Inter-ray relationships
Analyses reveal that in contrast to chimpanzees, the fourth and fifth metacarpals of gorillas are not significantly different in length (cf. Susman, 1979). However, within gorillas, the similarity in length is not produced by differential rates of growth, rather, the length of the fifth metacarpal is not significantly different than the length of the other metacarpals throughout ontogeny. In contrast, chimpanzees show a marked relative shortening of the fifth metacarpal compared to the other metacarpals at all sizes (Figure 8). The similarity in dimensions among metacarpals in gorillas and the variation among metacarpals in chimpanzees indicate that gorillas are possibly positioning individual rays differently than chimpanzees and orang-utans (Susman, 1974; Susman & Tuttle, 1976; Tuttle, 1967, 1969*a,b*, 1970, 1975, 1977, 1986; Tuttle & Basmajian, 1974; Tuttle & Beck, 1972; Tuttle & Cortright, 1988). Tuttle (1967, 1969*a,b,c*) noted that gorillas frequently use different manual rays in knuckle-walking postures to chimpanzees and suggested that gorilla adults use their

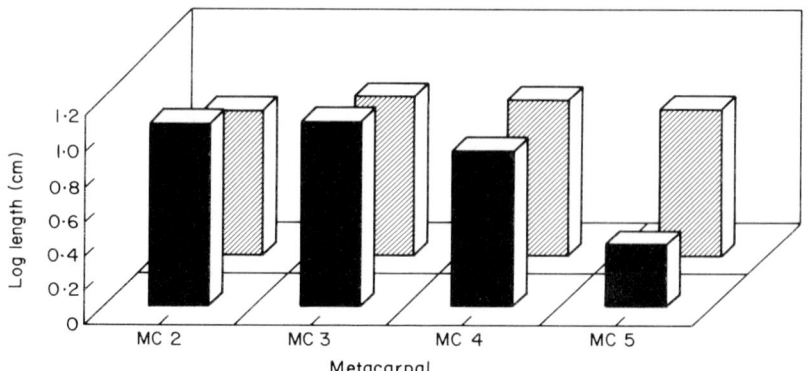

Figure 8. Schematic diagram of the relative length of chimpanzee and gorilla metacarpals during ontogeny. Metacarpal lengths are the *y*-intercept values for bivariate regressions of metacarpal lengths on pubic lengths. (■) *Pan troglodytes troglodytes*; (▨) *Gorilla gorilla gorilla*.

fifth digits to support body weight in knuckle-walking postures, while chimpanzees almost never use their fifth digits.

In gorillas, the evenness in length of metacarpals II–V straightens the line of articulation at the metacarpophalangeal joints and thereby increases the number of supports. This explanation was originally used by Etter (1973) to describe the similar phenomenon that distinguishes the more arboreal from the larger, more terrestrial cercopithecoid monkeys. From my own observations of knuckled hand postures of chimpanzees and gorillas (Inouye, 1989, 1992), this difference in locomotor hand posture probably occurs at all common postnatal sizes.

In addition to the length of metacarpals, metacarpal shaft and head dimensions are also undifferentiated among metacarpals for gorillas, while the fifth metacarpal (and usually the fourth) of chimpanzees is typically smaller than the remaining metacarpals in these dimensions. This also lends support to the hypothesis that gorillas use all of their metacarpals (excluding the first) to support their body weight during terrestrial hand postures, as gorillas have larger shaft and head dimensions for the fourth and fifth metacarpals relative to the second and third metacarpals than chimpanzees.

Conclusions

The use of ontogenetic analyses in morphological studies can be used as a powerful tool to provide insight into morphological differences among adult primates and other animals. This study demonstrates that many of the metacarpal and phalangeal characters that distinguish chimpanzee from gorilla adults are not significantly different after allometric adjustment. Most manual features, like other cranial and postcranial features (Shea, 1981, 1983*a*,*b*, 1986; Jungers & Susman, 1984), show the pervasive pattern of ontogenetic scaling between chimpanzees and gorillas; the relative proportional differences in the hand that are observed between adults are the result of differences in body size. These differences are what we would expect given the greater size of the latter.

Other features such as the lengths of the metacarpals and phalanges, phalangeal mediolateral midshaft widths and interray proportions, which were known to differ between chimpanzee and gorilla adults (Susman, 1979), are not underlain by a common growth pattern and consequently differ at common sizes. The "stout" metacarpals characteristic of adult gorillas and the "gracile" metacarpals of chimpanzees we now know are produced by the differential growth of metacarpal length with no differences in the growth of shaft diameters. The dissociations in the growth patterns between gorillas and chimpanzees for lengths and interray proportions imply adaptations for different shapes at common sizes. Thus, the shorter metacarpals and phalanges of gorillas, compared to chimpanzees, may be biomechanically adaptive for a greater frequency of terrestrial locomotion or a kinematically distinct pattern of knuckle-walking.

While differences in size do not explain all differences between chimpanzee and gorilla hands, it is important because size has a significant influence on the majority of morphological differences observed in adults that were previously thought to be functionally adaptive for each species at all sizes. By controlling for the effects of size, we can more accurately identify characters which are distinctive of, and perhaps functionally adaptive for, individual species. Differences in metacarpal and phalangeal lengths, phalangeal midshaft widths and in inter-ray proportions are perhaps the most important functional characters separating gorillas from chimpanzees. However, future studies on the functional and biomechanical properties of these features are needed before we can resolve competing hypotheses.

Acknowledgements

This work was supported by a National Science Foundation Dissertation Improvement award (NSF BNS 9000-964) and a Northwestern University Dissertation Year Grant (0100-510-152Y). I thank Drs Brian Shea, Jeff Meldrum, Bill Jungers and Russell Tuttle and two anonymous reviewers for their valuable comments and criticisms. I also thank Dr Matt Ravosa and Ann Gomez for inviting me to participate in this symposium. I gratefully acknowledge the support and assistance of Dr W. Van Neer, Mr D. R. Howlett, Dr R. W. Thorington and the helpful museum staffs at the Musee Royale De l'Afrique Centrale, The Powell Cotton Museum & Quex House and the National Museum of Natural History.

References

Brain, C. K., Churcher, C. S., Clark, J. D. *et al.* (1988). New evidence of early hominids, their culture and environment from the Swartkrans cave, South Africa. *S. Afr. J. Sci.* **84,** 828–835.
Bush, M. E., Lovejoy, C. O., Johanson, D. C. & Coppens, Y. (1982). Hominid carpal, metacarpal, and phalangeal bones recovered from the Hadar formation: 1974–1977. *Am. J. phys. Anthrop.* **57**(4), 651–677.
Clarke, M. R. B. (1980). The reduced major axis of a bivariate sample. *Biometrika* **67,** 441–446.
Currey, J. (1984). *The Mechanical Adaptations of Bones.* New Jersey: Princeton University Press.
Doran, D. M. (1991). The ontogeny of chimpanzee and pygmy chimpanzee locomotor behavior: a case study of morphological paedomorphism and its behavioral correlates. *Am. J. phys. Anthrop.* (suppl.) **12,** 69.
Doran, D. M. (1992). The ontogeny of chimpanzee and pygmy chimpanzee locomotor behavior: a case study of morphological paedomorphism and its behavioral correlates. *J. hum. Evol.* **23,** 139–157.
Erikson, G. E. (1963). Brachiation in new world monkeys and in anthropoid apes. *Symp. zool. Soc. Lond.* **10,** 135–163.
Etter, H. F. (1973). Terrestrial adaptations in the hands of Cercopithecinae. *Folia primat.* **20,** 331–350.
Fleagle, J. G. (1976). Locomotion and posture of the Malayan siamang and implications for hominoid evolution. *Folia primat.* **26,** 245–269.
Garn, S. M., Sulivan, T. M., Decker, S. A. & Hawthorne, V. M. (1991). Brief communication: on the optimum number of metacarpals for roentgenogrammetric measurement. *Am. J. phys. Anthrop.* **85,** 229–232.
Goldman, D., Rathna Giri, P. & O'Brien, S. J. (1987). A molecular phylogeny of the hominoid primates as indicated by two-dimensional protein electrophoresis. *Proc. natn. Acad. Sci. U.S.A.* **84,** 3307–3311.
Goodman, M., Tagle, D. A., Fitch, D. H. A. *et al.* (1990). Primate evolution at the DNA level and a classification of Hominoids. *J. molec. Evol.* **30,** 260–266.
Gould, S. J. (1966). Allometry and size in ontogeny and phylogeny. *Q. Biol. Rev.* **41,** 587–640.

Gould, S. J. (1975). Allometry in primates, with emphasis on scaling and the evolution of the brain. In (F. Szalay, Ed.) *Approaches to Primate Paleobiology, Contributions in Primatology*, vol. 5, pp. 244–292. New York: Karger, Basel.

Gould, S. J. (1977). *Ontogeny and Phylogeny*. Cambridge: Harvard University Press.

Gregory, W. K. (1928). Were the ancestors of man primitive brachiators? *Proc. Am. Phil. Soc.* **67,** 129–150.

Hasegawa, M., Kishino, H. & Yano, T. (1989). Estimation of branching dates among primates by molecular clocks of nuclear DNA which slowed down in Hominoidea. *J. hum. Evol.* **18,** 461–476.

Huxley, T. H. (1893). *Man's Place in Nature*. New York: D. Appleton and Co.

Inouye, S. E. (1989). Variability of knuckle-walking behavior in the African apes. *Am. J. phys. Anthrop.* **75,** 245.

Inouye, S. E. (1990). Variation in the presence and development of the dorsal ridge of the metacarpal head in African apes. *Am. J. phys. Anthrop.* **81,** 243.

Inouye, S. E. (1991a). Ontogeny and allometry in African ape fingers. In (A. Ehara *et al.*, Eds) *Primatology Today*, pp. 537–538. Tokyo: Elsevier.

Inouye, S. E. (1991b). Variation in the growth of African ape fingers. *Am. J. phys. Anthrop.* (suppl.) **12,** 96.

Inouye, S. E. (1992). Ontogeny of knuckle-walking hand postures in African apes. *Am. J. phys. Anthrop.* (suppl.) **14,** 93.

Jolicoeur, P. (1990). Bivariate allometry: interval estimation of the slopes of the ordinary and standardized normal major axes and structural relationship. *J. theor. Biol.* **144,** 275–285.

Jungers, W. L. (1984). Aspects of size and scaling in primate biology with special reference to the locomotor skeleton. *Yearb. phys. Anthop.* **27,** 73–97.

Jungers, W. L. & Susman, R. L. (1984). Body size and skeletal allometry in African apes. In (R. L. Susman, Ed.), *The Pygmy Chimpanzee: Evolutionary Biology and Behavior*, pp. 131–177. New York: Plenum Press.

Kishino, H. & Hasegawa, M. (1989). Evaluation of maximum likelihood estimate of the evolutionary tree topologies from DNA sequence data, and branching order in Hominoidea. *J. molec. Evol.* **29,** 170–179.

Koop, B. F., Tagle, D. A., Goodman, M. & Slightom, J. L. (1989). A molecular view of primate phylogeny and important systematic and evolutionary questions. *Mol. Biol. Evol.* **6,** 580–612.

Kuhry, B. & Marcus, L. F. (1977). Bivariate linear models in biometry. *Syst. Zool.* **26,** 201–209.

Lewis, O. J. (1969). The hominoid wrist joint. *Am. J. phys. Anthrop.* **30,** 251–269.

Lewis, O. J. (1972a). Osteological features characterizing the wrists of monkeys and apes, with a reconsideration of this region in *Dryopithecus* (Proconsul) *africanus*. *Am. J. phys. Anthrop.* **36,** 45–58.

Lewis, O. J. (1972b). Evolution of the hominoid wrist. In (R. H. Tuttle, Ed.) *Functional and Evolutionary Biology of Primates*, pp. 207–222. Chicago: Aldine-Atherton.

Lewis, O. J. (1973). The hominoid os capitatum with special reference to the fossil bones from Sterkfontein and Olduvai Gorge. *J. hum. Evol.* **2,** 1–11.

Lewis, O. J. (1977). Joint remodelling and the evolution of the human hand. *J. Anat.* **123,** 157–201.

Marzke, M. W. (1983). Joint function and grips of the *Australopithecus afarensis* hand, with special reference to the region of the capitate. *J. hum. Evol.* **12,** 197–211.

Marzke, W. M. & Marzke, R. F. (1987). The third metacarpal styloid process in humans: origin and functions. *Am. J. phys. Anthrop.* **73,** 415–432.

Marzke, W. M. & Shackley, M. S. (1986). Hominid hand use in the Pliocene and Pleistocene: evidence from experimental archaeology and comparative morphology. *J. hum. Evol.* **15,** 439–460.

McArdle, R. H. (1988). The structural relationship: regression in biology. *Can. J. Zool.* **66,** 2329–2339.

Meldrum, D. J. (1991). Kinematics of the cercopithecine foot on arboreal and terrestrial substrates with implications for the interpretation of hominid terrestrial adaptations. *Am. J. phys. Anthrop.* **84,** 173–189.

Midlo, C. (1934). Form of hand and foot in primates. *Am. J. phys. Anthrop.* **19,** 337–389.

Miyamoto, M., Slightom, J. L. & Goodman, M. (1987). Phylogenetic relations of humans and African apes from DNA sequences in the -globin region. *Science* **238,** 369–373.

Musgrave, J. H. (1969). A comparative study of the hand bones of Neanderthal man. *Hum. Biol.* **41,** 587–588.

Musgrave, J. H. (1970). An anatomical study of the hands of the Pleistocene and recent man. Ph.D. Dissertation, Churchill College, University of Cambridge.

Napier, J. H. Jr (1956). The prehensile movements of the human hand. *J. Bone Jt. Surg.* **38B,** 902–913.

Napier, J. H. Jr (1959). Fossil metacarpals from Swartkrans. *Fossil Mammals of Africa No. 17*. London: British Museum of Natural History.

Napier, J. H. Jr (1960a). Studies of the hands of living primates. *Proc. zool. Soc. Lond.* **134,** 647–656.

Napier, J. H. Jr (1960b). Prehensility and opposability in the hands of primates. *Symp. zool. Soc. Lond.* **134,** 647–657.

Napier, J. H. Jr (1962). Fossil hand bones from Olduvai Gorge. *Nature (Lond.)* **196,** 409–411.

Napier, J. H. Jr (1980). *Hands*. London: George Allen & Unwin.

Plotnick, R. E. (1989). Application of bootstrap methods to reduced major axis line fitting. *Syst. Zool.* **38,** 144–153.

Preuschoft, H. (1973). Functional anatomy of the upper extremity. In (H. Bourne, Ed.), *The Chimpanzee*, vol. 6, pp. 34–120. New York: Karger, Basel.

Rayner, J. M. V. (1985). Linear relations in biomechanics: the statistics of scaling functions. *J. Zool. Lond.* **206,** 415–439.

Ricklan, D. E. (1986). The differential frequency of preservation of early hominoid wrist and hand bones. *Hum. Evol.* **1,** 373–382.

Sarmiento, E. E. (1988). Anatomy of the Hominoid wrist joint: its evolutionary and functional implications. *Int. J. Primat.* **9**, 281–345.
Schultz, A. H. (1927). Studies on the growth of gorilla and other higher primates with special reference to a fetus of gorilla, preserved in the Carnegie Museum. *Mem. Carnegie Mus.* **11**, 1–88.
Schults, A. H. (1936). Characters common to higher primates ad characters specific for man. *Q. Rev. Biol.* **11**, 259–283, 425–455.
Schultz, A. H. (1940). Growth and development of the chimpanzee. *Carnegie Institution of Washington Publication 518, Contributions to Embryology* **28**, 1–63.
Schultz, A. H. (1956). Postembryonic age changes. In (H. Hofer, A. H. Schultz & D. Stark, Eds) *Primatologia*, vol. 1, pp. 887–964.
Schultz, A. H. (1963). Relations between the lengths of the main parts of the foot skeleton in primates. *Folia primat.* **1**, 150–171.
Shea, B. T. (1981). Relative growth of the limbs and trunk of the African apes. *Am. J. phys. Anthrop.* **59**, 179–202.
Shea, B. T. (1983a). Size and diet in the evolution of African ape craniodental form. *Folia primat.* **40**, 32–68.
Shea, B. T. (1983b). Phyletic size change and brain/body scaling: a consideration based on the African pongids and other primates. *Int. J. Primatol.* **4**, 33–62.
Shea, B. T. (1983c). Allometry and heterochrony in the African apes. *Am. J. phys. Anthrop.* **62**, 275–289.
Shea, B. T. (1985). Ontogenetic allometry and scaling: a discussion based on the growth and form of the skull in African apes. In (W. L. Jungers, Ed.), *Size and Scaling in Primate Biology*, pp. 175–205. New York: Plenum Press.
Shea, B. T. (1986). Scapula form and locomotion in chimpanzee evolution. *Am. J. phys. Anthrop.* **70**, 475–488.
Shea, B. T. (1988). Allometry. In (I. Tattersall, E. Delson & J. Van Couvering, Eds) *Encyclopedia of Human Evolution and Prehistory*, pp. 20–22. New York: Garland Publishing.
Smith, S. L. (1990). Variation in the first three rays of the hand in an evolutionary context. *Am. J. phys. Anthrop.* **81**, 297.
Stern, J. T. Jr & Susman, R. L. (1983). The locomotor anatomy of *Australopithecus afarensis*. *Am. J. phys. Anthrop.* **60**, 279–317.
Steudal, K. (1985). Allometric perspectives on fossil catarrhine morphology. In (W. L. Jungers, Ed.) *Size and Scaling in Primate Biology*, pp. 449–475. New York: Plenum Press.
Strasser, E. (1992). Hindlimb proportions, allometry, and biomechanics in Old World monkeys (Primates, Cercopithecidae). *Am. J. phys. Anthrop.* **87**, 187–213.
Strasser, E. (1991). Form and function of the cercopithecoid foot. *Am. J. phys. Anthrop.* (suppl.) **12**, 170.
Straus, W. L. (1940). The posture of the great ape hand in locomotion and its phylogenetic implications. *Am. J. phys. Anthrop.* **27**, 199–207.
Straus, W. L. (1949). The riddle of man's ancestry. *Q. Rev. Biol.* **24**, 200–223.
Susman, R. L. (1978). Functional morphology of hominoid fingers. In (D. J. Chivers & K. A. Joysey, Eds) *Recent Advances in Primatology*, vol. 3, pp. 77–80. New York: Academic Press.
Susman, R. L. (1979). Comparative and functional morphology of hominoid fingers. *Am. J. phys. Anthrop.* **50**, 215–236.
Susman, R. L. (1988a). Hand of *Paranthropus robustus* from Member 1, Swartkrans: fossil evidence for tool behavior. *Science* **240**, 781–784.
Susman, R. L. (1988b). New postcranial remains from Swartkrans and their bearing on the functional morphology and behavior of *Paranthropus robustus*. In (F. E. Grine, Ed.), *Evolutionary History of the Robust Australopithecines*, pp. 149–172. New York: Aldine de Gruyter.
Susman, R. L. (1989). New hominid fossils from the Swartkrans Formation (1979–1986 excavations): postcranial specimens. *Am. J. phys. Anthrop.* **79**, 451–474.
Susman, R. L. & Creel, N. (1979). Functional and morphological affinities of the subadult hand (O.H. 7) from Olduvai Gorge. *Am. J. phys. Anthrop.* **51**, 311–332.
Susman, R. L., Jungers, W. L. & Stern, J. T. Jr (1982). The functional morphology of the accessory interosseous muscle in the gibbon hand: determination of locomotor and manipulatory compromises. *J. Anat.* **134**, 111–120.
Susman, R. L. & Stern, J. T. Jr (1980). EMG of the interosseous and lumbrical muscles in the chimpanzee (*Pan troglodytes*) hand during locomotion. *Am. J. Anat.* **157**, 389–397.
Tsutakawa, R. K. & Hewett, J. E. (1977). Quick test for comparing two populations with bivariate data. *Biometrics*. **33**, 215–219.
Tuttle, R. H. (1967). Knuckle-walking and the evolution of hominoid hands. *Am. J. phys. Anthrop.* **26**, 171–206.
Tuttle, R. H. (1969a). Quantitative and functional studies on the hands of the anthropoidea. *J. Morph.* **128**, 309–364.
Tuttle, R. H. (1969b). Terrestrial trends in the hands of anthropoidea. In *Proceedings of the 2nd International Congress on Primatology*, vol. 2, pp. 192–200. Atlanta: Karger, Basel.
Tuttle, R. H. (1969c). Knuckle-walking and the problem of human origins. *Science* **166**, 953–961.
Tuttle, R. H. (1970). Postural, propulsive, and prehensile capabilities in the cheiridia of chimpanzees and other great apes. In (G. H. Bourne, Ed.), *The Chimpanzee*, vol. 2, pp. 167–253. Atlanta: Karger, Basel.
Tuttle, R. H. (1975). Parallelism, brachiation and hominoid phylogeny. In (W. P. Luckett & F. S. Szalay, Eds) *Phylogeny of the Primates*, pp. 447–480. New York: Plenum Press.

Tuttle, R. H. (1977). Naturalistic positional behavior of apes and models of hominid evolution. 1929–1976. In (G. H. Bourne, Ed.), *Progress in Ape Research*, pp. 277–296. New York: Academic Press.

Tuttle, R. H. (1986). *Apes of the World. Studies on the Lives of Great Apes and Gibbons. 1929–1985*. New Jersey: Noyes, Park Ridge.

Tuttle, R. H. & Basmajian, J. V. (1974). Electromyography of forearm musculature in Gorilla and problems related to knuckle-walking. In (F. A. Jenkins, Jr, Ed.), *Primate Locomotion*, pp. 293–347. New York: Academic Press.

Tuttle, R. H., Basmajian, J. V., Regenos, E. & Shine, G. (1972). Electromyography of knuckle-walking: results of four experiments of the forearm of *Pan gorilla*. *Am. J. phys. Anthrop.* **37,** 255–266.

Tuttle, R. H. & Beck, B. B. (1972). Knuckle walking hand postures in an orangutan (*Pongo pygmaeus*). *Nature* **236,** 33–44.

Tuttle, R. H. & Cortright, G. W. (1988). Positional behavior, adaptive complexes, and evolution. In (J. H. Schwartz, Ed.) *Orang-utan Biology*, pp. 311–330. Oxford: Oxford University Press.

Tuttle, R. H. & Watts, D. P. (1985). The positional behavior and adaptive complexes of *Pan gorilla*. In (S. Kondo, Ed.), *Primate Morphophysiology, Locomotor Analyses and Human Bipedalism*, pp. 261–288. Tokyo: University of Tokyo Press.

Weil, W. B. Jr (1962). Adjustment for size—a possible misuse of ratios. *Am. J. Clin. Nutr.* **11,** 249–252.

Wilkinson, L. (1990). *SYSTAT: The System for Statistics*. Evanston, IL: Systat, Inc.

Williams, S. A. & Goodman, M. (1989). A statistical test that supports a human/chimpanzee clade based on noncoding DNA sequence data. *Mol. Biol. Evol.* **6,** 325–330.

Appendix
Relative growth statistics for Ray III (regressed against humeral length)

			Metacarpal III		
Measurement Species	n	r	OLS slope (95% CI)	RMA[1] slope (95% CI)	y-intercept[2]
Metacarpal length					
Pp[3]	26	0·97	0·995(0·906–1·084)	1·026(0·919–1·134)	−1·014
Pt	95	0·90	0·828(0·758–0·898)	0·923(0·837–1·009)	−0·510
Ggg	96	0·89	0·900(0·823–0·977)	1·007(0·912–1·101)	−0·998
Ggb	22	0·96	0·940(0·836–1·044)	0·983(0·848–1·117)	−1·216
Pongo	36	0·86	0·885(0·731–1·039)	1·036(0·841–1·230)	−0·718
Mediolateral midshaft width					
Pp	28	0·94	0·920(0·806–1·034)	0·982(0·841–1·123)	−3·251
Pt	110	0·90	0·806(0·744–0·868)	0·893(0·817–0·968)	−2·767
Ggg	99	0·89	0·803(0·734–0·872)	0·900(0·816–0·984)	−2·806
Ggb	23	0·84	0·810(0·614–1·006)	0·964(0·716–1·211)	−2·813
Pongo	38	0·74	0·990(0·735–1·245)	1·339(1·012–1·665)	−3·689
Dorsopalmar midshaft width					
Pp	27	0·95	0·948(0·841–1·055)	1·000(0·867–1·132)	−3·257
Pt	117	0·83	0·877(0·785–0·969)	1·057(0·943–1·171)	−2·972
Ggg	111	0·79	0·958(0·838–1·078)	1·220(1·068–1·372)	−3·261
Ggb	22	0·93	1·131(0·962–1·300)	1·213(1·005–1·420)	−3·825
Pongo	38	0·72	0·948(0·688–1·208)	1·320(0·985–1·655)	−3·295
Mediolateral head width					
Pp	27	0·95	1·078(0·952–1·204)	1·140(0·986–1·295)	−3·295
Pt	96	0·85	0·976(0·871–1·081)	1·149(1·020–1·278)	−2·881
Ggg	96	0·85	0·982(0·878–1·086)	1·153(1·024–1·282)	−2·948
Ggb	22	0·91	0·999(0·817–1·181)	1·106(0·879–1·332)	−2·991
Pongo	36	0·80	1·075(0·844–1·306)	1·336(1·045–1·627)	−3·391
Dorsopalmar head width					
Pp	26	0·92	0·763(0·647–0·879)	0·833(0·689–0·976)	−2·061
Pt	96	0·87	0·862(0·778–0·946)	0·987(0·885–1·088)	−2·316

continued

Appendix (*Continued*)

Metacarpal III					
Measurement Species	n	r	OLS slope (95% CI)	RMA[1] slope (95% CI)	y-intercept[2]
Dorsopalmar head width					
Ggg	96	0.85	0.902(0.803–1.001)	1.066(0.945–1.188)	−2.573
Ggb	22	0.94	0.996(0.860–1.132)	1.057(0.890–1.225)	−2.871
Pongo	36	0.78	0.881(0.674–1.088)	1.134(0.870–1.398)	−2.469
Biepicondylar width					
Pp	25	0.92	1.078(0.910–1.246)	1.177(0.969–1.385)	−3.079
Pt	105	0.91	1.190(1.100–1.280)	1.312(1.201–1.422)	−3.376
Ggg	100	0.88	1.183(1.073–1.293)	1.352(1.216–1.487)	−3.488
Ggb	22	0.96	1.064(0.940–1.188)	1.112(0.959–1.265)	−3.130
Pongo	36	0.82	1.240(0.994–1.486)	1.507(1.194–1.820)	−3.750

Proximal phalanx III					
Phalangeal length					
Pp	22	0.95	0.842(0.737–0.947)	0.886(0.756–1.015)	−1.010
Pt	93	0.93	0.753(0.701–0.805)	0.810(0.747–0.874)	−0.658
Ggg	97	0.91	0.804(0.742–0.866)	0.883(0.807–0.959)	−1.095
Ggb	13	0.99	0.834(0.763–0.905)	0.845(0.756–0.934)	−1.233
Mediolateral midshaft width					
Pp	23	0.89	0.844(0.684–1.004)	0.944(0.747–1.142)	−2.655
Pt	108	0.83	0.674(0.600–0.748)	0.810(0.720–0.900)	−1.960
Ggg	107	0.89	0.772(0.707–0.837)	0.867(0.788–0.946)	−2.172
Ggb	14	0.95	1.007(0.828–1.186)	1.067(0.843–1.290)	−2.977
Dorsopalmar midshaft width					
Pp	23	0.88	1.048(0.839–1.257)	1.188(0.926–1.450)	−3.711
Pt	108	0.91	1.201(1.111–1.291)	1.322(1.213–1.431)	−4.052
Ggg	107	0.90	1.082(0.998–1.166)	1.199(1.096–1.301)	−3.709
Ggb	14	0.90	0.997(0.749–1.245)	1.110(0.796–1.423)	−3.446
Mediolateral basal with					
Pp	22	0.90	0.886(0.726–1.026)	0.980(0.780–1.179)	−2.489
Pt	103	0.91	0.919(0.847–0.991)	1.014(0.927–1.100)	−2.499
Ggg	105	0.94	0.900(0.845–0.955)	0.960(0.894–1.026)	−2.458
Ggb	14	0.95	0.925(0.764–1.086)	0.977(0.777–1.178)	−2.603
Dorsopalmar basal width					
Pp	22	0.91	0.910(0.748–1.072)	1.003(0.802–1.204)	−2.616
Pt	102	0.92	0.976(0.907–1.045)	1.057(0.975–1.139)	−2.777
Ggg	105	0.92	0.882(0.820–0.944)	0.957(0.883–1.031)	−2.571
Ggb	14	0.93	0.861(0.689–1.033)	0.924(0.709–1.139)	−2.563
Mediolateral trochlear width					
Pp	23	0.79	0.712(0.503–0.921)	0.906(0.637–1.175)	−2.191
Pt	109	0.89	0.739(0.677–0.801)	0.831(0.756–0.905)	−2.221
Ggg	106	0.90	0.775(0.715–0.835)	0.859(0.785–0.933)	−2.359
Ggb	14	0.96	0.818(0.699–0.937)	0.850(0.703–0.997)	−2.524

continued

Appendix (*Continued*)

Measurement Species	n	r	OLS slope (95% CI)	RMA[1] slope (95% CI)	y-intercept[2]
Proximal phalanx III					
Dorsopalmar trochlear width					
Pp	23	0.85	0.860(0.659–1.061)	1.013(0.761–1.266)	−2.801
Pt	109	0.92	0.904(0.841–0.967)	0.985(0.908–1.061)	−2.899
Ggg	106	0.86	0.853(0.769–0.937)	0.994(0.891–1.096)	−2.869
Ggb	14	0.96	0.893(0.767–1.019)	0.926(0.770–1.082)	−3.028
Biepicondylar width					
Pp	23	0.77	0.724(0.501–0.947)	0.938(0.651–1.226)	−2.226
Pt	108	0.873	0.759(0.690–0.828)	0.870(0.786–0.955)	−2.210
Ggg	106	0.90	0.917(0.842–0.992)	1.024(0.933–1.115)	−2.743
Ggb	14	0.92	0.817(0.635–0.999)	0.891(0.662–1.120)	−2.462
Middle phalanx III					
Phalangeal length					
Pt	79	0.89	0.780(0.701–0.859)	0.881(0.785–0.977)	−1.064
Ggg	93	0.48	0.720(0.489–0.951)	1.498(1.180–1.816)	−1.175
Mediolateral midshaft width					
Pt	100	0.80	0.669(0.584–0.754)	0.840(0.733–0.948)	−2.183
Ggg	102	0.81	0.761(0.667–0.855)	0.942(0.826–1.058)	−2.420
Dorsopalmar midshaft width					
Pt	100	0.90	1.036(0.952–1.120)	1.146(1.046–1.247)	−3.933
Ggg	102	0.88	0.980(0.891–1.069)	1.113(1.005–1.221)	−3.903
Mediolateral basal width					
Pt	97	0.91	0.984(0.905–1.063)	1.085(0.990–1.181)	−2.774
Ggg	98	0.90	0.931(0.854–1.008)	1.034(0.940–1.127)	−2.711
Dorsopalmar basal width					
Pt	96	0.91	1.055(0.975–1.135)	1.154(1.056–1.252)	−3.274
Ggg	98	0.89	0.899(0.819–0.979)	1.016(0.917–1.115)	−2.943
Mediolateral trochlear width					
Pt	100	0.87	0.723(0.653–0.793)	0.835(0.748–0.922)	−2.395
Ggg	102	0.89	0.745(0.682–0.808)	0.837(0.759–0.915)	−2.462
Dorsopalmar trochlear width					
Pt	100	0.91	0.931(0.857–1.005)	1.027(0.938–1.116)	−3.376
Ggg	102	0.86	0.808(0.728–0.888)	0.941(0.842–1.040)	−3.146
Biepicondylar width					
Pt	100	0.86	0.730(0.699–0.761)	0.846(0.757–0.936)	−2.386
Ggg	102	0.88	0.921(0.836–1.006)	1.050(0.947–1.154)	−3.018

[1]RMA = reduced major axis regression.
[2]y-intercept = y-intercept of OLS regressions.
[3]Pp = *Pan paniscus*; Pt = *Pan troglodytes troglodtyes*; Ggg = *Gorilla gorilla gorilla*; Ggb = *Gorilla gorilla beringei*; Pongo = *Pongo pygmaeus*.

Diane M. Doran
Department of Anatomical Sciences, Suny at Stony Brook, Stony Brook, New York 11794-8081, U.S.A.

Received 1 April 1991
Revision received 2 December 1991 and accepted 3 January 1992

Keywords: paedomorphism, positional behavior, heterochrony.

The ontogeny of chimpanzee and pygmy chimpanzee locomotor behavior: a case study of paedomorphism and its behavioral correlates

Pygmy chimpanzees (*Pan paniscus*) have been described as "paedomorphic" in comparison with common chimpanzees (*Pan troglodytes*) because some of their adult proportions and features are present in immature chimpanzees. However, interpretations of what this morphological paedomorphism means behaviorally have varied widely. This paper documents the changes in positional behavior that occur during the ontogeny of pygmy and common chimpanzees so as to assess whether the adult pygmy chimpanzee is more similar in its positional behavior to immature rather than adult chimpanzees.

Results of field studies indicate that both pygmy and common chimpanzees show significant and similar patterns of change in locomotor behavior during ontogeny. With increasing age, in both species, there is an increase in the frequency of quadrupedalism. In early infancy, chimpanzee positional behavior is clearly forelimb dominated, and clinging, climbing and arm-hanging are most common. Aided bipedalism is the earliest form of frequent hindlimb use. Only later is the infant able to coordinate forelimb and hindlimb use in quadrupedalism. As the frequency of quadrupedalism increases, the frequency of bipedalism decreases. By 2 years of age, common chimpanzee locomotion is primarily quadrupedal. In addition to changes in the type and frequency of locomotor activities performed, there are also ontogenetic changes in the type of substrates used and the locomotor activities performed on them.

Results also indicate that adult pygmy chimpanzee arboreal locomotor behavior differs from that of adult common chimpanzees and most closely resembles that of infant chimpanzees. Adult pygmy chimpanzees, like immature common chimpanzees, use more suspensory and quadrupedal behavior than adult common chimpanzees. Shea (1981) has demonstrated that heterochrony in African ape evolution has resulted in substantial morphological differentiation. The results of this study complement Shea's findings by demonstrating that behavioral differences are associated with the morphological patterns.

Introduction

Pygmy chimpanzees have been described as "paedomorphic" in comparison with common chimpanzees, because some of their adult proportions and features, such as skull size and shape, chest girth, arm span, scapula size and shape, and intralimb proportions, are present in juvenile common chimpanzees (Coolidge, 1933, 1984; Shea, 1984*a,b*, 1986; Jungers & Susman, 1984). As Shea (1984*b*:104) noted, "the proportions observed in adult common chimpanzees are those predicted if the growth patterns of the pygmy chimpanzee are simply extended to larger terminal size." Thus, many of the shape differences between the two species are size related and result from the simple extension or truncation of common growth allometries, i.e., they are ontogenetically scaled (Shea, 1984*a,b*; Jungers, 1984; Jungers & Susman, 1984).

Most important of these ontogenetically scaling features, in terms of a predictable locomotor difference, is the shape change in the scapula. In a study based on growth allometries, Shea (1986:485) stated that "adult *P. paniscus* scapulae are indistinguishable from those of

Send reprint requests to: Diane M. Doran, Department of Biological Anthropology and Anatomy, Duke University, Wheeler Bldg., 3705B Erwin Road, Durham, NC 27705, U.S.A.

adult and subadult *P. troglodytes* of comparable body size". Thus, the same morphological difference that distinguishes adult pygmy and common chimpanzees also serves to distinguish mature and immature common (and pygmy) chimpanzees.

The adult pygmy chimpanzee scapula is relatively more narrow (i.e., has a higher scapula index) and longer than that of common chimpanzee adults (Jungers & Susman, 1984; Shea, 1986). Historically, a functional correlation has been drawn between narrow scapulae and arm-swinging locomotion (Coolidge, 1933; Susman *et al.*, 1980). This has frequently been cited to predict that adult pygmy chimpanzees are more suspensory than their common chimpanzee counterpart (Coolidge, 1933; Frechkop, 1935; Roberts, 1974; Susman *et al.*, 1980). However, other researchers, based on allometric studies, have suggested that the shape changes are correlated with a change in overall body size rather than as a result of selection for a specific shape change. As a result of this proposed functional equivalence, they predict that the locomotor differences between the two species are likely to be minimal (McHenry & Corruccini, 1981; Shea, 1984a,b, 1986, 1988).

Underlying this interpretation of functional equivalence in ontogenetic scaling is an assumption that there is little behavioral change within a species during ontogeny, and by analogy, between ontogenetically scaled taxa such as pygmy and common chimpanzees (Shea, 1984a,b, 1986). However, Shea (1986) points out that shape changes during ontogeny may, in fact, be of locomotor significance, in which case, there should be "equally marked distinctions" in subadult and adult behavior as between ontogenetically scaled adults. Thus, if there are intraspecific differences in the type and frequency of locomotor activities performed during ontogeny, then one could predict that paedomorphism in morphology is associated with "paedomorphism" in behavior. In the case of pygmy and common chimpanzees, with increasing age (as measured by size) the scapula becomes less narrow. One could predict then, based on the correlation of narrow scapulae with arm-swinging, that with increasing age there is decreasing suspensory activity in both species. If this is true, then it follows that the paedomorphic pygmy chimpanzee adult (in regards to scapula shape), would be more similar in its locomotor behavior to immature rather than mature common chimpanzees, and would be more suspensory in its locomotor behavior. Clearly, before this issue can be resolved as regards pygmy and common chimpanzees, it must first be established what, if any, changes occur in the locomotor behavior of *P. troglodytes* during ontogeny.

Early qualitative field studies of primate behavior first suggested that locomotor behavior does change during development (Schaller, 1963; Goodall, 1968; Fossey, 1979). Ontogenetic changes in the type and frequency of positional behavior were further verified in quantitative studies among baboons (Rose, 1977), galagos (Crompton, 1983), macaques (Rawlins, 1976) and gorillas (Tuttle & Watts, 1985). To date, however, no one has specifically considered the ontogenetic changes in chimpanzee and pygmy chimpanzee locomotor and postural behavior (but see Goodall, 1968).

The purpose of this paper is two-fold: (1) to quantitatively assess the changes that occur in chimpanzee locomotor behavior and substrate use during development, and (2) to consider whether the pygmy chimpanzee acts as a "locomotor paedomorph", i.e., does the adult pygmy chimpanzee retain a more "juvenile" pattern of behavior in comparison with common chimpanzees?

Specific questions for both pygmy and common chimpanzees include:
(1) Do the locomotor profiles of pygmy and common chimpanzees change during ontogeny, so that with increasing age there is decreased suspensory behavior?
(2) Are there other age differences in locomotor behavior and substrate use?

(3) If there are age-related changes in locomotor behavior, do the same trends characterize the development of both species?
(4) Does the locomotor behavior of adult pygmy chimpanzees more closely resemble that of immature rather than mature chimpanzees, i.e., is the pygmy chimpanzee adult more suspensory in its locomotor behavior than common chimpanzee adults?

Materials and methods

Study site and sampling methods: Pan troglodytes

One non-provisioned, habituated community of *P. troglodytes* was observed for 430 hours during seven consecutive months, from March through September 1988, in the Tai National Park of the Ivory Coast. For a complete description of the study site and habituation process see Boesch & Boesch (Boesch, 1978; Boesch & Boesch, 1981, 1983, 1984, 1989).

At the time of this study, the Tai community included 70 chimpanzees, with seven adult males, 23 adult females, six adolescent females, zero adolescent males, 12 juveniles and 22 infants.

Exact ages of all individuals less than 9 years old are known from the long-term records at the Tai study site. Definitions of age classes used in this study are as follows:
(1) *Infant*—transported by mother during travel. The category infant is divided into 3 subclasses:
  (a) infant 1 (inf 1)—0–6 months of age.
  (b) infant 2 (inf 2)—6–24 months of age.
  (c) infant 3 (inf 3)—2 years until travel is independent of mother.
(2) *Juvenile*—travels and nests independently of mother, but accompanies mother during travel (if mother is still living).
(3) *Adolescent*—females are immature and nulliparous. There were no adolescent males at Tai at the time of this study.

Focal animal sampling (Altmann, 1974) was used and each focal animal was followed for an entire day, or as long as possible (mean length of focal animal sample = 259 min). Since this study is part of a larger study which addresses sex differences and seasonal variation in adult behavior, in addition to the age differences considered here, an attempt was made to sample male and female adults equally each month. The remaining time was divided among the other age classes.

Because of the fission–fusion nature of chimpanzee social organization, the choice of a focal animal was dependent upon which animals were present in the party actually located. Adult female chimpanzees travel more quietly (i.e., fewer vocalizations and drummings) and in smaller parties than adult males (Doran, 1989). As a result, it is virtually impossible to locate a specific individual female and her offspring. Thus, it could not be predicted in advance which female (and/or offspring) would serve as a focal animal on any given day.

It was decided in advance whether an adult or immature chimpanzee would serve as a focal animal that day. If an immature animal was to be followed, and more than one was present in the party located, then the individual of the age class that was least represented in the data was selected.

When a female had an infant of less than 1 year, both mother and infant served as focal animals. Infants of more than 1 year were sampled separately from their mothers.

The results of locomotor activity and substrate use during locomotion presented here are based on locomotor bout (with distance) sampling. For a complete description of methods used and data collected, see Doran (in press).

Table 1 Distribution of *Pan troglodytes* data by age class. Data are from focal animal sampling

Age class	No. of sightings per age class	No. of different individuals per age class	No. of chimps of each age class present at Tai	No. of locomotor bouts	Total time with each age class (1 min points)
Inf 1	4	2	8	41	958
Inf 2	8	5	9	198	687
Inf 3	3	3	5	83	992
Juvenile	12	6	12	690	4010
Adolescent	11	2	6	229	1721
Adult	48	15	30	1221	12,905

Study site and sampling methods: Pan paniscus

Pan paniscus was observed for a total of 220 hr during 10 consecutive months, from December 1986 through October 1987, in the Lomako Forest, Zaire. For a complete description of the study site see Badrian & Badrian (1977), Badrian & Malenky (1984), White (1986) and Thompson-Handler (1990).

Identical methods were used in the two study sites. However, individual ages were not known in the Lomako Forest, so the infant class was not subdivided into subclasses. In addition, animals in the Lomako Forest were less habituated than those in the Tai Forest, so lengths of focal animal samples were shorter. Finally, animals in the Lomako Forest were not followed as easily while travelling terrestrially, so all comparisons between the two chimpanzee species are based solely on arboreal travel.

Statistical analyses

G Tests of independence ($R \times C$ contingency tables) are used to determine whether the frequency of locomotor behavior or substrate use found in one age class is independent of the frequency of locomotor behavior or substrate use in another age class. The various statistics used are described in Sokal & Rohlf (1981). Symbols for significance used include: "*" = $P<0.05$, "**" = $P<0.01$, "***" = $P<0.001$ and NS = not significant. All G values have been corrected by William's correction factor in order to obtain a better approximation of the chi-square distribution. Unless otherwise stated, all tests are two-tailed.

Results

Age differences in the locomotor behavior of Pan troglodytes

The results of age class differences in locomotor behavior are from cross-sectional sampling of individuals of differing ages. Data on adult behavior represent 57·9% of the total data set. Of the remaining 42·1% of the data, adolescents account for 7·7%, juveniles 10·0% and infants 16·3% (Table 1). At least two different individuals were sampled from each age class. The data for each age class are based on a minimum of three separate sightings.

There are age class differences in the type and frequency of common chimpanzee locomotor activities performed (Figure 1). Adult chimpanzees primary mode of locomotion is knuckle-walking quadrupedalism. However, it takes several years and transitions in locomotor behavior before a chimpanzee engages regularly in this form of locomotion (Table 2).

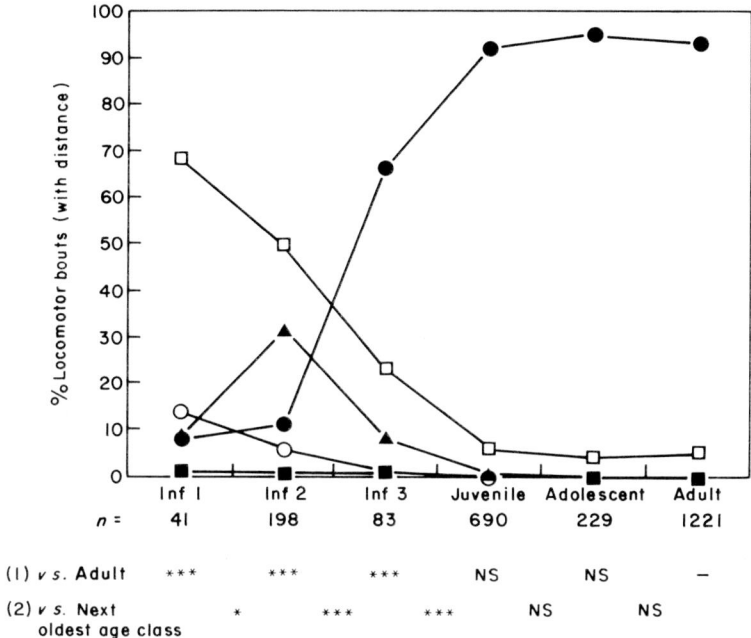

Figure 1. Age differences in the overall locomotor activity of *Pan troglodytes*. A plot of the percentage of locomotor bouts (with distance) spent in each locomotor category: (●) Quadrupedal; (□) climb/scramble; (▲) suspensory; (○) bipedal; (■) leap. $*P<0.05$, $***P<0.001$, NS = not significant.

Infants of less than 6 months (inf 1) differ significantly from adults in the frequency of locomotor activities performed (Figure 1; $G=96.3***$; d.f. $=4$). Infants of less than 6 months first engage in locomotor bouts independently from their mothers at approximately 5 months of age. In this first stage of locomotor development, locomotor activity consists primarily of climbing which occurs during locomotor play (Goodall, 1968). Infants of class 1 frequently climb up 3 or 4 feet on thin saplings (included in the substrate category liane) when the mother is resting within arm's reach. However, they do not climb down the sapling, but rather, descend by either dropping to the ground, or climbing onto the adjacent mother.

In addition to more frequent climbing, class 1 infants are also more suspensory than adults. These suspensory activities include: two-handed arm-hangs, one-handed arm-hangs, and one activity unique to infants, one-handed arm-hang twirls. This activity is used frequently during play and consists of the infant hanging by one arm, twisting his/her body 90° in one direction and then spinning around 180 degrees in the opposite direction.

Infants of less than 6 months were not observed using knuckle-walking quadrupedalism on the ground and only rarely engaged in palmigrade quadrupedalism. Infants of class 1 do not frequently move about independently on the ground. The few steps they take are usually bipedal and are aided by either grasping the mother or some branches in order to maintain balance.

Infants of class 2 (6–24 months) differ significantly in their locomotor activity from both class 1 infants and adults [G (inf 1 *vs.* inf 2) $=12.2*$, G (inf 2 *vs.* adult) $=546.0***$; d.f. $=4$]. This stage of locomotion is characterized by more frequent suspensory behavior than at any other time in a chimpanzee's life. In addition to the suspensory activities noted for class 1 infants, infants are also capable of arm-swinging along branches by an age of 12 months.

Table 2 Qualitative description of the ontogeny of *Pan troglodytes* locomotor behavior. Observations are based on focal animal observations of chimpanzees of known age

Age class	Description
Inf 1	
0–3 months:	No independent locomotor bouts, infant stays on or near mother at all times.
4–6 months:	Locomotor play (after Goodall, 1968) occurs near mother as she rests and includes frequent suspensory activity and aided bipedalism.
	Infant climbs up saplings short distances (3–4 feet) and arm-hangs and arm-hang twirls.
	Infant descends by either dropping to ground or climbing onto adjacent mother rather than climbing down sapling.
	Infant does not knuckle-walk and rarely uses palmigrade quadrupedalism. Infant moves short distances on the ground by aided bipedalism.
	Infant shows greater independence from mother in the trees than when on the ground.
Inf 2	
6–11 months:	The infant's frequency of suspensory behavior and climbing increases further. Knuckle-walking is still not observed.
12–15 months:	Suspensory behavior includes arm-swinging in addition to arm-hanging and twirling. Infant attempts to knuckle-walk.
	Uses some knuckle-galloping. Uneasy in quadrupedalism.
16–24 months:	Very independent in trees, remains 30–40 feet away from mother. Uses deliberate palmigrade quadrupedalism, frequent suspensory behavior, climbing and aided bipedalism.
	Continues to use small substrates. Attempts at climbing a larger trunk fails and infant falls. No knuckle-walking observed.
Inf 3	
2–3 years:	Infant knuckle-walks easily on ground and doesn't climb as much as younger infants. However these infants climb more and are more suspensory than adults. Infant is very independent (of mother) in trees. Infant is carried dorsally by mother except during precarious climbs when the infant is carried on ventrum.
3 years to juvenile:	Infant still is carried by mother during travel, but often ascends and descends trees alongside mother.
Juvenile	
	Adult-like in overall locomotion; a knuckle-walking quadruped. However, during arboreal locomotion, juveniles continue to engage in more suspensory behavior and less climbing than adults. Travels independently of mother.

The frequency of climbing in class 2 infants, although much greater than the frequency of climbing in adults, is less than that of class 1 infants. In addition, the frequency of bipedalism decreases in class 2 infants as the frequency of quadrupedalism begins to increase. At 12–15 months of age, infants were observed attempting unsuccessfully to knuckle-walk on the ground. However, they did manage to crutch-walk (after Goodall, 1968) a few steps. By 24 months of age, some knuckle-walking was observed, but crutch-walking was the more common form of quadrupedalism. Although use of quadrupedalism increases in class 2 infants, in comparison with class 1 infants, it is still used relatively infrequently in comparison with older chimpanzees.

Infants of class 3 differ significantly in their locomotor activities from class 2 infants and adults [G (inf 3 $vs.$ inf 2) = 79·3***, G (inf 3 $vs.$ adult) = 33·3***; d.f. = 4]. The transition from inf 2 to inf 3 includes a major shift to quadrupedalism, which is coincident with the infant's ability to knuckle-walk quadrupedally with ease at 29 months. In addition to an increased frequency of quadrupedalism, infants of class 3, in comparison with younger infants, also show a qualitative change in the type of quadrupedalism used. Prior to 2 years of age,

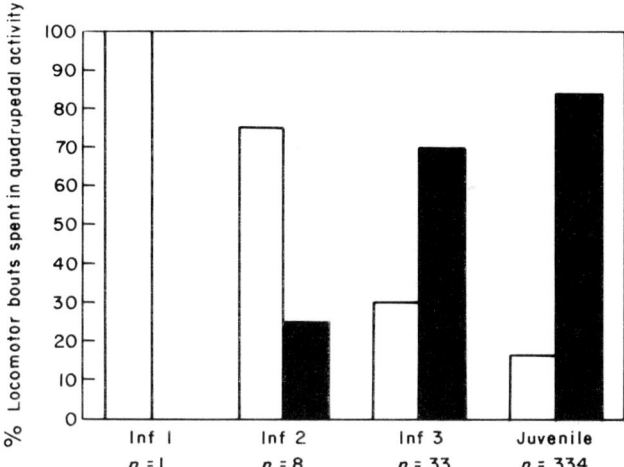

Figure 2. Age differences in immature *Pan troglodytes* quadrupedal locomotion. Data are from continuous locomotor bout sampling. (□) Palmigrade quadrupedalism; (■) knuckle-walking.

quadrupedalism is primarily palmigrade in nature, whereas, after 2 years of age, knuckle-walking quadrupedalism is the more frequent form (Figure 2).

Along with the dramatic increase in frequency of quadrupedalism by ages of 2 years and older, there is a resulting decrease in climbing, suspensory behavior and bipedalism in comparison with younger infants. However, in comparison with older chimpanzees, infants of class 3 show less frequent quadrupedalism and more frequent climbing, suspensory behavior and bipedalism.

There are no significant differences in the frequency of overall locomotor activities performed by adults, adolescents and juveniles [G (adult *vs.* adolescents) = 2·8 NS, G (adult *vs.* juvenile) = 2·3 NS, G (adolescent *vs.* juvenile) = 4·5 NS; d.f. = 4]. By the time a chimpanzee is old enough to travel separately from its mother (juvenile), the overall frequency of its locomotor behavior does not differ from that of adolescent and adults. Since the major means of travel for adult chimpanzees is terrestrial knuckle-walking (Doran, 1989), it is not unexpected that the major form of locomotion for juveniles is knuckle-walking quadrupedalism.

However, when the subset of arboreal locomotor data is extracted from the overall data set and considered, there are significant differences in the frequency of arboreal locomotor behaviors performed by adults and juveniles (Figure 3; $G = 12·9^*$; d.f. = 4). During arboreal locomotion, juveniles show a marked increase in the frequency of climbing in comparison to class 3 infants but still climb less frequently than adults. Juveniles are also more suspensory in their arboreal locomotor behavior than adults. There are no significant differences in the frequency of locomotor activities performed by adults and adolescents during arboreal locomotion ($G = 2·9$ NS; d.f. = 4). The same pattern trends that characterized *overall* locomotion of infants hold true in a consideration of infant *arboreal* locomotion.

Age differences in Pan troglodytes *substrate usage during locomotor activity*
Infants of less than 6 months perform the majority of their locomotor activities on lianes (category which also includes thin saplings) or on their mothers (Figure 4). They also use the ground relatively frequently, but rarely use vertical trunks, boughs or foliage.

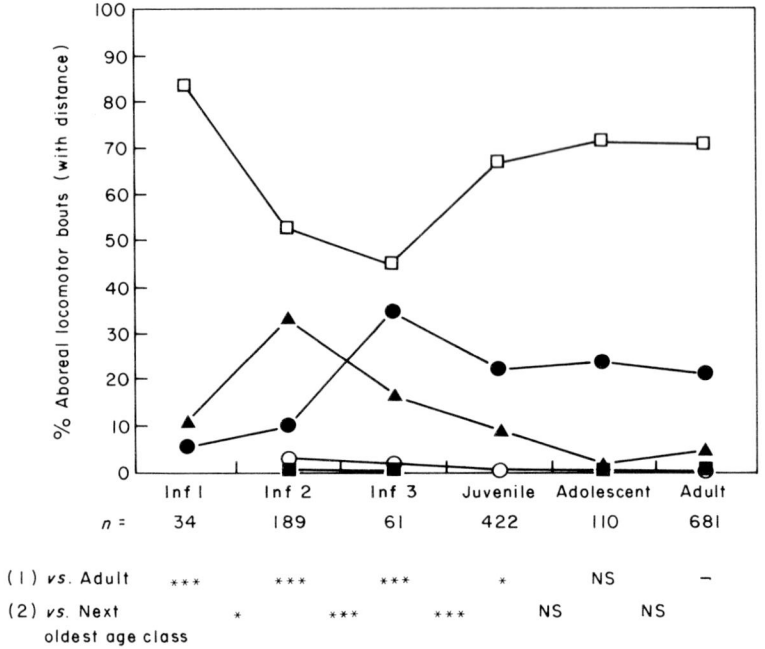

Figure 3. Age differences in the arboreal locomotor activity of *Pan troglodytes*. A plot of the percentage of arboreal locomotor bouts (with distance) spent in each locomotor category. (●) Quadrupedal; (□) climb/scramble; (▲) suspensory; (○) bipedal; (■) leap. *$P<0.05$, ***$P<0.001$, NS = not significant.

Class 2 infants differ significantly from class 1 infants in substrate use ($G=18.7**$; d.f. = 5). The older infants perform fewer locomotor activities on their mothers than do the younger infants. As a result, they perform more activities on other substrates, and in particular, there is a sharp increase in the use of branches. However, locomotor activities still occur on thin-diametered and relatively stable substrates. Infants of less than 2 years rarely use trunks, boughs or foliage during locomotor activities.

Infants of class 3 differ significantly from class 2 infants in substrate use ($G=110.2***$; d.f. = 6). Infants of greater than 2 years show a much greater diversity of substrate use than younger infants in that they begin to use trunks (although still in a limited fashion), boughs and foliage more frequently than their younger counterparts.

The transition in locomotor behavior made by an infant as it becomes a juvenile is also reflected in substrate use. Substrates used by juveniles during overall locomotion differ from those of class 3 infants ($G=70.7***$; d.f. = 6), but do not differ from those of adolescents or adults ($G=6.0$ NS). Since a juvenile adopts a nearly adult form of locomotion, it is not unexpected that juveniles show a marked increase in the frequency of ground use in comparison with infants.

Juveniles differ from adolescents and adults in the frequency of substrate usage when only above ground substrates are considered (Figure 5). Smaller bodied juveniles use trunks less frequently and lianes more frequently than adolescents and adults. This is most likely a factor of the juvenile's inability to climb larger vertical trunks. If a trunk is too large to climb (in relation to juvenile forelimb length), then juveniles ascend trees by climbing up adjacent lianes. Thus, the frequency of liane use is much higher in juveniles than in adolescents or adults.

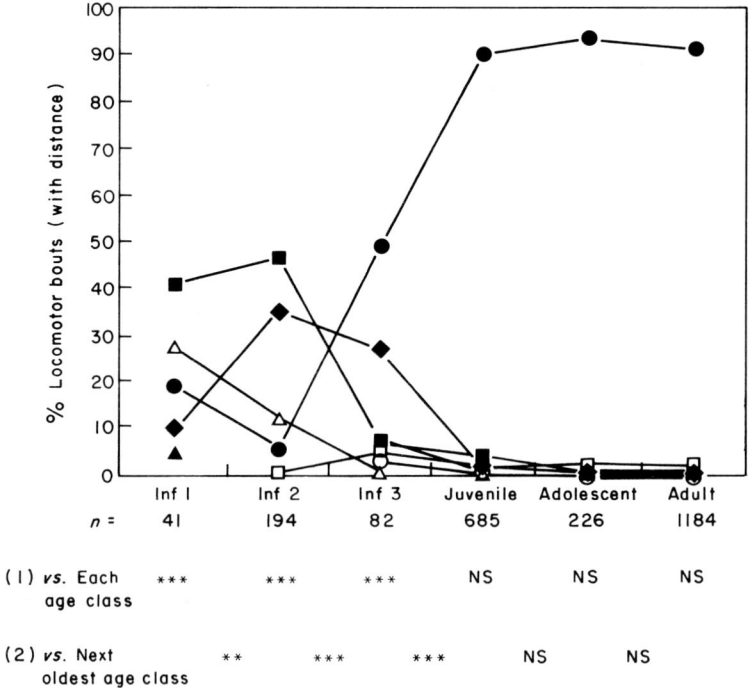

Figure 4. Age differences in *Pan troglodytes* substrate usage during locomotor activity. A plot of the percentage of locomotor bouts (with distance) spent on each substrate. (●) Ground; (□) trunk; (▲) bough; (◆) branch; (■) liane; (○) foliage; (△) chimpanzee (mother). $**P<0.01$, $***P<0.001$, NS = not significant.

Age differences in Pan troglodytes *locomotor activity performed on each substrate*

Figure 6 shows the freqency of locomotor activities performed on each substrate by each age class. There are no significant differences in the frequencies of locomotor activities performed on the ground by adults in comparison with adolescents, juveniles and class 3 infants [G (inf 3 vs. adult) = 1·8 NS, G (juvenile vs. adult) = 0·2 NS, G (adolescent vs. adult) = 0·2 NS; d.f. = 1]. The predominant locomotor activity while on the ground for each of these age groups is quadrupedalism.

However, class 1 and 2 infants differ significantly from adults in the frequencies of locomotor activities performed on the ground [G (inf 1 vs. adult) = 6·5*, G (inf 2 vs. adult) = 7·5*, d.f. = 2]. In testing whether the frequency of locomotor activities of one age class differs from that of the next oldest class (i.e., inf 1 vs. inf 2, inf 2 vs. inf 3 etc.), significant differences were found only in a comparison of infants class 2 and 3 [G (inf 1 vs. inf 2) = 1·7 NS, G (inf 2 vs. inf 3) = 12·7***, G (inf 3 vs. juvenile) = 0·1 NS, G (juvenile vs. adolescent) = 0·4 NS, G (adolescent vs. adult) = 0·2 NS; d.f. = 3]. During ontogeny, there are two different patterns of ground use by chimpanzees. Chimpanzees of less than 2 years are primarily bipedal (aided) when on the ground, whereas older chimpanzees are primarily quadrupedal.

There are no significant age class differences in the results of locomotor activities performed on vertical trunks. When on vertical trunks all age classes climb, however, infants and juveniles use trunks much less frequently than adults and adolescents. This points out a qualitative difference in the climbing of adult and immature chimpanzees. Infants climb frequently on small saplings during locomotor play. The majority of climbing by adults is to

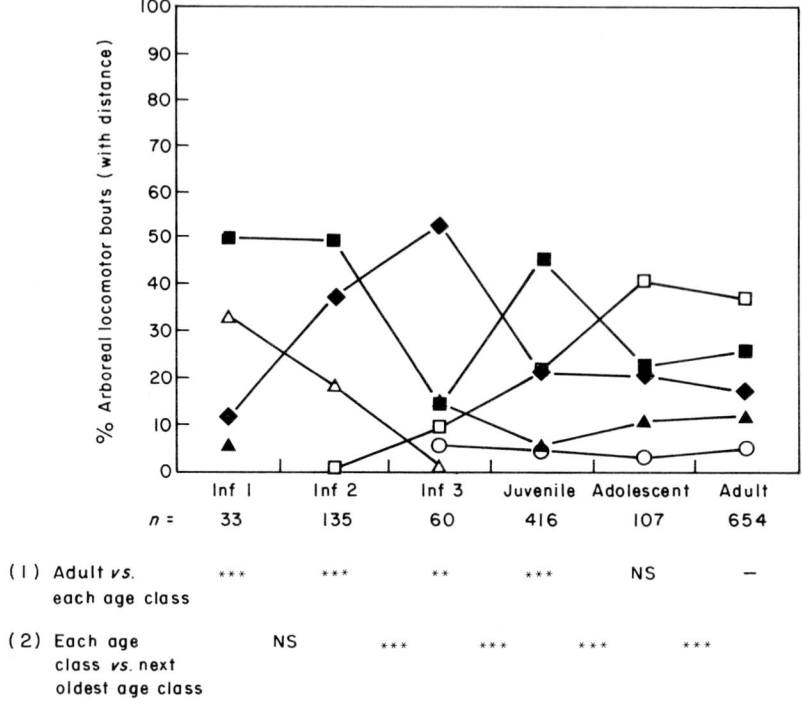

Figure 5. Age differences in *Pan troglodytes* substrate usage during arboreal locomotor activity. A plot of the percentage of locomotor bouts (with distance) spent on each substrate during arboreal locomotor activity. (●) Ground, (□) trunk, (▲) bough, (◆) branch, (■) liane, (○) foliage, (△) chimpanzee (mother). **$P<0.01$, ***$P<0.001$, NS = not significant.

ascend to or descend from feeding sites on large vertical trunks. Juveniles, like adults, also climb during ascents and descents from feeding trees. However, unlike adults, they tend to use lianes rather than large vertical trunks.

There are no significant age class differences in the type of locomotor activities performed on boughs. However, as discussed above, young infants use boughs less frequently than older chimpanzees. Class 2 infants are far more suspensory on branches than any other age class. Infants of less than 2 years differ significantly from adults in the frequencies of locomotor activities performed on lianes [G (inf 1 *vs.* adult) = 5·8*, G (inf 2 *vs.* adult) = 36·4***; d.f. = 1]. Although infants of class 1 and 2 climb frequently on lianes, they also use suspensory activities much more frequently than adults.

There are significant differences in the frequencies of locomotor behaviors performed by juveniles in foliage in comparison with adults ($G = 16·3**$; d.f. = 4). Although juveniles climb and scramble frequently in foliage, they also use suspensory activities more frequently than adults in foliage. The data are insufficient with regard to adolescent behavior in foliage. The small sample size does not allow it to be distinguished from either juveniles or adults [G (adolescent *vs.* juvenile) = 4·2 NS, G (adolescent *vs.* adult) = 3·0 NS].

Age differences in the arboreal locomotor behavior of Pan paniscus

The results of age class differences in pygmy chimpanzee behavior are also from cross-sectional sampling. Data on adult behavior represent 57·9% of the total data set. Of the

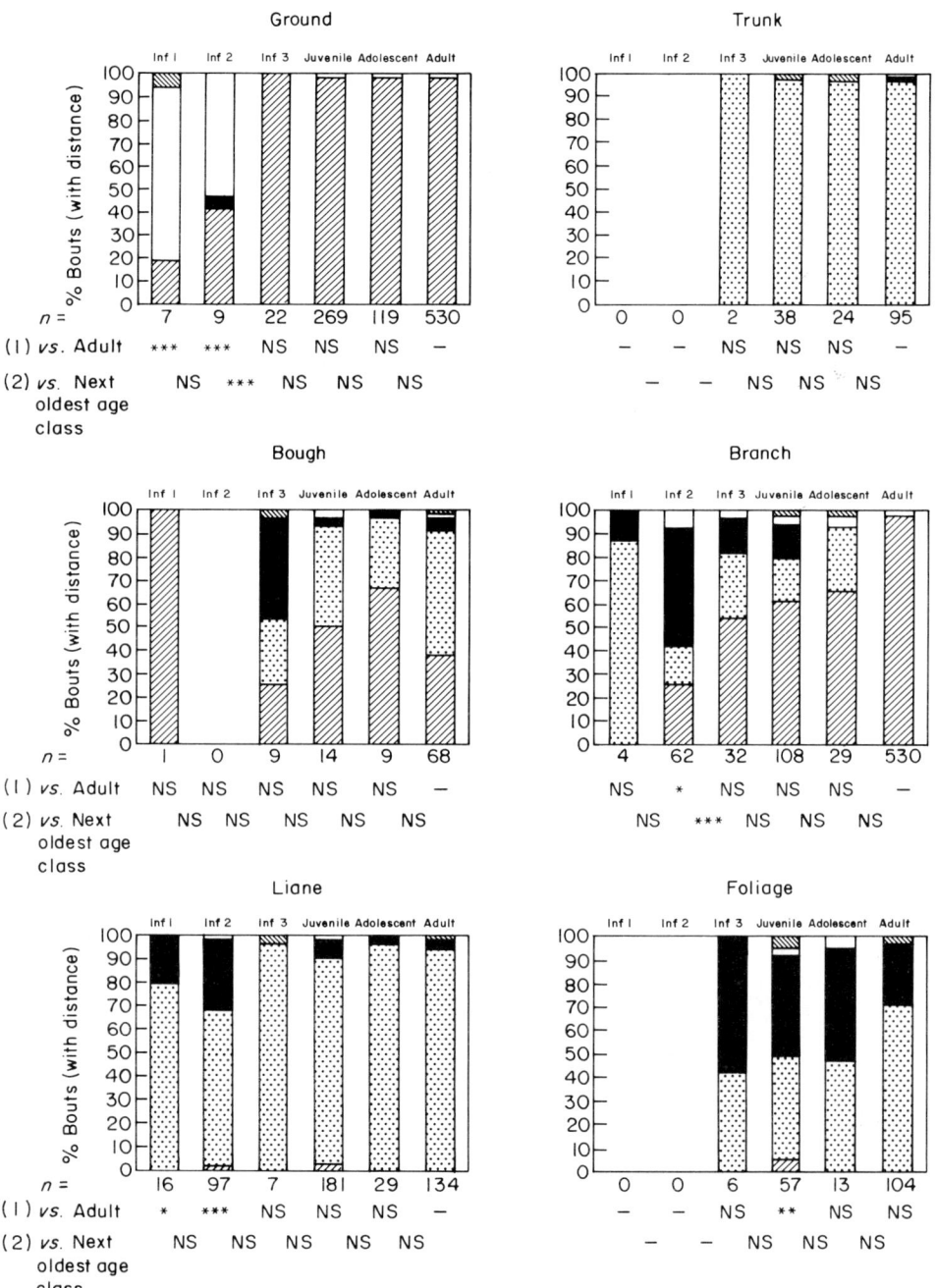

Figure 6. Age differences in *Pan troglodytes* locomotor activity performed on each substrate. A plot of the percentage of locomotor bouts (with distance) spent in each locomotor activity by a given age class on a given substrate. (▨) Leap; (□) bipedal; (■) suspensory; (▦) climb/scramble; (▧) quadrupedalism. $*P<0.05$, $**P<0.01$, $***P<0.001$, NS = not significant.

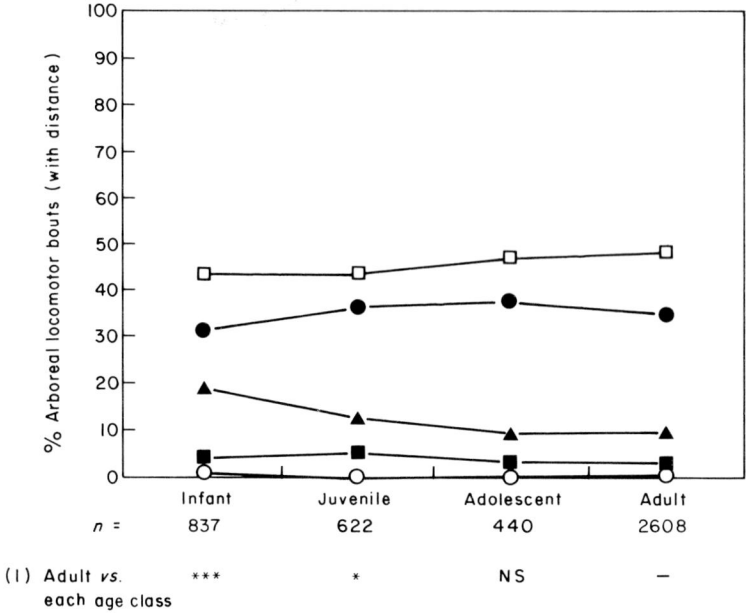

Figure 7. Age differences in the arboreal locomotor activity of *Pan paniscus*. A plot of the percentage of arboreal locomotor bouts (with distance) spent in each locomotor activity. (●) Quadrupedal; (□) climb/scramble; (▲) suspensory; (○) bipedal; (■) leap. $*P<0.05$, $***P<0.001$, NS = not significant.

remaining 42·1%, adolescents account for 9·8%, juveniles 13·8% and infants 18·6%. Exact ages of individual pygmy chimpanzee infants were not known, so all infants are included in the same age class.

As in the results of *P. troglodytes* age class differences in arboreal locomotion, there are no differences in the frequencies of pygmy chimpanzee adult and adolescent locomotor behavior (Figure 7; $G=2·2$ NS; d.f.=4). There are, however, differences in a comparison of the arboreal locomotor behavior of pygmy chimpanzee adults with juveniles and infants [G (adults *vs.* juvenile) = 13·2*, G (adults *vs.* inf) = 43·0***; d.f. = 4].

Infants differ from adults in using less quadrupedalism and more frequent suspensory behavior during arboreal locomotion.

One unexpected result is that there is no difference between adults and infants in the frequency of bipedalism used. This is probably a result of combining all infant classes (since older chimpanzee infants do not use bipedalism frequently) and because of very limited observations of pygmy chimpanzees while on the ground (since common chimpanzee infant bipedalism occurred primarily on the ground).

Juveniles differ from adults during arboreal locomotion in using less climbing and scrambling and more suspensory behavior.

A comparison of age class differences in the arboreal locomotor behavior of Pan troglodytes *and* Pan paniscus

Figure 8 shows a comparison of the age differences in arboreal locomotor activity of the two species. Within each species, the overall trends in locomotion are the same. With increasing age there is decreased suspensory behavior and increasing quadrupedalism. However, in spite of the similar overall trends in ontogeny, there are significant differences in the results

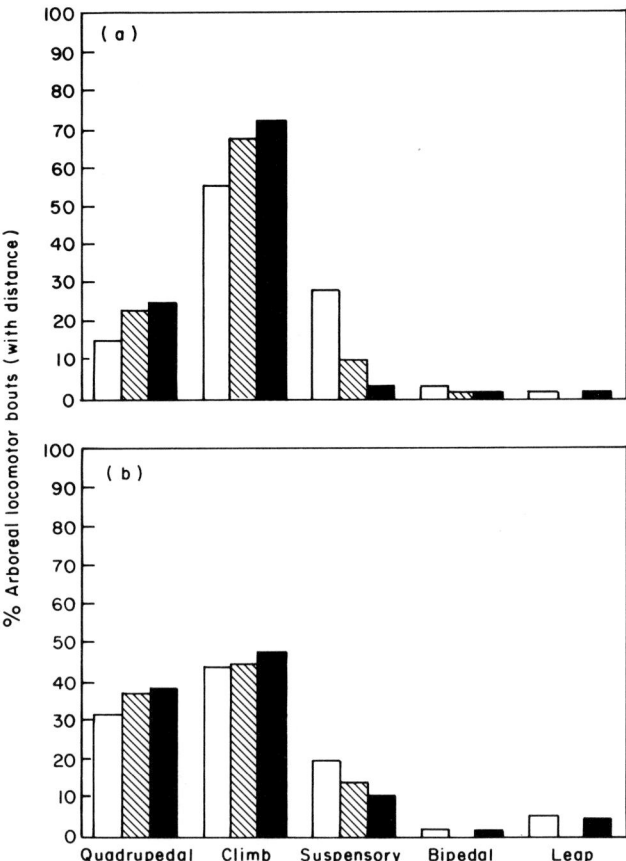

Figure 8. A comparison of age class differences in the arboreal locomotor behavior of (a) *Pan troglodytes* and (b) *Pan paniscus*. (a) (□) Infant, $n=284$; (▨) juvenile, $n=422$; (■) adolescent, $n=110$. (b) (□) Infant, $n=837$; (▨) juvenile, $n=622$; (■) adolescent, $n=440$. G (inf) $=49.7$***, G (juvenile) $=79.6$***, G (adolescent) $=25.8$***. ***$P<0.001$.

of interspecific comparisons (of each age class) of frequencies of locomotor activities [G (adolescents) $=25.8$***, G (juveniles) $=79.6$***, G (infants) $=49.7$***].

Pygmy chimpanzee infants, juveniles and adults (not included in Figure 8, but see Doran, 1989) all differ from their common chimpanzee counterparts in using more frequent quadrupedalism and less frequent quadrumanous climbing and scrambling. In addition, pygmy chimpanzee juveniles and adolescents are more suspensory than their common chimpanzee counterparts. The results do not indicate that pygmy chimpanzee infants are more suspensory than common chimpanzees. However, this may be related to the combining of all pygmy chimpanzee infants into one age class, since it is reported above that the frequency of suspensory behavior changes during infancy and it is not clear that pygmy chimpanzee infants of all ages were sampled evenly.

The pygmy chimpanzee as a "locomotor paedomorph"

The frequency of locomotor activities performed by adult pygmy chimpanzees differs significantly from those of adult, adolescent, juvenile, infant 1 and infant 2 common chimpanzees

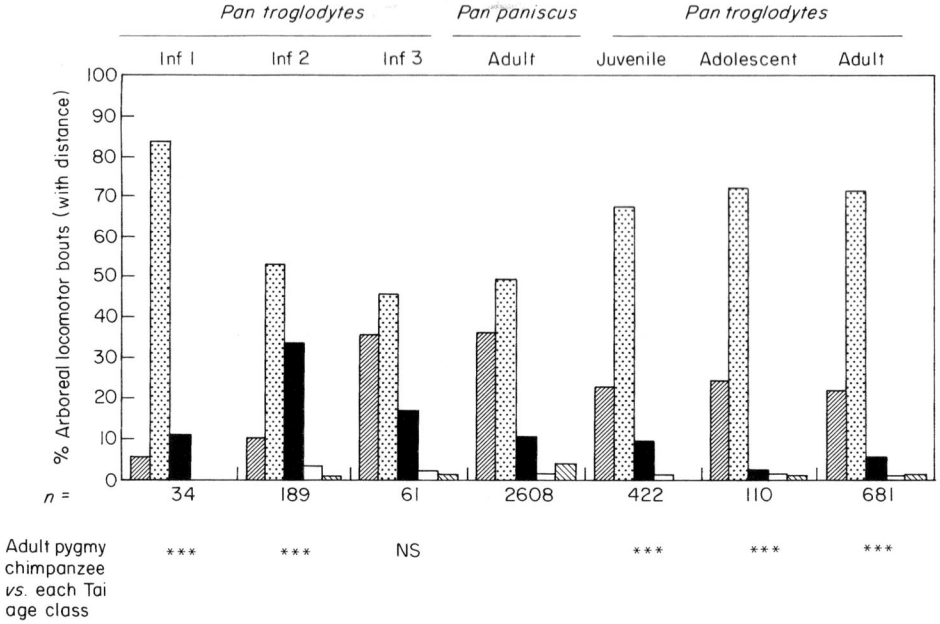

Figure 9. Results of adult *Pan paniscus* arboreal locomotor activity shown in comparison with age differences in the arboreal locomotion of *Pan troglodytes*. (▨) Quadrupedalism; (▦) climb/scramble; (■) suspensory; (□) bipedal; (▧) leap. **$P<0\cdot01$, ***$P<0\cdot001$, NS = not significant.

[Figure 9; G values for adult pygmy chimpanzees $vs.$: (adult common) $= 27\cdot7$***, (adolescent common) $= 27\cdot7$***, (juvenile common) $= 68\cdot8$***, (inf 2) $= 108\cdot5$***, (inf 1) $= 20\cdot4$***; d.f. $= 4$].

Adult pygmy chimpanzees do not differ significantly from common chimpanzee infant class 3 (infants greater than 2 years) in the frequency of locomotor activities performed. Both adult pygmy chimpanzees and class 3 common chimpanzee infants use more suspensory behavior and quadrupedalism, and less quadrumanous climbing and scrambling than adult common chimpanzees.

Discussion

The two primary aims of this paper are to determine whether there are positional behavior changes during the ontogeny of pygmy and common chimpanzees, and if so, whether the morphological paedomorphism of the adult pygmy chimpanzee is associated with a more "juvenile" pattern of behavior in comparison with common chimpanzees.

The results indicate that for both species there are significant changes in locomotor behavior during ontogeny. Moreover, the general trends in locomotor behavior during ontogeny are similar for both species.

With increasing age there is a decreased frequency of suspensory behavior and an increase in the frequency of quadrupedalism. In early infancy, chimpanzee positional behavior is clearly forelimb dominated, and clinging, climbing and arm-hanging are most common. Aided bipedalism is the earliest form of frequent hindlimb use. It is only through time that the infant is able to coordinate forelimb and hindlimb use in quadrupedalism. As the frequency of quadrupedalism increases, the frequency of bipedalism decreases. By 2 years of age, common chimpanzee locomotion is primarily quadrupedal in nature.

A comparison of the results of this study with those of Goodall (1968, 1969) show remarkably similar trends in the ontogeny of locomotor behavior, with two major differences.

The first difference is that these results show a different progression in the development of quadrupedalism. Goodall (1968, 1969) noted that, although there is individual variation, successful knuckle-walking occurs between 6–8 months of age. The results presented here indicate that infants become adept at palmigrade quadrupedalism before knuckle-walking, and that during that time, when on the ground, bipedalism is the primary form of locomotion. In this study, infants of less than 2 years rarely knuckle-walked.

The second difference in the results of the two studies is that Goodall stated that by 18 months of age, an individual proceeds in an adult fashion. The results of this study indicate that infants of greater than 2 years do, in fact, adopt a more adult-like form of locomotion than younger infants, but that they still use significantly less quadrupedalism and more climbing and suspensory activities than adults. Juveniles are also more suspensory than adults in arboreal locomotor behavior, although they are adult-like in their terrestrial locomotor profiles.

Thus, this paper demonstrates that chimpanzee positional behavior is not static during ontogeny, but rather, changes dramatically through time. Moreover, as was predicted, with increasing age there is decreasing suspensory activity.

The second aim of this study is to test whether the morphological paedomorphism of adult pygmy chimpanzees is associated with "behavioral paedomorphism", that is, whether adult pygmy chimpanzees are more suspensory in their locomotor behavior than adult common chimpanzees. The results presented here show that the arboreal locomotor behavior of adult pygmy chimpanzees differs from that of adult common chimpanzees and most closely resembles that of older infants. Adult pygmy chimpanzees, like immature common chimpanzees, use more suspensory and quadrupedal behavior than adult common chimpanzees.

The fact that an adult pygmy chimpanzee's locomotor activity is most similar to that of a 2–5 year old common chimpanzee infant is surprising. However, this is probably related to the paucity of terrestrial follows of pygmy chimpanzees. Clearly, like adult common chimpanzees, adult pygmy chimpanzees are knuckle-walking quadrupeds when on the ground. Most likely if adult pygmy chimpanzees were sampled more frequently when on the ground, their overall locomotor behavior pattern would most closely resemble that of juvenile rather than class 3 infants.

Another ontogenetic and interspecific difference in the results is an increasing frequency of quadrupedalism through time and the increased frequency of quadrupedalism of pygmy chimpanzees in comparison with common chimpanzees. These results were not predicted on the basis of morphological differences, but can best be understood in terms of body size differences. There is variation in the mean body weight of the three chimpanzee subspecies (Jungers & Susman, 1984). Although *P. t. schweinfurthii* does not differ in size from *P. paniscus*, both are significantly smaller than *P. t. troglodytes* (Jungers & Susman, 1984). Although the one available body weight for *P. t. verus* (the subspecies observed in this study) places it closer to that of *P. paniscus* and *P. t. schweinfurthii* than *P. t. troglodytes* (Jungers & Susman, 1984), it is highly likely that the mean body weight of *P. t. verus* probably exceeds that of *P. paniscus* (for further consideration of this issue, see Doran, 1989).

Predictions of primate locomotor differences resulting from differences in body size have been made by several authors (Napier, 1967; Cartmill, 1974; Cartmill & Milton, 1977; Fleagle, 1985). In a review of such predictions, Fleagle & Mittermeier (1980) observed that in a given arboreal habitat, larger bodied animals "see" relatively fewer discontinuities in the "arboreal highway", and there are relatively fewer substrates that are large enough to support their body weight relative to smaller bodied animals. As a result, they predicted that in a given arboreal habitat, larger bodied animals would leap less and climb and bridge more frequently than smaller animals. In addition, the larger animals would either use larger substrates than smaller animals or engage in relatively more frequent suspensory activity. Subsequent studies have lent support to these predictions by noting either intraspecific or interspecific differences in positional behavior or substrate use as a result of differences in body size (Mendel, 1976; Fleagle & Mittermeier, 1980; Galdikas & Teleki, 1981; Crompton, 1984; Tuttle & Watts, 1985; Sugardjito & van Hoof, 1986; Cant, 1987; Gautier-Hion, 1988; Hunt, 1989).

Thus, if it is assumed that *P. paniscus* is smaller than *P. t. verus*, one would predict that *P. paniscus* would engage in more quadrupedalism and leaping, and less climbing, scrambling and suspensory behavior than *P. t. verus* as a result of "seeing" relatively more large substrates and discontinuities in the habitat.

The results show that pygmy chimpanzees do, in fact, engage in more quadrupedalism and leaping, and less climbing and scrambling than their common chimpanzee counterparts as would be predicted on the basis of body size. However, pygmy chimpanzees show an increased frequency of suspensory behavior rather than the decreased frequency that would be predicted on the basis of body size. This result supports the prediction that, based on its longer and more narrow scapula, pygmy chimpanzees are more suspensory than common chimpanzees and thus "paedomorphic" in behavior.

An additional factor to consider is whether this interspecific difference is a true paedomorphic difference. Paedomorphism can result from either neoteny or rate or time hypomorphosis [see Shea (1984*b*) for a review]. The processes of rate and time hypomorphosis are characterized by ontogenetic scaling of proportions. Neoteny is characterized by an uncoupling of the ancestral size/shape relations.

In the case of pygmy and common chimpanzee locomotor behavior, it cannot be determined at present which developmental process, neoteny or hypomorphosis, is responsible for the retention of immature locomotor behavior. However, based on the ontogenetic scaling of scapular features and intralimb proportions, and on the similar locomotor trends during the ontogeny of both species, it seems most likely that either time or rate hypomorphosis is responsible. To further clarify the issue, further data are required, particularly on the locomotor behavior of pygmy chimpanzee infant classes 1, 2 and 3, in order to ascertain whether the ontogenetic locomotor trends of pygmy chimpanzees are parallel to, or dissociated from, those seen in common chimpanzees.

Shea (1984*a,b*) and Jungers & Susman (1984) demonstrated that many postcranial features of pygmy chimpanzees, such as chest girth, arm span, intralimb proportions and particularly scapula size and shape, scale ontogenetically in comparison with those of common chimpanzees. The results of this study support the hypothesis that these features reflect true behavioral differences within the ontogeny of each species, and between the adults of the two species.

Gould (1977:221) has stressed the importance of understanding ontogeny in studies of evolution, since "heterochrony—the temporal displacement of characters—is a pervasive phenomenon among evolutionary processes."

Shea (1981, 1983, 1984, 1986) has demonstrated that in African apes, and in pygmy chimpanzees in particular, both neoteny and time and rate hypermorphosis or hypomorphosis have resulted in substantial morphological differentiation. The results of this study complement those findings that "shuffling of developmental trajectories of various body regions may provide new morphological configurations" Shea (1983:522) by demonstrating that associated with the morphological patterns are true behavioral differences.

Summary

Locomotor activities vary among chimpanzees (*P. troglodytes*) of different ages. Independent locomotor activities occur by 5 months of age and consist initially of climbing, arm-hanging and twirling and (aided) bipedalism. As infants increase in age (6 months–2 years) suspensory behavior becomes more important. A major shift to quadrupedalism occurs in infants of greater than 2 years. Juveniles are adult-like in their locomotor behavior, with the exception of the juvenile's more frequent suspensory behavior during arboreal locomotor activities. There are no differences in the frequencies of locomotor behaviors performed by adult and adolescent chimpanzees in the Tai Forest.

In addition to age class differences in locomotor behavior, there are also age class differences in substrate use. Immature chimpanzees differ from adults both in the type of substrate used and in the type of locomotor activity performed on a given substrate. In early infancy (less than 6 months) infant chimpanzees use saplings or their mothers for most locomotor activities. Between 6 months and 2 years, infants are more independent of their mothers and use thin substrates during most activities. Infants of more than 2 years use a much wider range of substrates than younger infants. Juveniles use lianes more frequently and vertical trunks less frequently than adults. There are no differences in substrate use by adults and adolescents.

There are intraspecific differences in the arboreal locomotor behavior of chimpanzees. Pygmy chimpanzees show the same locomotor trends during ontogeny as common chimpanzees. With increasing age there is decreased suspensory behavior and increasing quadrupedalism. However, the end results for the two species differ greatly. Pygmy chimpanzees differ from common chimpanzees at every age by using more frequent quadrupedalism and less frequent quadrumanous climbing and scrambling. In comparison with common chimpanzee locomotor behavior, the frequency of locomotor behavior used by adult pygmy chimpanzees differs greatly from that of adults, and is most similar to (and does not differ statistically from) that of a class 3 infant (greater than 2 years). Thus, the adult pygmy chimpanzee has a more juvenilized pattern of positional behavior associated with its morphological paedomorphism in comparison with common chimpanzees.

Acknowledgements

I would like to thank Anne Gomez and Drs John Fleagle, Charles Janson, William Jungers, Matt Ravosa, Michael Rose, Brian Shea and Randall Susman for comments and discussion on this manuscript. I also thank Dr Matt Ravosa and Anne Gomez for the invitation to participate in the symposium. I am grateful to the Ministry of Scientific Research and the Station d'Ecologie Tropicale and its director, Dr Henri Dosso, for permission to conduct research in the Ivory Coast. I thank the Centre Suisse de Recherche Scientifique for logistical support during my stay in the Ivory Coast. I would also like to extend a special thanks to Drs

Christophe and Hedwige Boesch for friendship, guidance and permission to study the chimpanzees of Tai. I am grateful to the government of Zaire and its Institut de Recherche Scientifique for permission to study in the Lomako Forest. Special thanks go to Drs Nancy Thompson-Handler, Richard Malenky and Annette Lanjouw for my introduction to field work and to the pygmy chimpanzees. I gratefully acknowledge financial support from the Wenner-Gren Foundation, the Louis B. Leakey Foundation and the National Science Foundation (to R. L. Susman). Finally, I would like to acknowledge the debt I owe to the Department of Anatomical Sciences, SUNY at Stony Brook, and particularly Drs John Fleagle and Randall Susman for guidance during my graduate training.

References

Altmann, J. (1974). Observational study of behavior: sampling methods. *Behaviour* **49,** 227–267.
Badrian, A. & Badrian, N. (1977). Pygmy chimpanzees. *Oryx* **13,** 463–468.
Badrian, A. & Malenky, R. (1984). Feeding ecology of *Pan paniscus* in the Lomako Forest, Zaire. In (R. L. Susman, Ed.) *The Pygmy Chimpanzee: Evolutionary Biology and Behavior*, pp. 325–344. New York: Plenum Press.
Boesch, C. (1978). Nouvelles observations sur les chimpanzees de la foret de Tai (Cote D'Ivoire). *La Terre et la Vie* **32,** 195–201.
Boesch, C. & Boesch, H. (1981). Sex differences in the use of natural hammers by wild chimpanzees: a preliminary study. *J. hum. Evol.* **10,** 585–593.
Boesch, C. & Boesch, H. (1983). Optimization of nut-cracking with natural hammers by wild chimpanzees. *Behaviour* **3/4,** 265–286.
Boesch, C. & Boesch, H. (1984). Mental map in wild chimpanzees: an analysis of hammer transports for nut cracking. *Primates* **25,** 160–170.
Boesch, C. & Boesch, H. (1989). Hunting behavior of wild chimpanzees in the Tai National Park. *Am. J. phys. Anthrop.* **78,** 547–573.
Cant, J. G. (1987). Effects of sexual dimorphism in body size on feeding postural behavior of Sumatran orangutans (*Pongo pygmaeus*). *Am. J. phys. Anthrop.* **74,** 143–148.
Cartmill, M. (1974). Pads and claws in arboreal locomotion. In (F. A. Jenkins, Ed.) *Primate locomotion*. New York: Academic Press.
Cartmill, M. & Milton, K. (1977). The lorisiform wrist joint and the evolution of brachiating adaptations in Hominoidea. *Am. J. phys. Anthrop.* **47,** 249–272.
Coolidge, H. J. (1933). *Pan paniscus*: pygmy chimpanzee from the south of the Congo River. *Am. J. phys. Anthrop.* **18,** 1–59.
Coolidge, H. J. (1984). Historical remarks bearing on the discovery of *Pan paniscus*. In (R. L. Susman, Ed.) *The Pygmy Chimpanzee: Evolutionary Biology and Behavior*, pp. ix–xiii. New York: Plenum Press.
Crompton, R. H. (1983). Age differences in locomotion in the subtropical Galaginae. *Primates* **24,** 24–59.
Crompton, R. H. (1984). Foraging, habitat structure and locomotion in two species of galago. In (P. S. Rodman & J. G. H. Cant, Eds) *Adaptations for Foraging in Nonhuman Primates*. New York: Columbia University Press.
Doran, D. M. (1989). Chimpanzee and pygmy chimpanzee positional behavior: the influence of environment, body size, morphology, and ontogeny on locomotion and posture. Ph.D. Dissertation, SUNY at Stony Brook, U.S.A.
Doran, D. M. (in press). A comparison of instantaneous and locomotor bout sampling methods: a case study of adult male chimpanzee locomotor behavior and substrate use. *Am. J. phys. Anthrop.*
Fleagle, J. G. (1985). Size and adaptation in primates. In (W. L. Jungers, Ed.) *Size and Scaling in Primate Biology*, pp. 1–19. New York: Plenum Press.
Fleagle, J. G. & Mittermeier, R. A. (1980). Locomotor behavior, body size and comparative ecology of seven Surinam monkeys. *Am. J. phys. Anthrop.* **53,** 2011–2314.
Fossey, D. (1979). Development of the mountain gorilla (*Gorilla gorilla beringei*) through the first thirty six months. In (D. A. Hamburg & E. R. McCown, Eds) *The Great Apes*. Menlo Park, California: Benjamin Cummings.
Frechkop, S. (1935). Notes sur les mammiferes. XVII: A propos du chimpanze de la rive gauche du Congo. *Mus. R. Hist. nat. Belg. Bull.* **11,** 1–41.
Galdikas, B. M. & Teleki, G. (1981). Variations in subsistence activities of female and male Pongids: new perspectives on the origins of hominid labor division. *Curr. Anthrop.* **22,** 241–253.
Gautier-Hion, A. (1988). The diet and dietary habits of forest guenons. In (A. Gautier-Hion, F. Bourliere & J. P. Gautier, Eds) *A Primate Radiation: Evolutionary Radiation of African Guenons*. Cambridge: University Press.
Goodall, J. (1968). The behavior of free-living chimpanzees in the Gombe Stream, Reserve. *Anim. Behav. Monogr.* **1,** 161–311.
Goodall, J. (1969). Mother-offspring relationships in free-ranging chimpanzees. In (D. Morris, Ed.) *Primate Ethology*. Garden City, NY: Doubleday Anchor Books.

Gould, S. J. (1977). *Ontogeny and Phylogeny*. Cambridge, MA: Belknap Press of Harvard University Press.

Hunt, K. (1989). Positional behavior in *Pan troglodytes*. *Am. J. phys. Anthrop.* **78,** 242–243.

Jungers, W. L. (1984). Aspects of size and scaling in primate biology with special reference to the locomotor skeleton. *Yearb. phys. Anthrop.* **27,** 73–97.

Jungers, W. L. & Susman, R. L. (1984). Body size and skeletal anatomy in the African apes. In (R. L. Susman, Ed.) *The Pygmy Chimpanzee: Evolutionary Biology and Behavior*. New York: Plenum Press.

McHenry, H. M. & Corruccini, R. S. (1981). *Pan paniscus* and human evolution. *Am. J. phys. Anthrop.* **54,** 355–367.

Mendel, F. (1976). Postural and locomotor behavior of *Aloutta palliata* on various substrates. *Folia primat.* **26,** 36–53.

Napier, J. R. (1967). Evolutionary aspects of primate locomotion. *Am. J. phys. Anthrop.* **27,** 333–342.

Rawlins, R. (1976). Locomotor ontogeny in *Macaca mulatta*: I. Behavioral strategies and tactics. *Am. J. phys. Anthrop.* **44,** 201.

Roberts, D. (1974). Structure and function of the primate scapula. In (F. A. Jenkins, Ed.) *Primate Locomotion*, pp. 171–200. New York: Academic Press.

Rose, M. D. (1977). Positional behavior of olive baboons (*Papio anubis*) and its relationship to maintenance and social activities. *Primates* **18,** 59–116.

Schaller, G. (1963). *The Mountain Gorilla*. Chicago: University of Chicago Press.

Shea, B. T. (1981). Relative growth of the limbs and trunk in the African apes. *Am. J. phys. Anthrop.* **56,** 179–201.

Shea, B. T. (1983). Paedomorphism and neoteny in the pygmy chimpanzee. *Science* **222,** 521–522.

Shea, B. T. (1984*a*). Paedomorphism and neoteny in the pygmy chimpanzee. *Science* **222,** 521–522.

Shea, B. T. (1984*b*). An allometric perspective on the morphological and evolutionary relationships between pygmy (*Pan paniscus*) and common (*Pan troglodytes*) chimpanzees. In (R. L. Susman, Ed.) *The Pygmy Chimpanzee: Evolutionary Biology and Behavior*. New York: Plenum Press.

Shea, B. T. (1986). Scapula form and locomotion in chimpanzee evolution. *Am. J. phys. Anthrop.* **70,** 475–488.

Shea, B. T. (1988). Heterochrony in primates. In (M. L. McKinney, Ed.) *Heterochrony in Evolution*. New York: Plenum Press.

Sokal, R. A. & Rohlf, F. J. (1981). *Biometry*. New York: W. H. Freeman Co.

Sugardjito, J. & Van Hoof, J. (1986). Sex-age class differences in positional behavior of Sumatran orangutan (*Pongo pygmaeus abelii*) in the Gunung Leuser National Park, Indonesia. *Folia primat.* **47,** 14–25.

Susman, R. L., Badrian, N. L. & Badrian, A. J. (1980). Locomotor behavior of *Pan paniscus* in Zaire. *Am. J. phys. Anthrop.* **53,** 69–80.

Thompson-Handler, N. E. (1990). The pygmy chimpanzee: sociosexual behavior, reproductive biology and life history patterns. Ph.D. Dissertation, Yale University, U.S.A.

Tuttle, R. H. & Watts, D. P. (1985). The positional behavior and adaptive complexes of *Pan gorilla*. In (S. Kondo, Ed.) *Primate Morphophysiology, Locomotor Analyses and Human Bipedalism*. Tokyo: Tokyo University Press.

Vancata, V. (1982). Chimpanzee locomotion and implications for the origin of hominid bipedality. Man and his origins. *Anthropos/Brno.* **21,** 41–45.

White, F. J. (1986). Behavioral ecology of the pygmy chimpanzee. Ph.D. Dissertation, SUNY at Stony Brook, U.S.A.

D. Tab Rasmussen
Department of Anthropology, Washington University, St. Louis, Missouri, 63130, U.S.A.

Chia L. Tan
Department of Anthropology, University of California, Los Angeles, Los Angeles, CA 90024, U.S.A.

Received 21 June 1991
Revision received 20 December 1991 and accepted 16 January 1992

Keywords: primate behavior, ontogenetic trajectory, mother–infant interaction, suckling, weaning, body size, brain size.

The allometry of behavioral development: fitting sigmoid curves to ontogenetic data for use in interspecific allometric analyses

The behavior of young primates unfolds in a complex, non-linear fashion that has been difficult to quantify in a manner that allows precise interspecific comparison. As a consequence, allometric analyses have not been easily applied to patterns of behavioral ontogeny. Researchers studying the allometry of growth rates have approached the problem of comparing non-linear changes during development by fitting empirical data to sigmoid curves; coefficients obtained from these curves provide a precise basis for interspecific comparisons. We have explored similar methods that may be applied to behavioral data expressed as rates. Data on age-specific rates of suckling and physical independence were obtained from the literature for a range of primate species. These data are fitted to Gompertz models; the curves fitting each species' ontogenetic pattern can be precisely compared interspecifically while controlling for body size or other variables by use of standard allometric analyses. The resulting statistically significant allometric relationships yield allometric exponents for behavioral rate coefficients of -0.37 to -0.51, and for inflection points of 0.34–0.50, relative to adult body size and brain size. For the sample analysed here, 57–88% of the variation among species in the decrease of suckling and increase in physical independence is associated with variation in body and brain size, such that larger species have slower rates of behavioral change than small species. There are no statistically significant differences between prosimians and anthropoids in the ontogenetic variables examined in this study. Terrestrial primates attain physical independence earlier than arboreal primates when controlling for body size, reflecting an earlier age of locomotor independence in terrestrial primates. This method holds promise for: (1) more detailed interspecific studies of possible associations between behavioral ontogenetic patterns and diet, locomotion, social organization, morphological development and other key variables; (2) exploration of possible heterochronic effects in behavioral evolution; and (3) intraspecific study of the ontogenetic consequences of living under different conditions or in different habitats.

Journal of Human Evolution (1992) **23**, 159–181

Introduction

An enormous body of scientific literature exists on the development of behavior in young primates (see reviews by Pereira & Altmann, 1985; Baldwin, 1986; Nicolson, 1986; Walters, 1986). Despite the availability of this literature, there has been little effort to make precise, quantitative comparisons among primate species. The importance of making mathematically precise interspecific comparisons has become more and more evident as primatologists have come to fully appreciate the importance of scaling factors and size constraints on nearly all aspects of primate biology (see reviews by Fleagle, 1985; Harvey *et al.*, 1986). The early development of primate behavior is certainly constrained by growth rates, neurological or neuromuscular development, body size and metabolic rate, among other variables. Before many interesting questions about the adaptive significance of primate behavioral development can be fully addressed (e.g., what is the effect of different social organizations, mating systems, patterns of offspring care, diet, locomotion or habitat?) and the role of heterochrony in behavioral development can be explored, there must be an evaluation of the variance in behavioral ontogeny that might be attributable to size and scaling. In order to do this, precise interspecific comparisons among taxa are necessary.

Methods have been developed that allow fairly sophisticated investigations of the influences of size on physical growth, structural development and adult form (e.g., Shea, 1985; Jolicoeur & Pirlot, 1988; Leigh & Cheverud, 1991; Zullinger et al., 1984). However, these methods have not been systematically applied to the study of behavior and behavioral ontogeny, largely because the behavior of young primates unfolds in a complex, non-linear fashion that is difficult to characterize. The task of making quantitative comparisons among species boils down to three major problems: comparable behavioral data must be obtained from a variety of species, these data must adequately reflect the actual pattern of behavioral development and the data must be compared according to a standard mathematical currency. Previous interspecific work on behavioral development has made comparisons by one of two fairly simple methods. First, the absolute length of time that is required for a behavior to appear or disappear has been measured and compared. Second, an average rate of change may be obtained across a specified time interval bounded by two discrete end-points. The problem with these methods is that behavior does not often develop in discrete saltational events or usually change in a linear fashion. Both of these methods are therefore fairly crude abstractions of the actual observed behavior.

In studying the ontogeny of morphological systems, Alberch et al. (1979) have called the curvilinear ontogenetic path that an organism follows through a multidimensional space of time, shape and size an "ontogenetic trajectory". Such a trajectory can be projected onto a bivariate plot of some measure of size or shape as a function of time. Several non-linear models are available to describe the resulting projected trajectory (Alberch et al., 1979). For example, the physical growth of organisms has often been modeled by sigmoid functions such as the Gompertz model (Zullinger et al., 1984). Once the model has been fitted to data from several individuals or species, it is possible to make precise comparisons for use in allometric analyses or other studies.

Behavior, like morphology, develops along a multidimensional, curvilinear pathway or ontogenetic trajectory. Among the components that make up the behavioral ontogenetic trajectory are changes in the rate or frequency at which a behavior is expressed, in the quality and completeness of expression and in the contexts and social settings in which the behavior occurs. Projection of one or more of these variables onto a plot versus time should make the behavior amenable to analysis via non-linear regression in a way similar to how growth or structure have been examined. One of the most commonly reported measures of behavioral ontogeny is change in the frequency at which a behavior is expressed through time. An inspection of such data from a variety of species suggested to us that some primate behaviors might unfold in a sigmoid pattern basically similar to mammalian growth rates. By fitting a common class of sigmoid curves to behavioral developmental rates in a variety of primate species, a common currency for interspecific comparisons is available that is more precise than the use of developmental stages or average rates. The behavioral ontogenetic trajectory is thus abstracted to a useful and mathematically tractable *shape*. In this paper, we provide a preliminary exploration of such a method. Our results suggest that the study of "ontogeny curves" does yield statistically significant allometric relationships among species, and further study may prove useful in understanding allometric and adaptive aspects of behavioral development.

Materials and methods

Behavioral data

We chose to focus on the development of two infant behaviors: suckling and physical independence. These two behaviors, which are obviously correlated with each other, were chosen because they were represented in the literature by the greatest volume of work on the widest range of species. Physical independence was defined in many ways in the literature; we accepted those studies defining it as the lack of physical contact between infant and caretaker, again, because this was the most commonly used definition. In species with multiple caretakers, such as callitrichids, we used time out of contact with all caretakers as the measure of independence. Suckling was defined as being in suckling position; in most studies, there was no attempt to determine actual lactation on the part of the mother or ingestion of milk by the infant. The behavior of being "in suckling position" may differ significantly from actual ingestion of milk (Izard, 1987).

Data were accepted into our study sample if the recording method was continuous, instantaneous or one-zero, and if the period of data collection extended at least over most of infancy, from birth or shortly after birth until weaning. Because continuous and instantaneous sampling may provide reliable, unbiased estimates of activity budgets, the data obtained by these methods were pooled in a common analysis. One-zero methods do not provide a measure of time budgets and so these samples were treated separately (Altmann, 1974; Simpson & Simpson, 1977; Martin & Bateson, 1986).

Most studies in the literature presented suckling and physical contact in the form of bivariate plots, with time on the abscissa and a quantitative behavioral measure on the ordinate. The behavioral measure most often used was a frequency or proportion expressed as a percentage of the total activity budget. In some cases, unusual measures of behavior could be algebraically converted to a proportion of time if continuous or instantaneous recording methods were used; samples derived from one-zero recording cannot be converted. (Although rarely explicitly stated in the literature, we assumed calculations of activity budget incorporated only the active half of the total diel cycle, depending on whether the species was diurnal or nocturnal.) Few published studies presented the actual values of points on their bivariate plots. The values of plotted points were obtained by written request from the authors themselves, or in some cases in which authors were not contacted, by making enlarged copies of the plots and reading the ordinate values directly off the published figure by use of a set square.

Data on slow lorises (*Nycticebus coucang*), slender lorises (*Loris tardigradus*) and pottos (*Perodicticus potto*) were collected by instantaneous sampling at 5 min intervals of focal individuals housed at the Duke University Primate Center (DUPC). The methods are described in Rasmussen (1986). Table 1 provides a summary of all behavioral data included in our analyses.

Choice of models

For our sigmoid equation, we chose to use the Gompertz model, which has often been applied to growth situations (Alberch *et al.*, 1979). Zullinger *et al.* (1984) demonstrated that the Gompertz was the best choice among sigmoid equations for application to comparative studies of mammalian growth rates, and we therefore assumed that it would be a reasonable point of departure for our studies of behavioral rates. This does not imply that we expected a close correlation between physical growth and behavioral development, only that the

Table 1 Summary of behavioral data organized by species

ID[1]	Species	Behavior	Sampling	n^2	Reference
a	*Lemur macaco*	contact	contin	5	Harrington, 1978
b	*Lemur catta*	contact	contin	11	Gould, 1990
c	*Galago moholi*	suckle	contin	8	Doyle et al., 1969
d	*Nycticebus coucang*	contact	instan	3	Rasmussen, 1986
		suckle	instan	3	Rasmussen, 1986
e	*Loris tardigradus*	contact	instan	3	Rasmussen, 1986
		suckle	instan	3	Rasmussen, 1986
f	*Perodicticus potto*	contact	instan	1	Rasmussen, unpublished
		suckle	instan	1	Rasmussen, unpublished
A	*Saimiri sciureus*	contact	1/0	6	Kaplan, 1972
		suckle	1/0	6	Rosenblum, 1968
		suckle	1/0	6	Kaplan, 1972
B	*Cebus apella*	suckle	contin	1	Fragaszy, 1990
C	*Callicebus moloch*	contact	contin	2	Wright, 1984
D	*Aotus trivirgatus*	contact	contin	1	Wright, 1984
		contact	contin	10	Dixson & Fleming, 1981
		suckle	contin	10	Dixson & Fleming, 1981
E	*Alouatta seniculus*	contact	contin[3]	11	Mack, 1979
		suckle	contin[3]	11	Mack, 1979
F	*Callimico goeldii*	contact	instan[3]	2	Heltne et al., 1973
G	*Leontopithecus rosalia*	contact	contin	11	Hoage, 1978
H	*Saguinus oedipus*	contact	1/0	12	Cleveland & Snowdon, 1984
		suckle	1/0	12	Cleveland & Snowdon, 1984
I	*Saguinus mystax*	contact	instan	1	Heymann, 1990
J	*Saguinus fuscicollis*	contact	instan	2	Heymann & Sicchar, 1990
		contact	instan	5	Vogt et al., 1978
K	*Callithrix jacchus*	contact	contin	31	Ingram, 1977
		contact	contin	9	Locke-Hayden & Chalmers, 1983
L	*Cercopithecus aethiops*	contact	contin	35	Fairbanks & MacGuire, 1987
		contact	contin	15	Fairbanks, 1988
M	*Erythrocebus patas*	contact	1/0	7	Chism, 1986
		suckle	1/0	7	Chism, 1986
N	*Macaca silenus*	contact	instan	2	Kumar & Kurup, 1981
O	*Macaca nemestrina*	contact	contin	21	Reite & Short, 1980
		contact	contin	?	Bolwig, 1980
		contact	contin	10	Castell & Wilson, 1971
		suckle	contin	21	Reite & Short, 1980
		suckle	contin	?	Bolwig, 1980
P	*Macaca fascicularis*	contact	1/0	20	Nakamichi et al., 1990
		suckle	1/0	20	Nakamichi et al., 1990
Q	*Macaca fuscata*	contact	1/0	22	Murray & Murdoch, 1977
R	*Macaca mulatta*	contact	contin	20	Berman, 1980
		contact	contin	v	Johnson & Southwick, 1987
		contact	contin	44	Simpson et al., 1986
		suckle	contin	20	Berman, 1980
		suckle	contin	v	Johnson & Southwick, 1987
S	*Macaca arctoides*	contact	1/0	7	Rhine & Hendy-Neely, 1978
		suckle	1/0	7	Rhine & Hendy-Neely, 1978
T	*Papio cynocephalus*	contact	instan	12	Altmann, 1978
		contact	instan	11	Altmann, 1980
		suckle	instan	14	Rhine et al., 1985
		suckle	1/0	14	Rhine et al., 1985
U	*Papio hamadryas*	contact	contin	?	Bolwig, 1980
V	*Theropithecus gelada*	suckle	instan	10	Dunbar, 1984
W	*Presbytis entellus*	contact	contin	19	Dolhinow & Murphy, 1982
		contact	contin	19	Dolhinow & Krusko, 1984

Table 1 *Continued.*

ID[1]	Species	Behavior	Sampling	n^2	Reference
X	*Pongo pygmaeus*	contact	contin	4	Miller & Nadler, 1981
		suckle	contin	4	Miller & Nadler, 1981
Y	*Pan troglodytes*	contact	contin	4	Miller & Nadler, 1981
		suckle	contin	4	Miller & Nadler, 1981
Z	*Gorilla gorilla*	contact	contin	32	Fossey, 1979
		suckle	contin/instan[3]	14	Stewart, 1988

[1] The identifying letters in this column are used to label species in Figures 4 and 5.
[2] Sample size indicated here is the total number of individuals across the entire study period; sample size may vary over shorter time intervals. See original publication for details.
[3] Method of data presentation differs from others, requiring algebraic conversion to be comparable to activity budget data; see original publications for details. v = variable sample size; see the cited publication. Sampling: contin = continuous; instan = instantaneous.

Gompertz model has proven to be a flexible, useful model for application to biological development. We used a modified, two parameter version of the model used by Zullinger *et al.* (1984). Variation among specific Gompertz curves can therefore be characterized by these two parameters: (1) the *rate coefficient*, an instantaneous rate of change that reflects the steepness of the curve at any given point; and (2) the *inflection point*, which reflects the age of the maximum rate of behavioral change (Figure 1). The asymptote of our Gompertz model for application to the independence data was fixed at 100% and no starting point was designated. (In most primates, infant physical independence starts near 0% and rises to nearly 100% in adulthood.) The asymptote for the suckling data was set at 0% (no suckling) and the starting value at the maximum observed suckling rate (usually observed in the first week after birth). The resulting equation for independence, going from an initial low value to a final high value, is the following:

$$C_t = 100 e^{-e^{-K(t-I)}}$$

where I = age at the inflection point of the curve (days), K = behavioral rate coefficient (days^{-1}) and C_t = contact rate (proportion of time spent out of contact with caretaker) at age t (days). The coefficent K provides an estimate of the instantaneous rate of change in caretaker–infant contact rate at any age and thus is analogous to the slope of a linear regression; the coefficient I provides an age estimate for the time of maximum rate of change.

The inverse equation for suckling, starting with an initial high value and dropping to a low one, is:

$$S_t = m - m e^{-e^{-K(t-I)}}$$

where S_t = suckling rate (proportion of time spent in suckling position) at age t (days) and m = the maximum observed suckling rate (which is obtained from the empirical data).

Data were fitted to the models and the parameters K and I were estimated using the nonlinear procedure of SAS (Statistical Analysis System) with the Marquardt iteration method. A separate curve was determined for each published data set. For species represented by more than one data set, species values for the parameters are means of the individual sets.

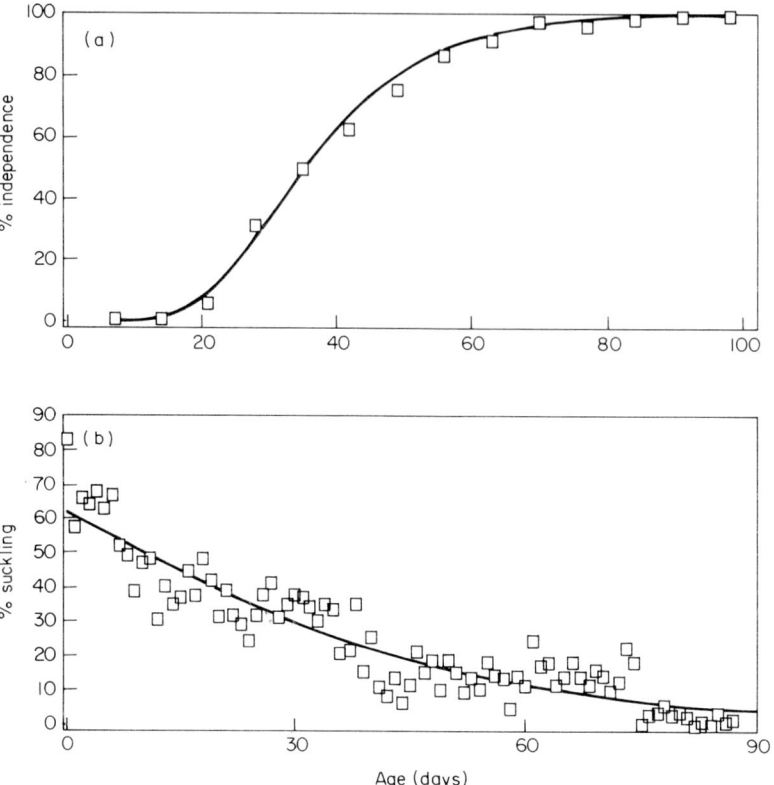

Figure 1. Examples showing the fit between behavioral data and Gompertz curves. (a) The attainment of physical independence in the common marmoset, *Callithrix jacchus* (Ingram, 1977). This plot demonstrates basic features of the Gompertz curve, including the inflection point, at 31·2 days and the rate coefficient, which is a relatively high value (0·0846), reflecting a relatively "curvy" curve, rather than a flat or gradually sloping curve. This is an unusually good-fitting curve, with minimal residual deviations and only a slight possibility of autocorrelation of residuals (there is a string of four consecutive negative residuals from about 40–60 days). (b) The reduction in suckling behavior in the Mohol lesser galago, *Galago moholi* (Doyle *et al.*, 1969). The inflection point, occurs at only 10·8 days and the rate coefficient, is 0·0578, resulting in a more gradual rate of change than the previous example. Autocorrelation of residuals is a significant problem between the ages of 60–90 days; the model cannot fit the empirical pattern in which the bushbabies attain a low level of suckling (10–20%) that is maintained fairly constantly from about 45–75 days, before abruptly dropping to near zero.

The goodness of fit of each equation was evaluated with respect to the residual sum of squares. Proper statistical interpretation of equations fit by non-linear regression require that observations are not correlated, which is rarely the case when using longitudinal behavioral data (Hoel, 1964). In particular, variance about the regression might not be random with respect to age, especially when only one or a few infants are sampled. Autocorrelation of residuals was analysed visually when plotted as a function of age. Patterns of variance for any species will also be influenced by the choice of time interval and the sample size within each interval. Most importantly, confidence in the initial behavioral data points themselves varies enormously with the sample size of individuals per species and the time observed per individual. Confidence intervals and hypothesis testing involving the direct comparison of separate curves will be reliable only to the extent that the above assumptions about correlation and independence are true.

Appendix 1 provides examples of SAS programs that can be used to fit data to the Gompertz curve and to graph the results.

Allometric analyses

Interspecific allometric analyses of log-transformed species means of the rate coefficient K and the inflection point I relative to adult body size and brain size were performed using least squares regression. Allometric exponents were determined using the straight-line relationship:

$$\log y = a \log x + \log b$$

where y is one of the behavioral rate coefficients (I or K), x is adult body weight or brain size, a is the slope and b is the intercept. The slope a becomes the allometric exponent when the equation is converted to the power law, while b is the allometric coefficient:

$$y = bx^a$$

Patterns of residual deviation from the least squares regressions were examined to evaluate whether or not differences existed between prosimians and anthropoids, and between arboreal and terrestrial species. Differences in residual distribution were analysed statistically by a non-parametric Mann–Whitney U test.

Body weight and brain size were obtained from the literature.

Results

Curve fitting

The results of the curve-fitting are presented in Tables 2 and 3. In general, the Gompertz equations fit the empirical data quite well, suggesting that this is an appropriate model for behavioral change. Standard errors averaged about 10–20% of the estimated parameter values. Analysis of the variance in the standard errors with respect to recording methods (continuous, instantaneous and one-zero) indicated that there was no significant difference in how well the model fit empirical data ($\alpha = 0.01$).

Figures 1–3 show examples of the fit between Gompertz curves and data. Analysis of the residuals about the regression suggest that in most cases autocorrelation is a minimal problem, especially when the curves are fit to mean data from several individuals. The distribution of residuals above and below predicted values is often fairly random in appearance (Figures 1a, 2b and 3). In other cases, there are short strings of consecutive positive or negative residuals that would suggest an inappropriate model or correlation of observations (Figures 1b and 2a). The greatest error often occurred at very young ages, a problem probably associated with regression leverage rather than a biological misfit to the model (investigation of regression methods weighted in favor of initial values might be useful). With misfits at young ages, autocorrelation of residuals was often observed throughout the curve.

Among the data sets with the poorest fit to the Gompertz equation is the one on parental contact in *Nycticebus coucang*, an example that is instructive when examined in more detail (Figure 2b). The contact data of *N. coucang* yielded the relatively widest standard error of I and also the lowest I of any species (Table 2). This surely reflects the "parking" behavior exhibited by this species, where mothers park their infants on branches beginning at a very young age while the mothers forage or travel (Charles-Dominique, 1977; Rasmussen, 1986).

Table 2 Parameters estimated by fitting the Gompertz model to contact data

Species	Rate ($\times 10^3$) K	SE	Inflection I	SE	n	Reference
Continuous and instantaneous samples						
L. macaco	46·3	5·5	84·2	2·0	19	Harrington, 1978
L. catta	28·7	2·1	49·9	1·8	16	Gould, 1990
N. coucang	12·4	2·9	8·3	14·6	19	Rasmussen, 1986
L. tardigradus	26·9	5·7	42·9	5·6	14	Rasmussen, 1986
P. potto	41·6	16·7	49·7	6·9	12	Rasmussen, unpublished
C. moloch	29·9	6·8	72·4	5·8	5	Wright, 1984
A. trivirgatus	22·4	2·3	91·3	2·7	6	Wright, 1984
	17·6	1·6	82·5	2·7	18	Dixson & Fleming, 1981
A. seniculus	14·9	1·3	110·3	4·3	9	Mack, 1979
C. goeldii	26·9	8·2	74·7	5·7	9	Heltne et al., 1973
L. rosalia	50·1	3·6	42·3	1·1	12	Hoage, 1978
S. mystax	45·4	5·5	42·7	1·8	11	Heymann & Sicchar, 1990
S. fuscicollis	22·2	6·8	70·5	10·3	7	Heymann, 1990
	61·3	4·7	55·9	1·1	20	Vogt et al., 1978
C. jacchus	84·6	1·8	31·2	0·2	14	Ingram, 1977
	107·9	12·7	18·7	0·9	7	Locke-Hayden & Chalmers, 1983
C. aethiops	15·7	1·1	51·6	3·0	6	Fairbanks & McGuire, 1987
	16·7	0·5	76·7	1·3	6	Fairbanks, 1988
M. silenus	55·2	8·0	49·4	1·2	8	Kumar & Kurup, 1981
M. nemestrina	10·0	1·2	127·3	5·2	5	Reite & Short, 1980
	15·2	2·7	43·9	5·4	15	Bolwig, 1980
	8·6	1·9	128·5	8·6	5	Castell & Wilson, 1971
M. mulatta	9·2	1·1	37·0	8·4	15	Berman, 1980
	15·9	2·2	68·3	3·9	7	Simpson et al., 1986
	21·5	1·8	53·3	1·6	11	Johnson & Southwick, 1987
P. cynocephalus	16·1	1·8	81·8	4·6	6	Altmann, 1978 (laissez-faire)
	10·9	1·9	110·1	7·4	6	Altmann, 1978 (restrictive)
	10·0	0·9	103·6	6·5	11	Altmann, 1980
P. hamadryas	20·4	1·6	96·4	1·9	15	Bolwig, 1980
P. entellus	15·1	2·1	69·8	4·3	10	Dolhinow & Krusko, 1984
	12·1	1·5	98·9	6·4	10	Dolhinow & Murphy, 1982
P. pygmaeus	8·4	1·0	206·2	9·1	12	Miller & Nadler, 1981
P. troglodytes	8·0	1·0	288·8	7·9	12	Miller & Nadler, 1981
G. gorilla	2·4	0·3	457·4	32·0	19	Fossey, 1979
One-zero sample						
S. sciureus	49·6	7·9	37·7	2·7	21	Kaplan, 1972
S. oedipus	13·3	1·7	85·9	4·4	20	Cleveland & Snowden, 1984
E. patas	14·0	1·1	66·8	3·3	21	Chism, 1986
M. fascicularis	10·2	1·9	128·8	11·4	7	Nakamichi et al., 1990
M. fuscata	23·4	2·8	45·6	2·5	6	Murray & Murdoch, 1977
M. arctoides	20·1	5·3	55·0	5·6	6	Rhine & Hendy-Neely, 1978

thus, physical independence in *N. coucang* follows a different ontogenetic trajectory than in other species, which does not fit well with the Gompertz model. In contrast, infants of another Asian lorisine, *Loris tardigradus*, are rarely parked early in life (Rasmussen, 1986) and this species is fitted well by the Gompertz model.

Intraspecific variation
The largest number of studies available for a single species are the three parental contact curves for the rhesus macaque, *Macaca mulatta* (Berman, 1980; Simpson et al., 1986; Johnson

Table 3 Parameters estimated by fitting the Gompertz model to suckling rate data

Species	Rate ($\times 10^3$) K	SE	Inflection I	SE	n	Reference
Continuous and instantaneous samples						
G. moholi	57·8	3·1	10·8	0·8	89	Doyle et al., 1969
N. coucang	27·0	4·9	12·6	5·9	29	Rasmussen, 1986
L. tardigradus	59·0	11·6	29·2	2·7	14	Rasmussen, 1986
P. potto	41·5	12·1	34·5	6·0	12	Rasmussen, unpublished
C. apella	44·6	10·6	19·0	4·2	11	Fragaszy, 1990
A. trivirgatus	24·6	4·4	64·6	5·1	6	Dixson & Fleming, 1981
M. nemestrina	10·5	3·1	116·6	12·0	5	Reite & Short, 1980
	17·1	2·8	44·5	4·6	15	Bolwig, 1980
M. mulatta	9·5	1·1	64·9	7·1	15	Berman, 1980
	23·5	2·5	46·5	2·1	11	Johnson & Southwick, 1987
P. cynocephalus	9·6	1·0	75·3	8·5	26	Rhine et al., 1985 (instan)
T. gelada	14·4	2·9	36·5	9·7	11	Dunbar, 1984
P. pygmaeus	8·8	1·4	108·8	13·1	12	Miller & Nadler 1981
P. troglodytes	6·7	1·4	90·2	21·8	12	Miller & Nadler, 1981
G. gorilla	2·4	0·6	640·9	76·4	7	Stewart, 1988
One-zero samples						
S. sciureus	137·5	42·2	27·7	1·8	6	Rosenblum, 1968
	11·1	3·3	37·9	15·4	21	Kaplan, 1972
S. oedipus	17·3	4·1	51·4	8·5	20	Cleveland & Snowden, 1984
E. patas	9·8	1·9	94·3	12·8	21	Chism, 1986
M. fascicularis	27·4	4·7	44·1	4·1	7	Nakamichi et al., 1990
M. arctoides	17·4	9·0	54·7	11·5	6	Rhine & Hendy-Neely, 1978
P. cynocephalus	9·8	0·8	134·5	6·2	26	Rhine et al., 1985 (1/0)

& Southwick, 1987) and the pig-tailed macaque, *M. nemestrina* (Reite & Short, 1980; Bolwig, 1980; Castell & Wilson, 1971). The three studies on each of these two macaque species yielded a fairly wide range of parameter estimates (in the case of *M. mulatta*, K values were 9·2, 15·9 and 21·5; I values were 37·0, 68·3 and 53·3), indicating that qualities of the study population and study methods probably contribute to notable variation in the best fitting Gompertz curve. Johnson and Southwick (1987) studied wild *M. mulatta* in India, Berman (1980) studied the Cayo Santiago population and Simpson et al. (1986) studied small captive groups. Thus, important nutritional and social variables are not controlled and these undoubtedly contribute to the intraspecific variation. The factors contributing to intraspecific variation cannot be addressed statistically without larger samples of studies from a given species.

In contrast, intraspecific comparison of data sets collected by the same researcher on the same social groups does not show as much unexplained variability. Altmann (1978) divided her baboon mothers into two groups, those that were permissive with their offspring (*"laissez-faire"*) and those that were "restrictive". The Gompertz curves fitted to these two classes shows that infant independence is gained at a greater rate (higher K) and the inflection point (I) occurs earlier in the permissive mothers than in restrictive mothers (Table 2), but the variation is less than that observed among the three studies of *M. mulatta*. The Altmann study suggests that ontogeny curves are sensitive to intraspecific behavioral differences when sampling methods and the study population are held constant.

Table 4 **Body weight and brain size for species included in the allometric analyses**

Species	Body weight[1]	Brain size[2]
L. macaco	2401[3]	23·3[4]
L. catta	2670[3]	—
G. moholi	160	5·3[5]
N. coucang	1268[5]	10·1[5]
L. tardigradus	194[5]	5·7[5]
P. potto	1150	14·0
S. sciureus	875	24·0
C. apella	2620	63·8[6]
C. moloch	1070	19·0
A. trivirgatus	1220	17·1
A. seniculus	6690	56·8[6]
C. goeldii	630[3]	11·0
L. rosalia	680	12·3[6]
S. oedipus	450	10·0
S. mystax	580[3]	—
S. fuscicollis	462[3]	8·7[6]
C. jacchus	310	7·6
C. aethiops	4100[7]	65·6[7]
E. patas	8500	108·0
M. silenus	—	—
M. nemestrina	8140[7]	99·9[7]
M. fascicularis	4160[7]	56·6[7]
M. fuscata	9550[8]	107·0
M. mulatta	6380[7]	84·0[7]
M. arctoides	7890[7]	101·1[7]
P. cynocephalus	17,575	201·0
P. hamadryas	16,650[3]	199·0[7]
T. gelada	15,350[3]	131·9[6]
P. entellus	15,000[7]	98·0[7]
P. pygmaeus	62,000[9]	350·0[7]
P. t. troglodytes	53,700	406·4[7]
G. g. beringei	128,450[3]	500·0

[1]From Jungers (1985) unless noted otherwise; weights for sexually dimorphic species are the means of male and female weights (in g).

[2]From Stephan et al. (1981) unless noted otherwise. Following Armstrong (1985) we used cranial capacity (cm^3) and brain weight (g) interchangeably; the specific gravity of brain tissue is slightly higher than that of water but cranial capacity includes volumes for cerebrospinal fluid, vessels and meninges.

[3]Fleagle (1988).

[4]From Stephan et al. (1981) but using brain size value of the closely related Lemur fulvus.

[5]Rasmussen & Izard (1988).

[6]M. J. Ravosa (unpublished data).

[7]S. Gauld, A. Nelson & A. Foley (unpublished data).

[8]Nakayama et al. (1971).

[9]Markham & Groves (1990); mean of male = 86,000, female = 38,000.

Allometry

The body weight and brain size data used in the allometric analyses are presented in Table 4. The results of the interspecific allometric analyses are provided in Table 5 and illustrated in Figures 4 and 5. The allometric analyses include only data collected by continuous or

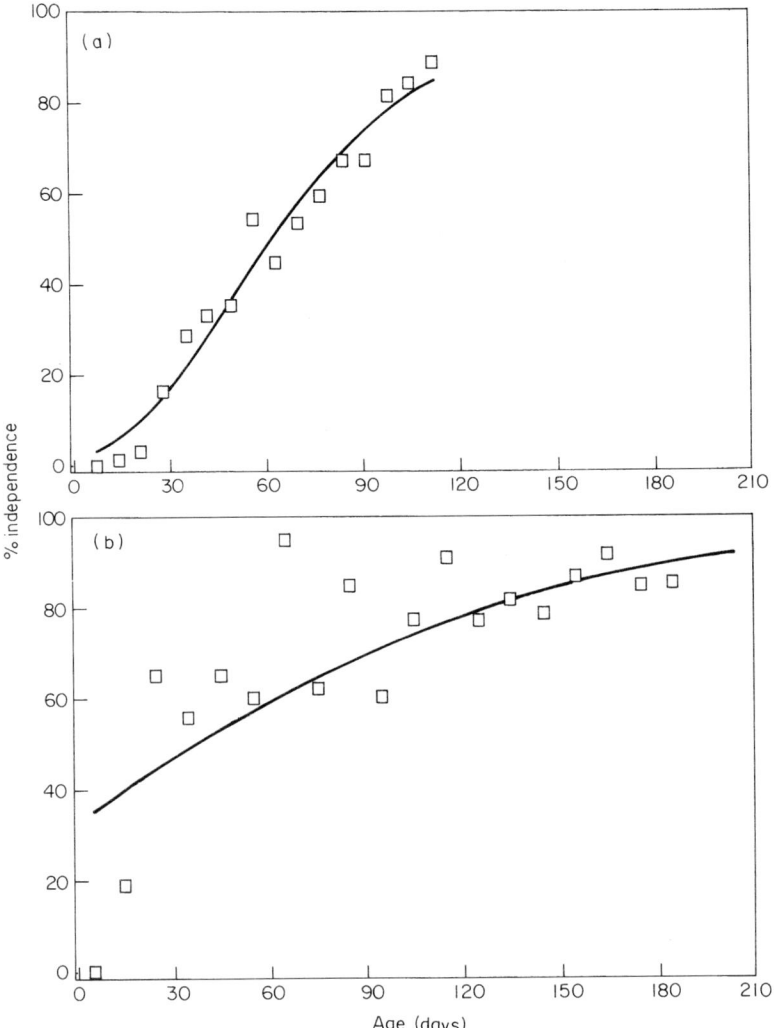

Figure 2. The attainment of physical independence in two primates drawn on the same axis scales, showing the fit between the behavioral data and the Gompertz curves. (a) Free-ranging ring-tailed lemurs, *Lemur catta*, at Berenty, Madagascar (Gould, 1991); a relatively good fit. (b) Captive slow lorises, *Nycticebus coucang*, at the Duke Primate Center (Rasmussen, 1986); the worst fitting curve in our sample. Beginning at a young age, infant *Nycticebus* are parked on branches and left alone while the mother forages, resulting in an ontogenetic trajectory that cannot be tracked easily by a Gompertz curve. The inflection point occurs very early and the rate coefficient is very low compared to primates of similar or larger sizes (e.g., *Lemur catta*).

instantaneous sampling (as proportion of time is the dependent variable). All of the allometric relationships are statistically significant at an alpha level of 0·01 or better, and a visual inspection of the figures suggests that the log–log relationships are approximately linear. The behavioral rate coefficient scaled to body and brain size by negative allometry in both the suckling and contact data ($-0·37$ to $-0·51$). The inflection point scaled to body and brain

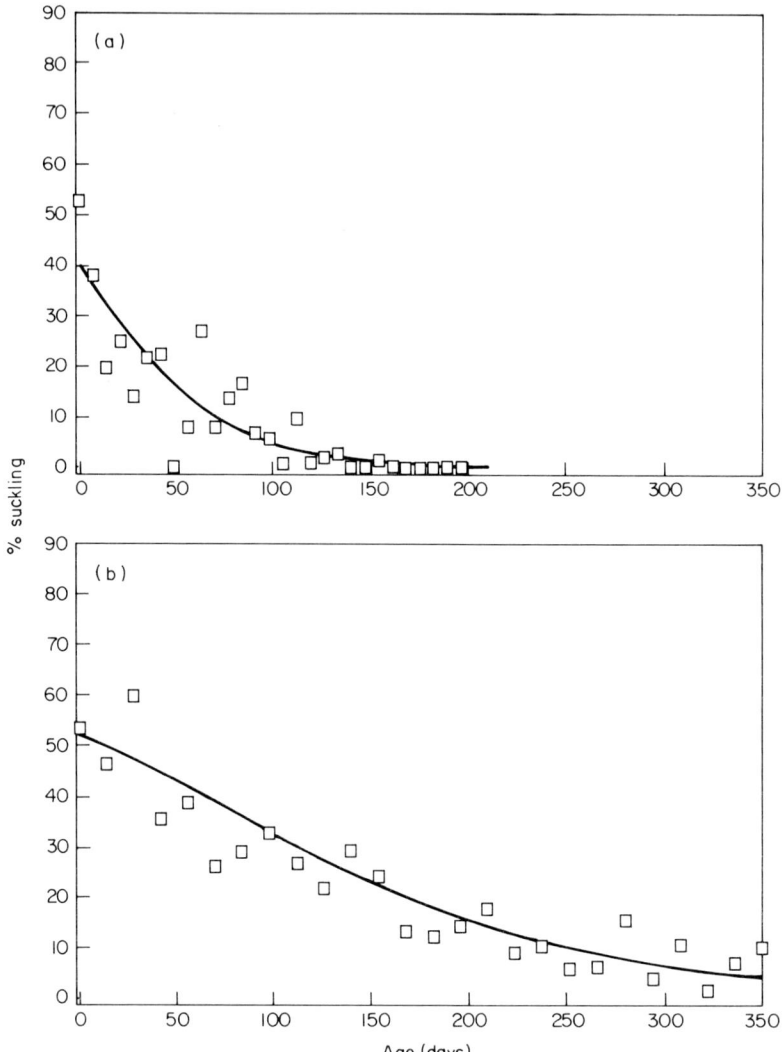

Figure 3. The decrease in suckling behavior in two primates species of radically different body sizes drawn on the same axis scales, showing the fit between the behavioral data and the Gompertz curves. (a) Captive slow lorises, *Nycticebus coucang* (Rasmussen, 1986); a curve with a low inflection point (12·6 days) and high rate coefficient (0·0270). (b) Captive savanna baboons, *Papio cynocephalus* (Rhine et al., 1985; instantaneous sample); a curve with a high inflection point (75·3 days) and a low rate coefficient (0·0096). Allometric analyses demonstrate that much of the disparity between these two species in their pattern of behavioral development can be accounted for by the difference in body size.

size by positive allometry in both suckling and contact data (0·34–0·50). On average, larger animals attain independence later and at a lower rate of change than smaller animals. The parameters I and K were weakly, inversely correlated with each other (Pearson correlation coefficient: independence, $r = -0.43$; suckling, $r = -0.46$).

Although the allometric relationships are statistically significant, the coefficients of determination (R^2) are often relatively low (Table 5). The two highest R^2 values were

Table 5 **Allometric exponents (a) with standard errors of the least squares estimate (SE), intercepts (b) with their standard errors (SE), F ratios, coefficients of determination (R^2) and sample sizes (n) for behavioral measures as functions of body weight, brain size and basal metabolic rate (continuous and instantaneous sampling only). All eight allometric relationships are statistically significant ($P<0.01$)**

Dependent variable	Independent variable						
	a	SE	b	SE	F	R^2	n
	Body weight						
Physical contact							
K	−0.37	0.05	6.0	0.4	47.5	0.70	22
I	0.34	0.07	1.5	0.5	26.4	0.57	22
Suckling position							
K	−0.41	0.05	6.4	0.4	80.4	0.88	13
I	0.40	0.10	0.5	0.8	19.4	0.64	13
	Brain size						
Physical contact							
K	−0.44	0.08	4.6	0.3	30.6	0.63	20
I	0.44	0.09	2.7	0.4	26.0	0.59	20
Suckling position							
K	−0.51	0.08	5.0	0.3	42.4	0.79	13
I	0.50	0.13	1.9	0.5	15.8	0.59	13

obtained for the suckling rate coefficient as a function of body size (0.88) and brain size (0.79). The lowest values of R^2 were obtained for the contact inflection points (0.57 and 0.59). The R^2s for K values were higher in both suckling and contact than for I values. In summary, size variables (body or brain) are associated with 63–88% of the variance in the rates of behavioral change for suckling and independence. Size variables are associated with roughly 57–64% of the variance in inflection points. More detailed analysis of the predictive value of body size versus brain size is not warranted with the current sample size given the demonstrated intraspecific variation and the imprecision concerning confidence intervals.

Sampling methods

In contrast to the above results based on the continuous and instantaneous samples, allometric analyses using the one-zero sample never attained an alpha level of 0.01 for K or I relative to body size with respect to either the contact data or the suckling data (brain size allometry was not examined in this smaller sample). Coefficients of determination (R^2) were also low (0.09–0.54), indicating that half the variance or less in the dependent variable was associated with body size.

Prosimians vs. anthropoids

To test the hypothesis that prosimians undergo behavioral development at a greater rate or at an earlier age than anthropoids, the pooled samples of prosimian residuals were

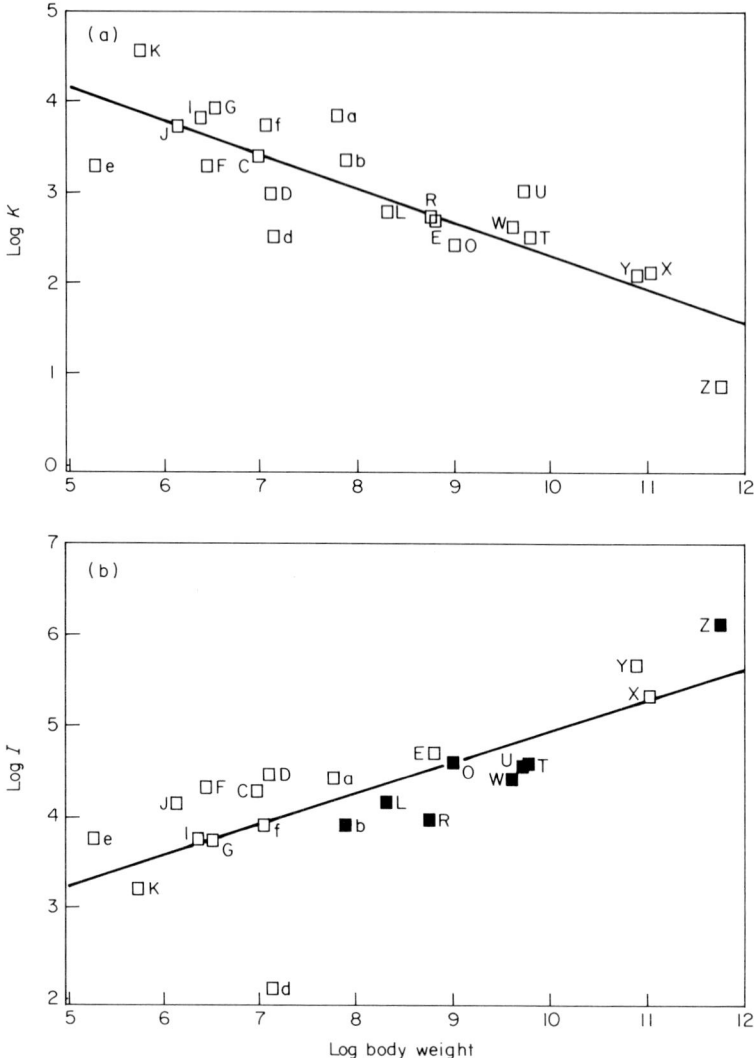

Figure 4. (a) Log–log linear regression of the contact rate coefficient (K) as a function of body weight, and (b) of the contact inflection point (I) as a function of body weight in 22 species of primates ranging in size from the slender loris, *Loris tardigradus* (e) to the gorilla, *G. gorilla* (Z). Each square represents the mean value of one species, identified by letters as in Table 1 (lower case = prosimians; capitals = anthropoids). In the regression of inflection points against size (b): terrestrial species (■); arboreal species (□). Notice the particularly low value for *Nycticebus coucang* (d) a species in which the offspring are "parked". Quantitative regression results are presented in Table 5.

compared to the pooled samples of anthropoid residuals for all allometric regressions. In no case was there a statistically significant difference between the suborders (Mann–Whitney U test). Thus, taking body size into account, there appears to be no difference between prosimians and anthropoids in the relative rates at which behavioral independence is attained.

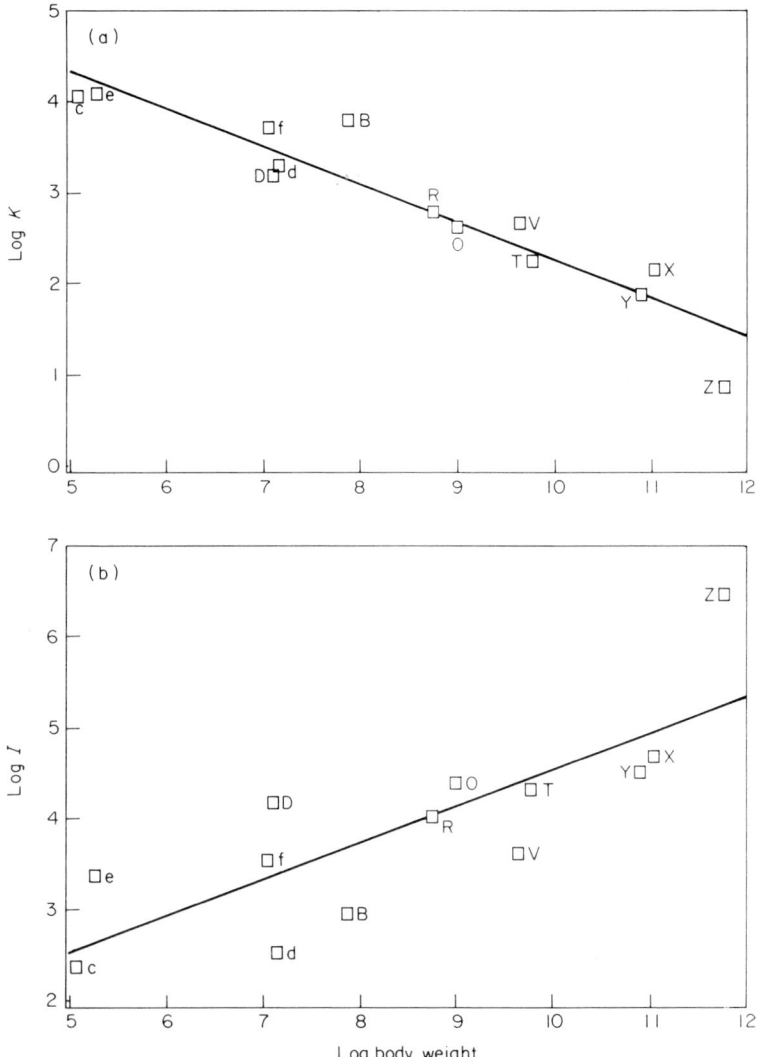

Figure 5. (a) Log–log linear regression of the suckling rate coefficient (K) as a function of body weight and (b) the suckling inflection point (I) as a function of body weight in 13 species of primates, ranging in size from the Mohol lesser galago, *Galago moholi* (c) to the gorilla, *G. gorilla* (Z). Each square represents the mean value for one species, identified by letters as in Table 1 (lower case = prosimians; capitals = anthropoids). Note that there is a much better fit of the rate coefficient to body size (a) than there is of the inflection point to body size (b), although both relationships are statistically significant. Quantitative regression results are presented in Table 5.

Arboreal vs. *terrestrial primates*

To test the hypothesis that terrestrial species attain behavioral independence earlier than arboreal species, the pooled samples of residuals from terrestrial primates were compared to the pooled samples of residuals from arboreal primates for all allometric regressions. (*Nycticebus* was removed from the analysis because of its unusual "parking" behavior.) Forcing primates into the exclusive categories of either arboreal or terrestrial obscures the fact that many

Table 6 Residual values of the contact inflection point derived from allometric regressions of contact behavior as a function of body weight and brain size; arranged in rank order of increasing residuals for the body weight regression

Species	Locomotion[1]	Body	Brain
M. mulatta	T	−0.61	−0.70
C. jacchus	A	−0.42	−0.53
P. entellus	T	−0.41	−0.29
L. catta	T	−0.40	—
P. hamadryas	T	−0.31	−0.43
P. cynocephalus	T	−0.30	−0.41
C. aethiops	T	−0.28	−0.41
P. potto	A	−0.14	−0.08
L. rosalia	A	−0.13	−0.19
S. mystax	A	−0.08	—
M. nemestrina	T	−0.05	−0.13
P. pygmaeus	A	0.04	0.11
A. seniculus	A	0.11	0.18
L. macaco	A	0.16	0.25
C. moloch	A	0.26	0.18
L. tardigradus	A	0.27	0.12
S. fuscicollis	A	0.39	0.34
A. trivirgatus	A	0.40	0.40
P. troglodytes	A	0.42	0.39
C. goeldii	A	0.46	0.42
G. gorilla	A	0.61	0.77

[1]T = largely terrestrial; A = largely arboreal.

species do both (e.g., *Pan troglodytes*). Data on the percentages of time spent in different forms of locomotion would be preferable to the abstract categories used here, but such precise information is not presently available for all taxa and interspecific variation among studies is difficult to interpret. For now, the analysis presented here relies on categorical data, but the comparative use of ontogeny curves is well suited for use with continuous behavioral variables when these are available.

There was no significant difference between arboreal and terrestrial primates in the rate coefficient for contact data or suckling data, nor was there a significant difference when comparing residuals of suckling inflection points. In contrast, terrestrial primates did differ from arboreal species in their contact inflection points (Figure 4b; Table 6; Mann–Whitney U two-tailed tests, relative to body size: $n_1 = 8, n_2 = 13, U = 22, P \leqslant 0.05$; relative to brain size: $n_1 = 7; n_2 = 12, U = 18, P \leqslant 0.05$). Terrestrial primates had relatively earlier inflection points for their sizes than did arboreal species, indicating that physical independence in the former was attained at a younger age.

Discussion

The results described here demonstrate that allometric analyses can be successfully applied to patterns of behavioral development. In this study, changes in the frequency of suckling and of physical independence were compared across taxa using coefficients obtained from

Gompertz curves that had been fitted to empirical data. The Gompertz model fit behavioral data closely and without any extreme directional bias. Changes in the rate of suckling and physical independence were then modeled against body size and brain size. In the sample of species examined here, 57–88% of the variation in the rate at which suckling and independence develop can be accounted for by body or brain size. Not surprisingly, in larger species, the rate of behavioral change is lower and spread out over a longer time interval than in smaller species. This indicates that the longer periods of infant dependence observed in the great apes, for example, may be largely explicable in terms of large body and brain size. By itself, this rather intuitive result is not very interesting. However, by formalizing the relationships between behavior and size in the form of an explicit allometric analysis, the precise allometric exponents and patterns of variation among species can be analysed in detail.

Although statistically significant relationships exist between behavioral development and body or brain size, the unexplained variance was greater than that typically observed in allometric analyses of structural variables, or of life history variables such as gestation length, age at sexual maturity or interbirth interval (Harvey et al., 1986). There was notable variation among three independent studies of a single species, *M. mulatta*, suggesting that details of methodology, environment or other sources of variation heavily influence the conformation of the Gompertz curve. It follows from this that slight interspecific variation must be interpreted with caution. The variance in behavioral ontogeny that is not accounted for by sampling methodology, body or brain size, may reflect adaptive differences among species not associated tightly with size variables, phenotypic differences associated with local environmental conditions, the effects of phylogenetic constraints or the greater flexibility and variability in behavior than in life history characters. One advantage of the curve-fitting methods described here is that they provide a tool to precisely investigate and evaluate these hypothetical sources of variation.

A broader analysis using a larger sample of primate species, or using multiple studies of a given species, will be necessary to investigate sources of variation in greater detail. The compilation of a larger data set from the published literature would be facilitated if there was greater standardization among researchers in how behavior is measured. Currently, there is great variation in sampling method, recording method, recording interval, sample size, mathematical manipulation and abstraction of the raw data, a data summarization, and finally, data presentation in a publication. Often, unusual methods are justifiable and necessary given the particular research goals. However, basic descriptive data on changes in behavior through time should be standardized, unless data of a specific form are required to evaluate particular hypotheses being addressed by the researcher.

Data collected by continuous and instantaneous sampling seem to be closely comparable to each other and the estimated Gompertz parameters obtained from both sampling methods fall closely around a common allometric regression. The compatibility of continuous and instantaneous sampling is not unexpected given the fact that these provide relatively unbiased estimates of activity budgets (Altmann, 1974; Simpson & Simpson, 1977; Martin & Bateson, 1986). One-zero sampling has been criticized for weaknesses in its statistical properties and in its usefulness for interspecific comparisons (Altmann, 1974). In contrast, other researchers have commented that one-zero sampling is valuable for characterizing patterns of ontogenetic change within species, or in evaluating behaviors that are expressed irregularly (Rhine & Flanigan, 1978; Rhine et al., 1985; Tyler, 1979). Our results suggest that one-zero sampling may be valuable for describing ontogenetic changes within a study population; Gompertz models fit one-zero data very well (Tables 2 and 3). However, the variance among one-zero samples is too great to yield statistically significant allometric results, at least for the small sample analysed here. This indicates that there is an additional source of variation that cannot be accounted for by size variables; the additional source is presumably the sampling method itself. Even if one-zero samples were consistent among themselves, they would still not be directly comparable to continuous or instantaneous samples, because proportion of arbitrary intervals is an entirely different variable than proportion of time (and one that is difficult to justify biologically). We urge researchers to utilize focal animal or point sampling whenever possible, to make their data available for broader interspecific comparative studies.

We chose to use Gompertz curves because the behavioral data we wanted to compare appeared to parallel or resemble growth trajectories. An explicit comparison of Gompertz curves to other sigmoid models may prove useful. Other behaviors exhibit a very different pattern of development through time that will not follow a sigmoid pattern. For example, play behavior occurs at low frequencies very early in development and during adulthood, but at high frequencies during late infancy and declining through juvenility and adolescence. Clearly, a different class of curves would be required to model this behavior and others that also fail to conform closely to a sigmoid curve. However, once an appropriate class of curves is found, these behaviors could be subjected to interspecific comparisons using methods otherwise generally similar to those described here.

One area of evolutionary research that may be enhanced by the use of precise interspecific comparisons of behavioral ontogeny is the study of residual variation not accounted for by the interspecific allometric regressions. For example, we have shown that terrestrial primates attain physical independence earlier than arboreal primates, as reflected by relatively low contact inflection points. The difference between the residuals of terrestrial and arboreal primates is statistically significant when analysed with respect to body or brain size (despite the fact that the terrestrial gorilla had the highest residuals of any primate in the sample for this behavior). We interpret the ontogenetic differences between terrestrial and arboreal species to be an adaptation reflecting the different risks of falling that inexperienced infants are exposed to during early exploring forays away from their caretakers. Interestingly, despite the difference between arboreal and terrestrial primates in the timing of independence, the rate of development (indicated by K) does not differ.

Summary

The behavior of young primates unfolds in a complex, non-linear fashion that has been difficult to quantify in a manner that allows precise interspecific comparisons. As a consequence, allometric analyses have not been easily applied to patterns of behavioral ontogeny. We have explored statistical methods that utilize non-linear regression of behavioral data expressed as rates, followed by allometric analysis of parameters derived from the non-linear models. Data on age-specific rates of suckling and physical independence were obtained from the literature for a range of primate species, and these were fitted by Gompertz curves. Patterns of behavioral ontogeny, as expressed by a rate coefficient, K, and an inflection point, I, were precisely compared interspecifically while controlling for adult body and brain size. The resulting allometric relationships yielded allometric exponents for behavioral rate coefficients of -0.37 to -0.51 and for inflection points of $0.34-0.50$, relative to adult body and brain size. For the sample analysed here, 57–88% of the variation among species in the decrease of suckling and the increase in physical independence is associated with variation in body and brain size, such that larger species have slower rates of behavioral change. There are no statistically significant differences between prosimians and anthropoids in the ontogenetic variables examined in this study. Terrestrial species attain physical independence earlier than arboreal species when controlling for body size, reflecting an earlier age of locomotor independence in terrestrial primates.

With a more complete sample of primate species, and with greater control of methodological variation, it might prove possible to test additional adaptive and taxonomic hypotheses concerning behavioral development. The method should also prove applicable in illuminating environmental effects on intra specific variation. Furthermore, with better samples it may also prove possible to study heterochronic processes in the evolution of behavioral ontogeny. The use of ontogeny curves will also be applicable to a broad range of other quantitative, comparative techniques apart from classic allometric methods, including methods designed to address the problem of phylogenetic constraints (e.g., Felsenstein, 1985; Harvey & Pagel, 1991). We conclude that such studies are currently limited by the low diversity and uneven quality of available, quantitative behavioral data, not by the absence of effective comparative methods.

Acknowledgements

We thank the following researchers for graciously sharing with us their data on primate behavioral development: C. M. Berman, J. Chism, D. M. Fragaszy, L. Gould, E. W. Heymann, R. D. Nadler, M. Nakamichi and R. J. Rhine. We thank S. Gauld and M. J. Ravosa for generously providing us with unpublished data on primate body weights and brain sizes. We are indebted to M. E. Pereira, C. van Schaik, F. White, A. Gomez, M. J. Ravosa, J. B. Silk and R. Boyd for reading earlier versions of this paper and offering their expert suggestions. M. Tyler provided valuable help in preparing the illustrations. Behavioral and morphological data on lorisines were collected with the assistance and cooperation of the excellent technical staff of the Duke University Primate Center. Finally, we would again like to acknowledge the many researchers whose published data were utilized in this study, as cited in Table 1.

References

Alberch, P., Gould, S. J., Oster, G. F. & Wake, D. B. (1979). Size and shape in ontogeny and phylogeny. *Paleobiology* **5,** 296–317.
Altmann, J. (1974). Observational study of behavior: sampling methods. *Behaviour* **49,** 227–265.
Altmann, J. (1978). Infant dependence in yellow baboons. In (G. M. Burghardt & M. Bekoff, Eds) *The Development of Behavior: Comparative and Evolutionary Aspects*, pp. 253–277. New York: Garland.
Altmann, J. (1980). *Baboon Mothers and Infants*. Cambridge, MA: Harvard University Press.
Armstrong, E. (1985). Relative brain size in monkeys and prosimians. *Am. J. phys. Anthrop.* **66,** 263–273.
Baldwin, J. D. (1986). Behavior in infancy: exploration and play. In (G. Mitchell & J. Erwin, Eds) *Comparative Primate Biology, Volume 2, Part A, Behavior, Conservation & Ecology*, pp. 295–326. New York: Alan R. Liss.
Berman, C. M. (1980). Mother-infant relationships among free-ranging rhesus monkeys on Cayo Santiago: a comparison with captive pairs. *Anim. Behav.* **28,** 860–873.
Bolwig, N. (1980). Early social development and emancipation of *Macaca nemestrina* and species of *Papio*. *Primates* **21,** 357–375.
Castell, R. & Wilson, C. (1971). Influence of spatial environment on development of mother–infant interaction in pigtail monkeys. *Behaviour* **39,** 202–211.
Charles-Dominique, P. (1977). *Ecology and Behaviour of Nocturnal Primates*. New York: Columbia University Press.
Chism, J. (1986). Development and mother–infant relations among captive patas monkeys. *Int. J. Primatol.* **7,** 49–81.

Cleveland, J. & Snowdon, T. (1984). Social development during the first twenty weeks in the cotton-top tamarin (*Saguinus o. oedipus*). *Anim. Behav.* **32,** 432–444.

Dixson, A. F. & Fleming, D. (1981). Parental behaviour and infant development in owl monkeys (*Aotus trivirgatus griseimembra*). *J. Zool. Lond.* **194,** 25–39.

Dolhinow, P. & Krusko, N. (1984). Langur monkey females and infants: the female's point of view. In (M. F. Small, Ed.) *Female Primates: Studies by Women Primatologists*, pp. 37–57. New York: Alan R. Liss.

Dolhinow, P. & Murphy, G. (1982). Langur monkey (*Presbytis entellus*) development: the first 3 months of life. *Folia primat.* **39,** 305–331.

Doyle, G. A., Andersson, A. & Bearder, S. K. (1969). Maternal behaviour in the lesser bushbaby (*Galago senegalensis moholi*) under semi-natural conditions. *Folia primat.* **11,** 215–238.

Dunbar, R. I. M. (1984). *Reproductive Decisions: An Economic Analysis of Gelada Baboon Social Strategies.* Princeton, NJ: Princeton University Press.

Fairbanks, L. A. (1988). Vervet monkey grandmothers: effects on mother–infant relationships. *Behaviour* **104,** 176–188.

Fairbanks, L. A. & McGuire, M. T. (1987). Mother–infant relationships in vervet monkeys: response to new adult males. *Int. J. Primatol.* **8,** 351–366.

Felsenstein, J. (1985). Phylogenies and the comparative method. *Am. Nat.* **125,** 1–15.

Fleagle, J. G. (1985). Size and adaptations in primates. In (W. L. Jungers, Ed.) *Size and Scaling in Primate Biology*, pp. 1–19. New York: Plenum Press.

Fleagle, J. G. (1988). *Primate Adaptations & Evolution.* New York: Academic Press.

Fossey, D. (1979). Development of the mountain gorilla (*Gorilla gorilla beringei*): the first thirty-six months. In (D. A. Hamburg & E. R. McCown, Eds) *The Great Apes*. Menlo Park, CA: Benjamin/Cummings Publ. Co.

Fragaszy, D. M. (1990). Early behavioral development in capuchins (*Cebus*). *Folia primat.* **54,** 119–128.

Gould, L. (1990). The social development of free-ranging infant *Lemur catta* at Berenty Reserve, Madagascar. *Int. J. Primatol.* **11,** 297–337.

Harrington, J. E. (1978). Development of behavior in *Lemur macaco* in the first nineteen weeks. *Folia primat.* **29,** 107–128.

Harvey, P. H., Martin, R. D. & Clutton-Brock, T. H. (1986). Life histories in comparative perspective. In (B. B. Smuts, D. L. Cheney, R. M. Seyfarth, R. W. Wrangham & T. T. Struhsaker, Eds) *Primate Societies*, pp. 181–196. Chicago: University of Chicago Press.

Harvey, P. H. & Pagel, M. D. (1991). *The Comparative Method in Evolutionary Biology*. Oxford: Oxford University Press.

Heltne, P. G., Turner, D. C. & Wolhandler, J. (1973). Maternal and paternal periods in the development of infant *Callimico goeldii*. *Am. J. phys. Anthrop.* **38,** 555–560.

Hershkovitz, P. (1977). *Living New World Monkeys (Platyrrhini), with an Introduction to the Primates, Volume 1*. Chicago: University of Chicago Press.

Heymann, E. W. (1990). Social behaviour and infant carrying in a group of moustached tamarins, *Saguinus mystax* (Primates: Platyrrhini: Callitrichidae) on Padre Isla, Peruvian Amazonia. *Primates* **31,** 183–196.

Heymann, E. W. & Sicchar, L. A. (1990). Estudio ecologico del pichico barba blanca, *Saguinus mystax mystax* y del pichico comun, *Saguinus fuscicollis nigrifrons* (Primates; Callitrichidae) en un galpon al aire libre – resultados preliminares. *La Primatologia en el Peru, Investigaciones Primatologicas (1973–1985)*, pp. 359–381. Lima: Proyecto Peruano de Primatologia.

Hoage, R. J. (1978). Parental care in *Leontopithecus rosalia rosalia*: sex and age differences in carrying behavior and the role of prior experience. In (D. G. Kleiman, Ed.) *The Biology and Conservation of the Callitrichidae*, pp. 293–305. Washington, DC: Smithsonian Institution Press.

Hoel, P. G. (1964). Methods for comparing growth type curves. *Biometrics* **20,** 859–872.

Ingram, J. C. (1977). Interactions between parents and infants, and the development of independence in the common marmoset (*Callithrix jacchus*). *Anim. Behav.* **25,** 811–827.

Izard, M. K. (1987). Lactation length in three species of *Galago*. *Am. J. Primat.* **13,** 73–76.

Johnson, R. L. & Southwick, C. H. (1987). Ecological constraints on the development of infant independence in rhesus. *Am. J. Primatol.* **13,** 103–118.

Jolicoeur, P. & Pirlot, P. (1988). Asymptotic growth and complex allometry of the brain and body in the white rat. *Grow. Dev. Ag.* **52,** 3–10.

Jungers, W. L. (1985). Body size and scaling of limb proportions in primates. In (W. L. Jungers, Ed.) *Size and Scaling in Primate Biology*, pp. 345–382. New York: Plenum Press.

Kaplan, J. (1972). Differences in the mother–infant relations of squirrel monkeys housed in social and restricted environments. *Devol. Psychobiol.* **5,** 45–52.

Kumar, A. & Kurup, G. U. (1981). Infant development in the lion-tailed macaque, *Macaca silenus* (Linnaeus): the first eight weeks. *Primates* **2,** 512–522.

Leigh, S. R. & Cheverud, J. M. (1991). Sexual dimorphism in the baboon facial skeleton. *Am. J. phys. Anthrop.* **84,** 193–208.

Locke-Hayden, J. & Chalmers, N. R. (1983). The development of infant–caregiver relationships in captive common marmosets (*Callithrix jacchus*). *Int. J. Primatol.* **4,** 63–81.

Mack, D. (1979). Growth and development of infant red howling monkeys (*Alouatta seniculus*) in a free ranging population. In (J. F. Eisenberg, Ed.) *Vertebrate Ecology in the Northern Neotropics*, pp. 127–136. Washington, DC: Smithsonian Institution Press.

Markham, R. & Groves, C. P. (1990). Brief communication: weights of wild orang-utans. *Am. J. phys. Anthrop.* **81,** 1–3.

Martin, P. & Bateson, P. (1986). *Measuring Behaviour*. Cambridge: Cambridge University Press.

Miller, L. C. & Nadler, R. D. (1981). Mother-infant relations and infant development in captive chimpanzees and orang-utans. *Int. J. Primatol.* **2,** 247–261.

Murray, R. D. & Murdoch, K. M. (1977). Mother-infant dyad behavior in the Oregon troop of Japanese macaques. *Primates* **18,** 815–824.

Nakamichi, M., Cho, F. & Mirami, T. (1990). Mother-infant interactions of wild-born, individually-caged cynomolgus monkeys (*Macaca fascicularis*) during the first 14 weeks of infant life. *Primates* **31,** 213–224.

Nakayama, T., Hori, T., Nagasaka, T., Tokura, H. & Tadaki, E. (1971). Thermal and metabolic responses in the Japanese monkey at temperatures of 5–38°C. *J. appl. Physiol.* **31,** 332–337.

Nicolson, N. (1986). Infants, mothers, and other females. In (B. B. Smuts, D. L. Cheney, R. M. Seyfarth, R. W. Wrangham, T. T. Struhsaker, Eds) *Primate Societies*, pp. 330–342. Chicago: University of Chicago Press.

Pereira, M. E. & Altmann, J. (1985). Development of social behavior in free-living nonhuman primates. In (E. S. Watts, Ed.) *Nonhuman Primate Models for Human Growth and Development*, pp. 217–309. New York: Alan R. Liss.

Rasmussen, D. T. (1986). Life history and behavior of slow lorises and slender lorises. Ph.D. Dissertation, Duke University.

Rasmussen, D. T. & Izard, M. K. (1988). Scaling of growth and life history traits relative to body size, brain size, and metabolic rate in lorises and galagos (Lorisidae, Primates). *Am. J. phys. Anthrop.* **76,** 357–367.

Reite, M. & Short, R. (1980). A biobehavioral developmental profile (BDP) for the pigtailed monkey. *Devol. Psychobiol.* **13,** 243–285.

Rhine, R. J. & Flanigan, M. (1978). An empirical comparison of one-zero, focal-animal and instantaneous methods of sampling spontaneous primate social behavior. *Primates* **19,** 353–361.

Rhine, R. J. & Hendy-Neely, H. (1978). Social development of stumptail macaques (*Macaca arctoides*): synchrony of changes in mother–infant interactions and individual behaviors during the first 60 days of life. *Primates* **19,** 681–692.

Rhine, R. J., Norton, G. W., Wynn, G. M. & Wynn, R. D. (1985). Weaning of free-ranging infant baboons (*Papio cynocephalus*) as indicated by one-zero and instantaneous sampling of feeding. *Int. J. Primatol.* **6,** 491–499.

Rosenblum, L. A. (1968). Mother-infant relations and early behavioral development in the squirrel monkey. In (L. A. Rosenblum & R. W. Cooper, Eds) *The Squirrel Monkey*, pp. 207–233. New York: Academic Press.

Shea, B. T. (1985). Ontogenetic allometry and scaling: a discussion based on the growth and form of the skull in African apes. In (W. L. Jungers. Ed.) *Size and Scaling in Primate Biology*, pp. 175–206. New York: Plenum Press.

Simpson, M. J. A. & Simpson, A. E. (1977). One-zero and scan methods for sampling behavior. *Anim. Behav.* **25,** 726–731.

Simpson, M. J. A., Simpson, A. E. & Howe, S. (1986). Changes in the rhesus mother–infant relationship through the first four months of life. *Anim. Behav.* **34,** 1528–1539.

Stephan, H., Frahm, H. & Baron, G. (1981). New and revised data on volumes of brain structures in insectivores and primates. *Folia primat.* **35,** 1–29.

Stewart, K. J. (1988). Suckling and lactational anoestrus in wild gorillas (*Gorilla gorilla*). *J. Reprod. Fert.* **83,** 627–634.

Tyler, S. (1979). Time sampling: a matter of convention. *Anim. Behav.* **27,** 801–810.

Vogt, J. L., Carlson, H. & Mendez, E. (1978). Social behavior of a marmoset (*Saguinus fuscicollis*). Group I: parental care and infant development. *Primates* **19,** 715–726.

Walters, J. R. (1986). Transition to adulthood. In (B. B. Smuts, D. L. Cheney, R. M. Seyfarth, R. W. Wrangham, T. T. Struhsaker, Eds) *Primate Societies*, pp. 358–369. Chicago: University of Chicago Press.

Wright, P. C. (1984). Biparental care in *Aotus trivirgatus* and *Callicebus moloch*. In (M. F. Small, Ed.) *Female Primates: Studies by Women Primatologists*, pp. 59–75. New York: Alan R. Liss.

Zullinger, E. M., Ricklefs, R. E., Redford, K. H. & Mace, G. M. (1984). Fitting sigmoidal equations to mammalian growth curves. *J. Mammal.* **65,** 607–636.

Appendix 1

The following are sample SAS programs for fitting Gompertz curves to ontogenetic data. Specifically, the examples here fit curves to both contact data and suckling data of *Loris tardigradus* (Rasmussen, 1986). The data set is called "Loris" and it contains three variables:

days (age of the infant), *contact* (proportion of time spent in contact with caretaker) and *suckle* (proportion of time spent in suckling position on the mother):

```
data loris;
   input days contact suckle;
cards;
     5     0    96·9
    15     0    77·2
    25     0    83·5
    35   29·8   50·1
    45   58·9   14·7
    55   70·5   17·3
    65   59·0   33·3
    75   61·5    0
   115   67·3    0
   125   82·7    0
   135   76·9    0
   145   94·2    0
   155   94·9    0
   165  100      0
;
run;
```

The following program will fit the contact data with a Gompertz curve. Using the non-linear procedure of SAS ("proc nonlin"), starting values for the parameters I and K are defined and the model is spelt out, along with first derivatives with respect to the two parameters. An output data set called "curvoutl" is constructed for plotting of the results, or for further analysis. Finally, an example of two simple plots is provided, one for the Gompertz curve (in this case, a fairly tight fit) and one for the residuals (in this case, showing some apparent autocorrelation).

```
data contact;
   set loris;
   V = contact;
proc nlin best = 10 method = marquardt;
   parameters
   K = ·01 to ·02 by ·002
   I = 10 to 100 by 10;
   E = − EXP( − K*(days − I));
   model V = 100*EXP(E);
   der.K = (I-days)*(100*EXP(E))*E;
   der.I = 100*K*(EXP(E))*E;
   output out = curvoutl p = predictl r = residl;
proc plot;
   plot V*days = 'x' predictl*days = 'p'/overlay;
   plot residl*days = 'x'/vref = 0;
run;
```

The following program for the suckling data is similar to the one for contact data shown above; the key difference is the direction of the curve which goes from a high value to a low one, requiring minor modifications of the equation and the specification of an asymptote that will differ among species (which is set empirically at the maximum observed value). Unlike the previous example, autocorrelation is not a problem for these suckling data.

```
data suckle;
   set loris;
   V = suckle;
   A = 96·9;
proc nlin best = 10 method = marquardt;
   parameters
   K = ·01 to ·02 by ·002
   I = 10 to 100 by 10;
   E = − EXP( − K*(days − I));
   model V = A − (A*EXP(E));
   der.K = (I-days)*( − A*EXP(E))*E;
   der.I = − A*K*(EXP(E))*E;
   output out = curvout2 p = predict2 r = resid2;
proc plot;
   plot V*days = 'x' predict2*days = 'p'/overlay;
   plot resid2*days = '*'/vref = 0;
run;
```

Anne D. Yoder
Duke University Medical Center, Department of Biological Anthropology and Anatomy, Durham, North Carolina 27710, U.S.A.

Received 7 June 1991
Revision received 7 February 1992 and accepted 18 February 1992

Keywords: homology, polarity, parsimony, strepsirhini, ascending pharyngeal artery.

The applications and limitations of ontogenetic comparisons for phylogeny reconstruction: the case of the strepsirhine internal carotid artery

Ontogeny and phylogeny together determine organismal form and consequently, the two should be reciprocally illuminating. Ontogeny contributes valuable information for phylogenetic studies, not because it is a window into phylogeny, but because ontogenetic comparisons provide data that is easily incorporated into the framework of systematic investigation. Ontogenetic data allow independent tests of character homology and polarity, increase knowledge of character complexity, and can also provide new characters for phylogenetic analysis.

Cheriogaleid and lorisiform primates share a unique condition of the internal carotid artery. A large medial branch of the internal carotid, the ascending pharyngeal artery, bypasses the bulla to enter the cranial cavity through the foramen lacerum. This character has been defined by primate systematists as the primary synapomorphy of a cheirogaleid–lorisiform clade. To test the hypotheses of homology and polarity, the development of the internal carotid artery in a cheirogaleid, *Microcebus murinus*, was compared with that of a lorisiform, *Galago senegalensis senegalensis*. The comparison revealed that the ontogeny of this character is nearly identical in these two strepsirhine primates, thus supporting the hypothesis of homology. Also, a new character was identified that, if interpreted as derived, adds further support to the hypothesis that cheirogaleids and lorisiforms constitute a monophyletic clade. The same developmental sequences do not contribute to the determination of polarity for the ascending pharyngeal artery. Consequently, this study cannot distinguish between hypotheses of synapomorphy or symplesiomorphy. Outgroup comparisons, however, strongly support the hypothesis that the ascending pharyngeal artery is a derived character and therefore indicative of cheirogaleid–lorisiform monophyly.

Journal of Human Evolution (1992) **23**, 183–195

Introduction

The connection between ontogeny and phylogeny continues to be the subject of investigation and speculation. For centuries, philosophers and biologists have observed a correlation between patterns of individual development and patterns among adults of related taxa. With the advent of Darwin's theory of decent with modification, an obvious association between individual development (ontogeny) and adult morphology (phylogeny) was illuminated: descriptions of ontogenies and phylogenies are both descriptions of morphological change over time. An important discrepancy exists between ontogeny and phylogeny, however. The time over which morphology changes in phylogeny is vast and beyond the scope of human observation. That for ontogeny, on the other hand, is conveniently brief and can provide a ready source of data for biological investigation, thus increasing the appeal of ontogenetic comparisons. It is also true that there are many instances (e.g., developing vertebrate embryos) in which patterns seen in ontogeny seem to mirror those of phylogeny. These two circumstances together have presented biologists with a tantalizing possibility: perhaps investigations of ontogeny can illuminate otherwise hidden patterns of phylogeny.

This concept was formalized in Haeckel's (1866) now notorious phrase "ontogeny recapitulates phylogeny". The theory of recapitulation has inspired a profusion of theoretical and empirical investigations into the connection between ontogeny and phylogeny (e.g., de Beer, 1958; Gould, 1977). But despite all of this investigation, the relationship between ontogeny and phylogeny remains elusive (Alberch et al., 1979; Fink, 1982; Alberch, 1985; de Queiroz, 1985; Kluge, 1985, 1988; Kluge & Strauss, 1985; Northcutt, 1990). The only definite conclusion that has been reached is that Haeckel's theory of recapitulation is a gross oversimplification of reality.

Perhaps, rather than viewing the two as separate, the patterns of ontogeny and phylogeny will be more profitably considered as inextricable parts of a whole, the whole being organismal evolution. Ontogeny continuously modifies phylogeny just as phylogeny modifies ontogeny. Ontogeny and phylogeny together determine organismal form and consequently, the two should be reciprocally illuminating. Hennig's (1966) phylogenetic systematics offers investigators a procedure for incorporating ontogenetic data (or any other type of life-history data) into phylogenetic analysis. Hennig's methodology is explicit: in order to reconstruct the phylogenetic relationship among a group of organisms, the investigator must compare homologous traits and determine their relative evolutionary recency (i.e., their polarity). Lastly, these characters are arranged on a tree that minimizes the number of evolutionary transitions (i.e., the most parsimonious tree). Thus, if ontogeny is to contribute useful information for interpreting evolutionary relationships among organisms, it should do so within the framework of phylogenetic systematics.

This paper will describe a case in which the ontogeny of a single morphological character was examined and compared in two taxa in order to investigate a phylogenetic question. The results of the ontogenetic comparison are then discussed as they relate to the concepts of homology, polarity, and parsimony.

The case of the strepsirhine internal carotid artery

Background information
The strepsirhine primates (lemuriforms and lorisiforms) constitute a monophyletic clade (Hill, 1953; Charles-Dominique & Martin, 1970). The lemuriforms (here used as a generic term for Malagasy primates) are confined to the island of Madagascar, whereas the lorisiforms are distributed across Africa and parts of Asia. Due to their geographic isolation, primatologists have generally considered the lemuriforms to be the product of a single adaptive radiation on Madagascar. Likewise, the lorisiforms have also been regarded as a monophyletic group.

The belief that lemuriforms and lorisiforms formed two separate monophyletic groups held until the early 1970s, despite the fact that there are no known morphological synapomorphies to unite the lemuriforms. More recently, studies have indicated that one family of lemuriform primates, the Cheirogaleidae, actually shares some of the more distinctive lorisiform characteristics and that these are derived (Szalay & Katz, 1973; Cartmill, 1975). This is especially true of the basicranium. A morphocline exists for the cheirogaleid basicranium such that some members of the family appear more lemur-like, whereas others (particularly the genus *Allocebus*) are virtually indistinguishable from a lorisiform. This had led to the hypothesis that cheirogaleids and lorisiforms together form a monophyletic clade. But this presents a biogeographical problem. Madagascar separated from the African mainland approximately 150 million years ago (m.y.a.) but the first primates do not appear in the fossil record until approximately 60 m.y.a. Thus, the hypothesis that cheirogaleids and lorisiforms are monophyletic requires that there were at least two primate migrations across the Mozambique Channel (which is 300 miles wide at present). Consequently, primatologists are currently

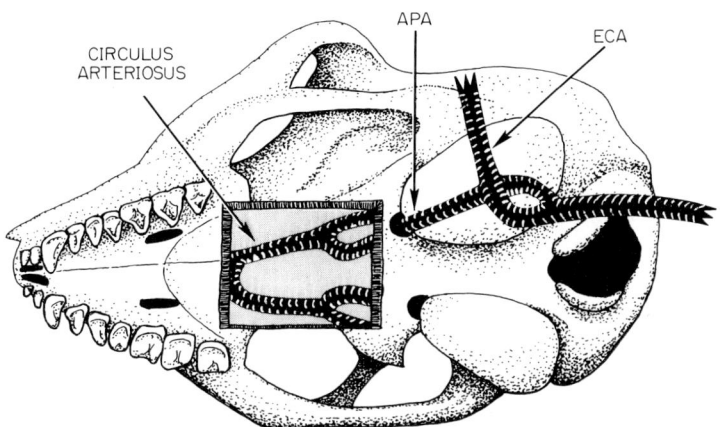

Figure 1. An illustration of the cranial arterial pattern shared by cheirogaleids and lorisiforms. The course of the ascending pharyngeal artery is extracranial until it passes medial to the auditory bulla to enter the cranial cavity through the foramen lacerum. After passing through this foramen, it joins the circulus arteriosus (indicated by the cutout). ECA = external carotid artery; APA = ascending pharyngeal artery.

unsure whether cheirogaleids are more closely related to other Malagasy lemuriforms or to the Afro–Asian lorisiforms.

Although it is true that there are a number of characters shared by some cheirogaleids and all lorisiforms, there is only one character that is uniformly possessed by all cheirogaleids and all lorisiforms: an enlarged ascending pharyngeal artery that is the major contributor of cranial blood. Thus, the hypothesis of cheirogaleid–lorisiform monophyly rests on the synapomorphic status of this character. In both groups, the promontory and stapedial arteries are reduced or completely involuted and are replaced by an ascending pharyngeal artery that runs along the pharyngeal roof, medial to the auditory bulla, to enter the cranial cavity through the foramen lacerum (Figure 1). Indeed, the adult cranial arterial morphology of cheirogaleids is virtually identical to that of lorisiforms except that a carotid rete mirabile has been described for lorisiforms but not for cheirogaleids. The carotid rete is a complex weave of arteries and veins associated with the internal carotid artery, which although present in a number of mammals, is unique to lorisiforms among the primates. Non-cheirogaleid lemuriforms, on the other hand, have also lost the promontory artery but show an enlarged stapedial artery. This leaves cranial arterial blood to be supplied solely by the vertebral arteries (Figure 2).

There are three possible hypotheses to explain the similarity in cranial arterial patterns shared by cheirogaleids and lorisiforms: (1) it is a synapomorphy for a cheirogaleid–lorisiform clade, (2) it is a symplesiomorphy and cannot resolve relationships within the Strepsirhini, or (3) lemuriforms and cheirogaleids are monophyletic and it is a parallelism. Because an enlarged ascending pharyngeal artery is unique within the primate clade, outgroup analysis indicates that it is a derived character. The lemuriform character of an enlarged stapedial artery is found in several fossil primate taxa (Gregory, 1920) and is therefore more likely to be primitive for strepsirhines. Given these assumptions, the three competing hypotheses can be illustrated and their relative degrees of parsimony compared (Figure 3). Hypothesis 1 (Figure 3a), in which cheirogaleids and lorisiforms are together monophyletic, requires only one character-state change to explain cranial arterial patterns among the strepsirhines. Hypothesis 2 (Figure 3b) requires that the ascending pharyngeal

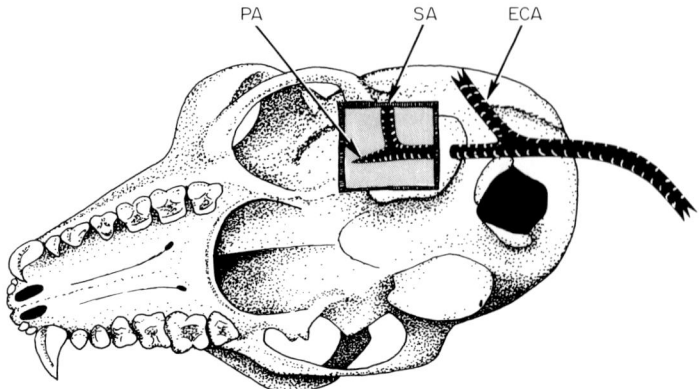

Figure 2. An illustration of the typical non-cheirogaleid lemuriform arterial pattern. The stapedial artery is enlarged and the promontory artery is reduced or involuted. PA = promontory artery; SA = stapedial artery; ECA = external carotid artery.

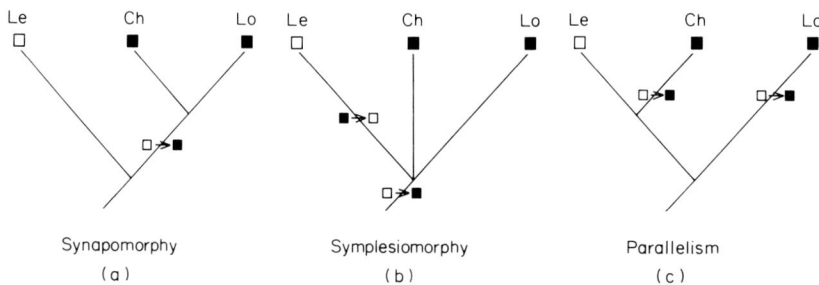

Figure 3. These three cladograms illustrate the conflicting hypotheses that may explain the shared cranial arterial pattern seen in cheirogaleids and lorisiforms. Given that fossil outgroups show an enlarged stapedial artery: (a) synapomorphy requires only one character-state change; (b) symplesiomorphy requires one gain and one loss; and (c) parallelism requires two gains. (□) Lemuriform condition of an enlarged stapedial artery; (■) cheirogaleid–lorisiform condition of an enlarged ascending pharyngeal artery.

artery was gained once, prior to the diversification of living strepsirhines, and then lost in non-cheirogaleid lemuriforms, thus indicating two character-state changes. Hypothesis 3 (Figure 3c) also requires two character-state changes, one gain in the cheirogaleids and one in the lorisiforms. Thus, although parsimony supports the hypothesis of synapomorphy, it cannot distinguish between the hypotheses of symplesiomorphy or parallelism.

Development of the strepsirhine internal carotid artery
Because the arterial pattern is the primary character arguing for the monophyly of cheirogaleids and lorisiforms, independent tests of character homology and polarity are desirable and ontogenetic comparisons offer an ideal source of data for such tests. The pattern of arterial development in *Galago* has been described by Butler (1980, 1983). Butler's study revealed that there are three distinct stages in the development of the ascending pharyngeal artery in this lorisiform (Figure 4). In the first stage, the primary blood supply to the brain is

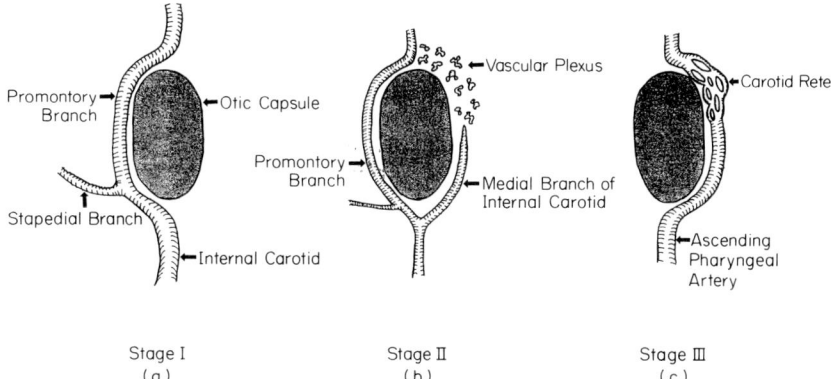

Figure 4. The three stages of *Galago* arterial development are illustrated following Butler (1983). Lateral is to the left of the figure and anterior is to the top of the figure. (a) In Stage I there are large promontory and stapedial arteries. (b) In Stage II, the promontory and stapedial arteries begin to diminish in size. Concurrently, undifferentiated vascular spaces appear on the ventral surface of the otic capsule and a medial branch of the internal carotid has appeared. (c) In Stage III, the adult morphology is completely formed. The promontory and stapedial arteries have disappeared and the ascending pharyngeal artery and carotid rete are the primary supply of cranial arterial blood.

via the promontory artery. This vessel runs forward on the roof of the pharynx, lateral to the developing otic capsule. Before entering the braincase, it gives off a large stapedial artery (Figure 4a). In the second stage, the promontory and stapedial arteries begin to diminish in size. Undifferentiated vascular spaces which will eventually form the carotid rete appear on the ventral aspect of the otic capsule. A new branch of the internal carotid arises from the commencement of the promontory artery and runs medial to the otic capsule to join the caudal end of the vascular spaces (Figure 4b). In the third and final stage of development, the adult arterial pattern is completely formed. The promontory and stapedial arteries have involuted leaving only the ascending pharyngeal artery to supply the fully formed arterial rete and the circulus arteriosus (Figure 4c). Thus, Butler's study revealed that the lorisiform ascending pharyngeal artery is a compound vessel formed from the commencement of the promontory artery, the artery to the rete, the rete, and the terminal intracranial part of the promontory artery.

A comparison of the patterns of arterial development in a cheirogaleid (*Microcebus murinus*) and a lorisiform (*Galago senegalensis senegalensis*) was undertaken to enrich the morphological detail of the arterial characters in order to test the hypotheses of homology and polarity (Yoder, 1991). Here histologically prepared specimens of corresponding developmental stages for both taxa are compared. These specimens reside in the Carnegie Laboratory of Embryology; the *Microcebus* specimens are part of the Bluntschli collection and the *Galago* specimens are those prepared by Butler. Consequently, it was possible to make direct comparisons between Butler's *Galago* specimens and similarly prepared specimens of a cheirogaleid. The arterial morphology of *Microcebus* was reconstructed using a computer-assisted, three-dimensional reconstruction program (Jandel Scientific, San Rafael, CA) (Figure 5). At each developmental stage, *Microcebus* carotid morphology and structural relationships show remarkable similarities to *Galago*. In the first stage of arterial development, a well-developed promontory artery passes lateral to the otic capsule and gives off a moderate sized stapedial artery. There is no sign of the carotid rete or of an artery to the rete (Figure 5a). The second stage of *Microcebus* development is also identical to that of the *Galago*.

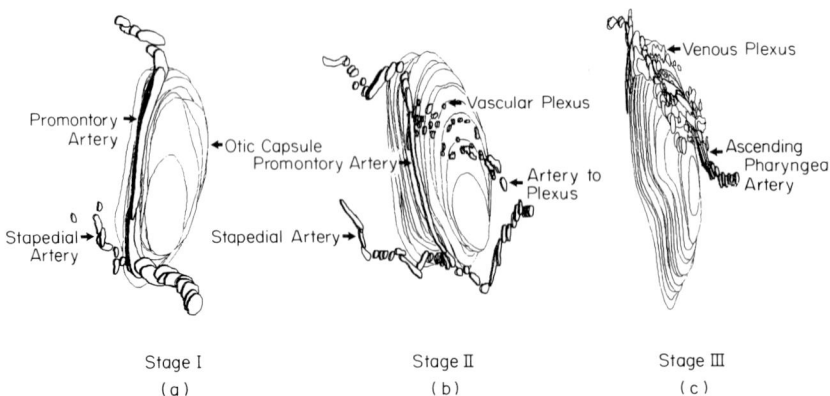

Figure 5. The three stages of *Microcebus* arterial development are illustrated in these 3-dimensional reconstructions. Lateral is to the left of the figure and anterior is to the top of the figure. (a) In Stage I there are large promontory and large stapedial arteries. (b) In Stage II, the promontory and stapedial arteries begin to diminish in size. Concurrently, undifferentiated vascular spaces appear on the ventral surface of the otic capsule and a medial branch of the internal carotid artery has appeared. (c) In Stage III, the adult morphology is completely formed. The promontory and stapedial arteries have virtually disappeared leaving the ascending pharyngeal artery to supply cranial blood. The vascular plexus has become a primarily venous plexus.

Vascular spaces appear on the ventral aspect of the otic capsule to form a sponge-like collection of undifferentiated vessels. The promontory and stapedial arteries are considerably narrowed and a medial branch of the internal carotid joins the caudal end of the vascular spaces (Figure 5b). It is only at the third stage of arterial development that *Microcebus* differs from *Galago*. The promontory artery has virtually disappeared, leaving only the ascending pharyngeal artery to convey internal carotid blood to the cerebral arterial circle. The only difference is that the vascular spaces that first appeared in the previous stage have been incorporated into a primarily venous plexus rather than a primarily arterial plexus as in lorisiforms (Figure 5c).

Discussion

This ontogenetic study of the strepsirhine internal carotid artery demonstrates that the ascending pharyngeal artery goes through similar developmental pathways in both cheirogaleids and lorisiforms. But how much phylogenetic information can be extrapolated from these ontogenetic patterns? To answer this question, ontogenetic data and their relation to the concepts of homology, polarity and parsimony must be considered in turn.

Homology

Characters that are phylogenetically informative are by definition homologous. Although the definitive test of character homology is congruence (Hennig, 1966; Wiley, 1981; Fink, 1982; Kluge & Strauss, 1985), there are a number of commonly used *a priori* tests to detect evolutionary homology: shared function, details of structure, topological relationships and ontogenetic patterns.

There is a fundamental reason why ontogeny should be a powerful test of homology. Ontogeny is the "translation or elaboration" of the stored genetic information that is passed from generation to generation; it "is therefore a manifestation of homology" (Roth, 1988:10).

This manifestation of genetic continuity thus provides a definable link of *process* between ontogeny and homology. Kluge (1988:75) warns, however, that there is a general "failure to distinguish between ontogeny as an explanation of homology and ontogeny as a criterion involved in its recognition." But unlike Kluge, I believe that the two can be complementary. A process that explains a pattern endows that pattern with deeper significance.

Darwin (1872:466) declared that "community in embryonic structure reveals community of descent; but dissimilarity ... does not prove discommunity of descent." This statement should be examined in two parts: first, the claim is made that similarity of development demonstrates homology; second, it is asserted that dissimilarity of development does not disprove homology. The implications are that a comparison of ontogenies can potentially support a hypothesis of homology but cannot falsify that hypothesis. Yet, a number of workers claim that divergent developmental pathways are usually indicative of convergent evolution (e.g., Van Valen, 1982; Roth, 1984; Creighton & Strauss, 1986). The question remains, can ontogenetic comparisons falsify a hypothesis of homology? If so, ontogeny becomes a powerful tool for phylogeny reconstruction.

Most workers agree that when it is demonstrated that adult characters in two or more taxa share details of a complex developmental pathway, it is persuasive evidence in favor of homology (Rieppel, 1979; Roth, 1984, 1988; Mishler, 1988). One classic example in which ontogeny reveals the homology of otherwise dissimilar adult structures is the case of the mammalian middle ear ossicles and the bony components of the reptilian jaw articulation. The middle ear ossicles of adult mammals bear very little resemblance to the hyomandibular, quadrate and articular bones of the reptilian jaw. Yet, we recognize them as homologous structures due to (among other evidence) their similar embryological development. Thus, in this case, ontogeny supports the hypothesis of homology.

But what about falsification? Darwin claimed that dissimilarity in development does not necessarily falsify a hypothesis of homology, yet Van Valen (1965) is generally credited with the exclusion of the treeshrews from the primate order due to their very different pattern of auditory bulla ontogeny. In another example, Gegenbaur (1873, as discussed by Wiley, 1981) proposed that the pectoral girdle of gnathostomes is the serial homologue of the gill arches based on its relative anatomical position. Balinsky (1970) demonstrated, however, that this structure develops from lateral plate mesoderm rather than from the neural crest cells from which the gill arches develop. Thus, it is generally accepted that the pectoral girdle of gnathostomes is not a modified gill arch.

These examples suggest that ontogeny can falsify a hypothesis of homology. But unfortunately for the systematist, there are at least as many examples in which it cannot. There are a number of cases in which ontogeny disagrees with hypotheses of homology based on adult structure, the fossil record, and phylogeny (Roth, 1988; Striedter & Northcutt, 1991). Several of the more decisive examples involve the comparison of urodele development with that of other tetrapods. It is virtually certain, based on comparative anatomy and the fossil record, that urodeles and other tetrapods are monophyletic. Yet the mode of development of the limbs and neural cord of salamanders is profoundly different than that observed for all other tetrapods (de Beer, 1971; Jarvik, 1980). The mode of limb development is so different, in fact, that Jarvik (1980) has argued that urodeles originated independently from the eutetrapods, and that the tetrapods are therefore diphyletic in origin. If we are to accept the traditional view that all tetrapods are monophyletic, then we must admit that the neural cord and limbs of salamanders and other tetrapods are homologous, despite their very different modes of development.

Wiley (1981) describes the recognition of homology as a two-step process: first, characters are subjected to *a priori* morphological tests of homology; second, the various hypotheses of homology are tested one against the other in a maximum parsimony analysis. The morphological tests improve the overall quality of a phylogenetic study by "weeding out" those characters that are phylogenetically uninformative. Two of the tests that he advocates most highly are (1) the criterion of position and (2) the criterion of special similarity.

The comparison of adult morphology in cheirogaleids and lorisiforms provides the necessary information to analyse the criterion of position. Other than its association with the carotid rete, the ascending pharyngeal artery of lorisiforms resembles that of cheirogaleids in all details of position and relationship to other structures. But adult morphology, in this case, is inadequate to test special similarity (i.e., character complexity). The comparison of arterial ontogeny, however, has demonstrated that the ascending pharyngeal artery is actually a complex array of anatomical components. In both taxa, the ascending pharyngeal artery is a compound vessel formed from the proximal end of the promontory artery, an artery to a vascular plexus, a vascular plexus and the terminal intracranial part of the promontory artery. Clearly, the adult ascending pharyngeal artery satisfies the criterion of special similarity.

Wiley (1981:132) further states that the criterion of special similarity "may either reinforce or falsify homologues based on positional criteria." Consequently, by Wiley's standards, a comparison of arterial development in *Microcebus* and *Galago* could have falsified the hypothesis of homology, not because ontogenetic comparisons reveal essential evolutionary truths, but because it could have contradicted the observation of positional similarity. All known *a priori* tests of homology support the hypothesis that the ascending pharyngeal artery of cheirogaleids and lorisiforms is an homologous structure. The arteries in both taxa share the same function, exhibit the same positional relationships, show special similarity and demonstrate nearly identical ontogenetic patterns. Ontogeny thus gives convincing support for the hypothesis of homology. Nonetheless, the possibility of parallelism should be discussed.

Parallelism is often defined as non-homologous similarity of adult structures due to common inherited genetic factors among related taxa (Simpson, 1945; Fink, 1982; Sluys, 1989). Consequently, structures or features that are the result of parallelism should resemble each other in every detail of structure and ontogeny because they are the consequence of identical genetic and epigenetic mechanisms. If true, the conclusion reached by both Fink (1982) and Roth (1984) seems inescapable: the distinction between homology and parallelism is difficult to detect and it is doubtful that ontogenetic comparisons of a single feature can resolve the problem.

Several papers in the recent literature attempt to refine the concept of parallelism (Saether, 1983; Gosliner & Ghiselin, 1984; Sluys, 1989). In particular, Sluys makes a distinction between phylogenetically uninformative parallelism and "underlying synapomorphy." He recommends two tests whereby the two can be distinguished: (1) character complexity and (2) outgroup distribution. A morphologically complex structure that appears frequently within a clade, but not in outgroups, is probably not the result of independent parallel selection. But Sluys is able to postulate underlying synapomorphy only in situations where a phylogenetic framework has been previously deduced from the distribution of strict synapomorphies. Given that this framework does not currently exist for the strepsirhines (work in progress), what then is the utility of Sluys' two tests for the case of the ascending pharyngeal artery?

The ontogenetic study discussed in this paper has demonstrated that the ascending pharyngeal artery is a morphologically complex character. Furthermore, it appears only within the strepsirhines. Thus, this character passes both of Sluys' tests, indicating that it is not a phylogenetically uninformative parallelism. Rather, the ascending pharyngeal artery is indicative of a close genetic affinity between cheirogaleids and lorisiforms. Therefore, the best working hypothesis, prior to a comprehensive parsimony analysis, is that the ascending pharyngeal artery is homologous in cheirogaleids and lorisiforms. This, however, leaves open the question of polarity.

Polarity

Systematists currently use a number of tests to determine character polarity: geological precedence, commonness within the study group, outgroup distributions and ontogenetic comparisons. The way in which workers have attempted to use ontogenetic data to polarize characters is surely related to the concept of recapitulation. Hennig (1966) emphasized the usefulness of developmental sequences for evaluating character polarity—the earlier a character appears in ontogeny, the more primitive it is. Yet, de Beer (1958) and others (Gould, 1977; Alberch et al., 1979; Mabee, 1987, 1989) have repeatedly demonstrated that the phenomenon of heterochrony can alter and rearrange developmental sequences in such a way that they do not recapitulate phylogenetic sequences. Direct observation of ontogeny fails to correctly polarize characters under conditions in which deletion or substitution have affected the developmental program (Brooks & Wiley, 1985; O'Grady, 1985).

In one example, Rieppel (1979) illustrates the potential for error when one assumes that early ontogenetic stages represent the more primitive condition than later ontogenetic stages. All modern adult reptiles, except for snakes, have tropibasic skulls (the trabeculae cranii are completely fused). Yet, most reptiles pass through an ontogenetic stage in which they have the platybasic condition (the trabeculae cranii are unfused). If one were to accept ontogenetic precedence as the best test of polarity, the conclusion would be that a platybasic skull is primitive for reptiles. But outgroup comparison, ingroup commonness, and geological precedence all indicate that the adult tropibasic condition is primitive. Rieppel (1979:59) thus concludes that "the reptile skull goes through an initial platybasic condition because an ancestral skull . . . also went through an initial platybasic condition during ontogeny".

Given that ontogenetic character precedence has been shown to be unreliable as a test of polarity, some workers have attempted to find alternative ontogenetic methods. Indeed, the most extreme claims for the power of developmental sequences in determining character polarity have been presented in the past 10 years (Nelson, 1978; Nelson & Platnick, 1984; Rosen, 1982; Bonde, 1984). In particular, Nelson (1978) and Nelson & Platnick's (1984) "direct method" of assessing character polarity has aroused a storm of controversy. These authors claim that in the comparison of ontogeny between two species, characters that appear in both are primitive whereas characters that appear in only one are derived. In other words, generality indicates primitiveness and thus Nelson's formulation of the biogenetic law is clearly derived from von Baer's second law (1828)—the less general features of an organism develop from the more general. De Queiroz (1985:280) claims that the sequence of ontogenetic transformations is irrelevant; "generality is the critical factor."

But generality is not always a reliable guide to primitiveness either. Hinchliffe and Griffiths (1983) reveal that there is no archetypal phase of prechondrogenic limb formation in developing tetrapods. On the contrary, the early patterns of limb formation are nearly as specialized as the adult limb skeletons. This presumably makes it impossible to observe a general (interpreted as primitive) pattern of tetrapod limb structure.

It seems then that ontogeny falls short as a dependable test of character polarity, whether one employs ontogenetic character precedence or generality as the criterion. Brooks and Wiley (1985) assert that ontogeny does not correctly resolve questions of character polarity where outgroup comparisons fail, and outgroup comparisons sometimes serve where ontogeny fails. Nonetheless, the ontogenetic criterion for polarity determination continues to receive a great deal of attention in the literature (e.g. Kraus, 1988; Mabee, 1989; Wheeler, 1990). Furthermore, both Kraus (1988) and Wheeler (1990) found the ontogenetic criterion (whether precedence or generality) to be as effective as the outgroup criterion for polarizing their data. Clearly, the issue remains open to debate. But even with uncertainty as to its effectiveness, the primary strength of the ontogenetic criterion is that it can be used independently of the outgroup criterion and the two can consequently be used to check one another.

In his discussion of *Galago* arterial development, Butler (1980) claims that lorisiforms pass through an ontogenetic stage that resembles the adult lemuriform condition. He regards this as evidence that the lemuriform pattern is primitive and the lorisiform pattern derived based on the recapitulationist concepts of ontogenetic character precedence and generality. And if this were true, the ontogenetic data would indeed support the evidence from outgroup comparisons and geological precedence. In fact, however, Butler is oversimplifying the ontogenetic data.

The ontogenetic stage that Butler equates with the adult lemuriform condition is stage I (Figure 4a). At this stage, in both *Galago* and *Microcebus*, there is a large promontory and a large stapedial artery. This is in fact significantly different than the adult lemuriform morphology. Adult lemurs do have a large stapedial artery but the promontory artery is either weak or completely involuted (Bugge, 1974; Conroy & Wible, 1978). Stage I is actually closer to the primitive eutherian condition (Wible, 1984) and does not resemble the adult arterial morphology for any living primate.

Unfortunately, there is no published description of early lemuriform arterial development nor is there (to my knowledge) a collection of sufficiently early lemuriform embryos (Carnegie Stage 18 through very early fetus) to permit such a description. Consequently, any discussion of arterial development of non-cheirogaleid Malagasy primates must be a matter of speculation. It is probable that lemuriforms also start cranial arterial development with a pattern indistinguishable from Stage I. The remainder of arterial development presumably involves a steady increase in the diameter of the stapedial artery and a steady decrease in the diameter of the promontory artery. In the adult lemur, cranial arterial blood is supplied from the vertebral artery alone (Bugge, 1974).

Thus, ontogeny offers no additional information pertinent to the determination of character polarity for the ascending pharyngeal artery. Outgroup comparisons and geological precedence (i.e., fossil evidence) both support the hypothesis that the condition observed in cheirogaleids and lorisiforms is derived. Ontogeny neither supports nor contradicts this hypothesis.

Parsimony, congruence and other applications
Although it is true that ontogeny should not be the sole criterion for deciding issues of homology and polarity, it is a powerful tool when used in combination with other tests. Each time that a hypothesis of character polarity or homology is tested, and the hypothesis is

supported or not, the systematist builds an impression of character credibility. Swofford & Olsen (1990:451) define parsimony as the notion that "simpler hypotheses are preferable to more complicated ones and ... *ad hoc* hypotheses should be avoided whenever possible." Consequently, if adult structure, the fossil record and ontogenetic information agree on a hypothesis of character homology, then it is most parsimonious to accept the hypothesis of homology. Likewise, if outgroup and ontogenetic information agree on a particular hypothesis of character polarity, then it is most parsimonious to accept that hypothesis.

Perhaps the most important contribution that ontogenetic analysis can make to a systematic study is that it can so deeply enrich appreciation of character complexity. The examination of ontogeny can often provide a level of morphological detail that is not available in the comparison of adult forms. This not only increases the accuracy of *a priori* judgements of character homology, it can also provide new characters for phylogenetic analysis. Indeed, it has been recommended that the ontogenetic transformations themselves should be utilized as a source of phylogenetic characters (Kluge, 1985, 1988; Kluge & Strauss, 1985; de Queiroz, 1985; Creighton & Strauss, 1986; Mishler, 1988).

An important result of the ontogenetic comparison of the cheirogaleid and lorisiform ascending pharyngeal artery is the revelation of an important new character: the appearance of a vascular plexus ventral to the otic capsule. Prior to this study, a vascular plexus associated with the ascending pharyngeal artery had never been described for a cheirogaleid. Although it is true that the same plexus in lorisiforms develops into primarily arterial rather than venous channels (although it is a combination of both), the vascular plexus is another character that cheirogaleids and lorisiforms share and lemuriforms do not. (Again, speculation on lemuriform ontogeny is called upon. But it is virtually certain that lemuriforms do not possess such a plexus at any time in development. In both cheirogaleids and lorisiforms, the plexus develops in conjunction with a medial, extracranial branch of the internal carotid artery. Lemuriforms have no such artery.) If the arterial characters are used to construct a cladogram, the most parsimonious distribution of these characters gives support to the hypothesis that cheirogaleids and lorisiforms form a monophyletic clade (Figure 6).

Conclusions

There are three significant results of this comparative ontogenetic study of the strepsirhine internal carotid artery: (1) the hypothesis that the ascending pharyngeal artery is homologous in cheirogaleids and lorisiforms is strongly supported, (2) the hypothesis that this is a derived character is neither supported nor falsified, and (3) a new character that is uniquely shared by cheirogaleids and lorisiforms, the extracranial vascular plexus, was discovered. These three results taken together support the hypotheses of synapomorphy and symplesiomorphy equally well.

But before this arterial character is left to a strict analysis of congruence, the issue of character weighting should be discussed. Swofford & Olsen (1990) rightly point out that all characters are not equally informative for phylogeny reconstruction and therefore a system for determining the relative strength of characters should be useful. Neff (1986) argues for a "rational basis" for character weighting. In essence, she is calling for a weighting system that is based on our knowledge of a character. Systematists should be asking not "How reliable is this character?" but "How reliable is the hypothesis of homology for this character?" As a character is repeatedly subjected to tests of polarity and homology, the confidence in that

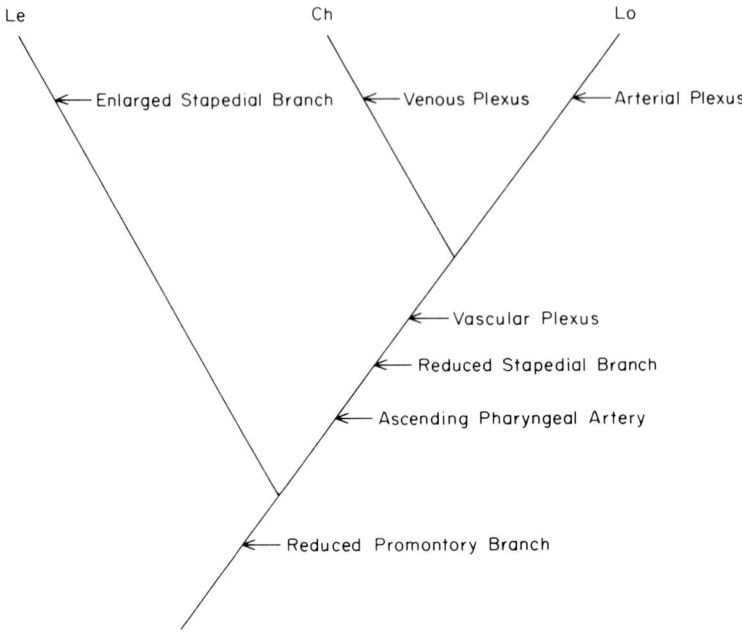

Figure 6. A cladogram constructed solely of developmental and adult arterial characters supports the hypothesis that cheirogaleids and lorisiforms are monophyletic and lemuriforms are their sister taxon.

character will either increase or decrease. "It is rational to weight more heavily those sources of information, those data, that are most internally consistent and most extensive and complete" (Neff, 1986:123).

Unfortunately, Neff does not recommend any operational steps whereby character weight information can be incorporated into a systematic study. By weighting characters prior to a Wagner parsimony analysis, the systematist is relying on the assumption that the weighted characters carry more phylogenetic information than non-weighted characters. The problem rests on the fact that it is not practical to subject every character in a phylogenetic study to the same degree of testing that has been described here for the ascending pharyngeal artery. Consequently a character that receives low weight may in fact be as informative as those with high weights, but for reasons of practicality, the phase of morphological testing was not as exhaustive. I would therefore suggest that unless every character is analysed equally, *a priori* character weighting is not valid. It is often the case, however, that a single most-parsimonious tree does not result from a phylogenetic analysis. In this case, if it is discovered that certain of the trees support the homology of the "valued" characters and others do not, the systematist has a "rational basis" for preferring the tree or trees that support the weighted character(s).

In conclusion, the ontogenetic comparison of the ascending pharyngeal artery in a cheirogaleid and a lorisiform has demonstrated that it is a morphologically complex character with a nearly identical pattern of development in both taxa. Thus, this character passes all known *a priori* tests of homology. Consequently, if in the final analysis, one must choose between a phylogenetic hypothesis that supports the homology of this character, and another that does not, one has a "rational basis" for selecting the hypothesis of homology.

Acknowledgements

For access to the Butler *Galago* specimens and the Bluntschli *Microcebus* specimens, I would like to thank Dr R. O'Rahilly and the staff of the California Regional Primate Center–University of California, Davis. Mr Jeff Rowland was extremely helpful in teaching me to master the art of photomicroscopy. Dr Matt Cartmill and Dr Kathleen Smith have provided support and encouragement throughout this project. The quality of this manuscript was significantly improved thanks to the comments of M. Cartmill, B. Mishler, M. Ravosa, K. Smith and three anonymous reviewers. Financial support was provided by a National Science Foundation, Dissertation Improvement Grant, a Sigma Xi Grant-in-Aid-of-Research and the Department of Biological Anthropology and Anatomy, Duke University Medical Center.

References

Alberch, P. (1985). Problems with the interpretation of developmental sequences. *Syst. Zool.* **34**, 46–58.
Alberch, P., Gould, S. J., Oster, G. F. & Wake, D. B. (1979). Size and shape in ontogeny and phylogeny. *Paleobiology* **5**, 296–317.
Baer von, K. E. (1828). *Entwicklungsgeschichte der Thiere: Beobachtung und Reflexion.* Konigsberg: Borntrger.
Balinsky, B. I. (1970). *An Introduction to Embryology.* Philadelphia: W.B. Saunders Co.
Bonde, N. (1984). Primitive features and ontogeny and phylogenetic reconstructions. *Vidensk. Meddr dansk naturh. Foren.* **145**, 219–236.
Brooks, D. R. & Wiley, E. O. (1985). Theories and methods in different approaches to phylogenetic systematics. *Cladistics* **1**, 1–11.
Bugge, J. (1974). The cephalic arterial system in insectivores, primates, rodents and lagomorphs, with special reference to the systematic classification. *Acta anat.* **87** (Suppl. 62), 1–160.
Butler, H. (1980). The homologies of the lorisoid internal carotid artery system. *Int. J. Primatol.* **1**, 333–343.
Butler, H. (1983). The embryology of the lesser galago (*Galago senegalensis*). *Contrib. Primatol.* **19**, 1–156.
Cartmill, M. (1975). Strepsirhine basicranial structures and the affinities of the Cheirogaleidae. In (W. P. Luckett & F. Szalay, Eds) *Phylogeny of the Primates: a Multidisciplinary Approach*, pp. 313–354. New York: Plenum Press.
Charles-Dominique, P. & Martin, R. D. (1970). Evolution of lorises and lemurs. *Nature* **227**, 257–260.
Conroy, G. C. & Wible, J. R. (1978). Middle ear morphology of *Lemur variegatus*: some implications for primate paleontology. *Folia primat.* **29**, 81–85.
Creighton, G. K. & Strauss, R. E. (1986). Comparative patterns of growth and development in cricetine rodents and the evolution of ontogeny. *Evolution* **40**, 94–106.
Darwin, C. (1872). *On the Origin of Species, with Additions and Corrections.* London: Murray.
de Beer, G. (1958). *Embryos and Ancestors*, 3rd edn. London: Oxford University Press.
de Beer, G. (1971). *Homology, an Unsolved Problem.* London: Oxford University Press.
de Queiroz, K. (1985). The ontogenetic method for determining character polarity and its relevance to phylogenetic systematics. *Syst. Zool.* **34**, 280–299.
Fink, W. L. (1982). The conceptual relationship between ontogeny and phylogeny. *Paleobiology* **8**, 254–264.
Gegenbaur, C. (1873). Ueber das Archipterygium. *Jenaische Z. Med. Nat. Wiss.* **7**, 131–141.
Gosliner, T. M. & Ghiselin, M. T. (1984). Parallel evolution in opisthobranch gastropods and its implications for phylogenetic methodology. *Syst. Zool.* **33**, 255–274.
Gould, S. J. (1977). *Ontogeny and Phylogeny.* Cambridge: Harvard University Press.
Gregory, W. K. (1920). On the structure and relations of *Notharctus*, an American Eocene primate. *Mem. Am. Mus. nat. Hist.*, n.s. **3**, 51–243.
Haeckel, E. (1866). *Generelle Morphologie der Organismen: Allgemeine Grundzuge der organischen Formen-Wissenschaft, mechanisch begrundet durch die von Charles Darwin reformirte Descendenz-Theorie*, 2 vols. Berlin: Georg Reimer.
Hennig, W. (1966). *Phylogenetic Systematics.* Urbana: University of Illinois Press.
Hill, W. C. O. (1953). *Primates, Vol. I, Strepsirhini.* Edinburgh: Edinburgh University Press.
Hinchliffe, J. R. & Griffiths, P. J. (1983). The prechondrogenic patterns in tetrapod limb development and their phylogenetic significance. In (B. C. Goodwin, N. Holder & C. C. Wylie, Eds) *Development and Evolution*, pp. 99–121. London: Cambridge University Press.
Jarvik, E. (1980). *Basic Structure and Evolution of Vertebrates*, 2 vols. London: Academic Press.
Kluge, A. (1985). Ontogeny and phylogenetic systematics. *Cladistics* **1**, 13–27.
Kluge, A. (1988). The characterization of ontogeny. In (C. J. Humphries, Ed.) *Ontogeny and Systematics*, pp. 57–81. New York: Columbia University Press.

Kluge, A. G. & Strauss, R. E. (1985). Ontogeny and systematics. *Ann. Rev. Ecol. Syst.* **16,** 247–268.
Kraus, F. (1988). An empirical evaluation of the use of the ontogenetic polarization criterion in phylogenetic inference. *Syst. Zool.* **37,** 106–141.
Mabee, P. M. (1987). Phylogenetic change and ontogenetic interpretation in the family Centrarchidae (Perciformes: Centrarchidae). Ph.D. Dissertation, Duke University.
Mabee, P. M. (1989). An empirical rejection of the ontogenetic polarity criterion. *Cladistics* **5,** 409–416.
Mishler, B. D. (1988). Relationships between ontogeny and phylogeny, with reference to bryophytes. In (C. J. Humphries, Ed.) *Ontogeny and Systematics*, pp. 117–136. New York: Columbia University Press.
Neff, N. (1986). A rational basis for a priori character weighting. *Syst. Zool.* **35,** 110–123.
Nelson, G. (1978). Ontogeny, phylogeny, paleontology, and the biogenetic law. *Syst. Zool.* **27,** 348–352.
Nelson, G. & Platnick, N. I. (1984). Systematics and evolution. In (M. W. Ho & P. T. Saunders, Eds) *Beyond Neo-Darwinism*, pp. 143–158. New York: Academic Press.
Northcutt, R. G. (1990). Ontogeny and phylogeny: a re-evaluation of conceptual relationships and some applications. *Brain, Behav. Evol.* **36,** 116–140.
O'Grady, R. T. (1985). Ontogenetic sequences and the phylogenetics of parasitic flatworm life cycles. *Cladistics* **1,** 159–170.
Rieppel, O. (1979). Ontogeny and the recognition and primitive character states. *Z. Zool. Syst. Evolutionsforsch* **17,** 57–61.
Rosen, D. E. (1982). Do current theories of evolution satisfy the basic requirements of explanation? *Syst. Zool.* **31,** 76–85.
Roth, V. L. (1984). On homology. *Biol. J. Linn. Soc.* **22,** 13–29.
Roth, V. L. (1988). The biological basis of homology. In (C. J. Humphries, Ed.) *Ontogeny and Systematics*, pp. 1–26. New York: Columbia University Press.
Saether, O. A. (1983). The canalized evolutionary potential: inconsistencies in phylogenetic reasoning. *Syst. Zool.* **32,** 343–359.
Simpson, G. G. (1945). The principles of classification and a classification of mammals. *Bull. Am. Mus. nat. Hist.* **85,** 1–350.
Sluys, R. (1989). Rampant parallelism: an appraisal of the use of nonuniversal derived character states in phylogenetic reconstruction. *Syst. Zool.* **38,** 350–370.
Striedter, G. F. & Northcutt, R. G. (1991). Biological hierarchies and the concept of homology. *Brain, Behav. Evol.* **38,** 177–189.
Swofford, D. L. & Olsen, G. J. (1990). Phylogeny reconstruction. In (D. M. Hillis & C. Moritz, Eds) *Molecular Systematics*, pp. 411–501. Sunderland, Massachusetts: Sinauer Associates.
Szalay, F. S. & Katz, C. C. (1973). Phylogeny of lemurs, galagos and lorises. *Folia primat.* **19,** 88–103.
Van Valen, L. (1965). Tree shrews, primates, and fossils. *Evolution* **19,** 137–151.
Van Valen, L. (1982). Homology and causes. *J. Morph.* **173,** 305–312.
Wheeler, Q. D. (1990). Ontogeny and character phylogeny. *Cladistics* **6,** 225–268.
Wible, J. R. (1984). The ontogeny and phylogeny of the mammalian cranial arterial pattern. Ph.D. Dissertation, Duke University.
Wiley, E. O. (1981). *Phylogenetics: the theory and practice of phylogenetic systematics*. New York: John Wiley and Sons.
Yoder, A. D. (1991). A comparison of the early developmental stages of the internal carotid artery in a cheirogaleid and a lorisiform: implications for phylogeny. *Am. J. phys. Anthrop.* **84,** 187.

Matthew J. Ravosa
Duke University Medical Center, Department of Biological Anthropology & Anatomy, Durham, North Carolina 27710, U.S.A.

Received 1 June 1991
Revision received 16 January 1992 and accepted 28 January 1992

Keywords: allometry, heterochrony, ontogeny, skull form, lemurs, Madagascar.

Allometry and heterochrony in extant and extinct Malagasy primates

Measurements of the skull and dentition were obtained for three groups of closely-related Malagasy lemurs: *Propithecus verreauxi* and *P. diadema*, *Hapalemur griseus* and *H. simus*, and *Varecia variegata* and subfossil *Pachylemur insignis*. In *Propithecus*, ontogenetic series for the larger *P. diadema* and smaller *P. verreauxi* were compared to evaluate whether species-level differences in skull form result from the differential extension of common patterns of relative growth. In the other two cases, mostly adult *H. simus* data and adult *Pachylemur* data were compared to ontogenetic series for their smaller sister taxa, respectively *H. griseus* and *Varecia*, to similarly infer whether morphological differences between these taxa result from the ontogenetic scaling of cranial proportions. First, analyses of the data indicate that cranial proportions for both species of *Propithecus* are ontogenetically scaled. As such, *P. diadema* apparently attains larger overall size primarily by growing at a faster rate, but not for a longer duration, than *P. verreauxi*. Second, analyses of the *Hapalemur* data suggest that facial proportions, but not mandibular dimensions, in both species are ontogenetically scaled. It is inferred that differences in patterns of relative growth for the mandibular corpus and symphysis of bamboo lemurs are correlated with differences in the loading regime of the lower jaw due to variation in the mechanical properties of their diets. Finally, analyses of the data indicate that most cranial proportions for *Varecia* and *Pachylemur* are not ontogenetically scaled, thus supporting claims for the generic separation of these taxa. Additional consideration of these comparisons and examples from the literature illustrate potential differences in the effects of selection for body size increases versus decreases, and selection for greater inter- versus intra-specific body size variation, on postcanine dental allometry in ontogenetically-scaled taxa. Possible differences in the development process underlying phyletic size change are also discussed.

Journal of Human Evolution (1992) **23**, 197–217

Introduction

Recent allometric and heterochronic approaches have contributed much to our understanding of the evolutionary biology of primates. One benefit of such approaches is to facilitate a more direct investigation of the developmental processes underlying phyletic size change and ecogeographic size variation among closely related taxa. For instance, Shea (1983*a–c*, 1984) demonstrates that the vast majority of differences in the adult cranial morphology of gorillas, chimpanzees and pygmy chimpanzees result from the differential extension of common patterns of relative growth to different adult body sizes. That is, the three species of African apes are *ontogenetically scaled* (Figure 1).

Given a pervasive pattern of ontogenetic scaling between a larger and smaller species, there are two ways for the larger species to have increased its body size (phyletic gigantism) relative to the ancestral or primitive condition of the smaller species (Shea, 1983*a*, 1986, 1988).* First, members of the larger species can grow faster via *rate hypermorphosis*. That is, in comparisons at the same non-adult ages as those of the smaller species, individuals of the larger species develop progressively larger craniofacial measures but attain maturity at the

---

*First off, this scenario assumes that the smaller species' adult body size and ontogenetic trajectory are viewed as the ancestral pattern, which is often, but not always the case. Second, this scenario assumes no dissociation of allometric trajectories between species as occurs in cases of "acceleration". Lastly, this scenario does not necessarily assume that both species have the same neonatal size, just that absolute differences in neonatal body size between species will be much less than species differences in adult body size.

0047–2484/92/080197+21 $08.00/0

© 1992 Academic Press Limited

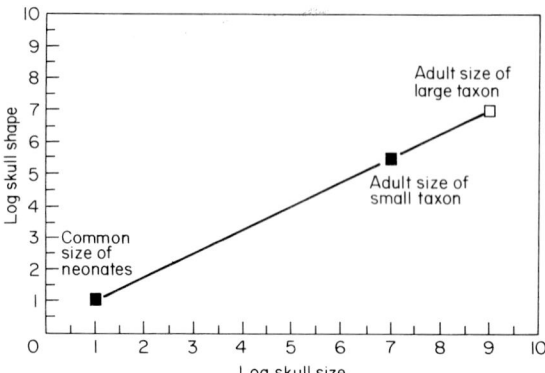

Figure 1. Ontogenetic scaling and phyletic gigantism. Growth trajectories for a measure of skull shape versus a measure of skull size for a smaller-bodied species (■, ancestral pattern) and a larger-bodied species (□, phyletic giant). Both species share similar growth allometries, so that any morphological differences between adults of each species are due to the differential elongation of common patterns of relative growth.

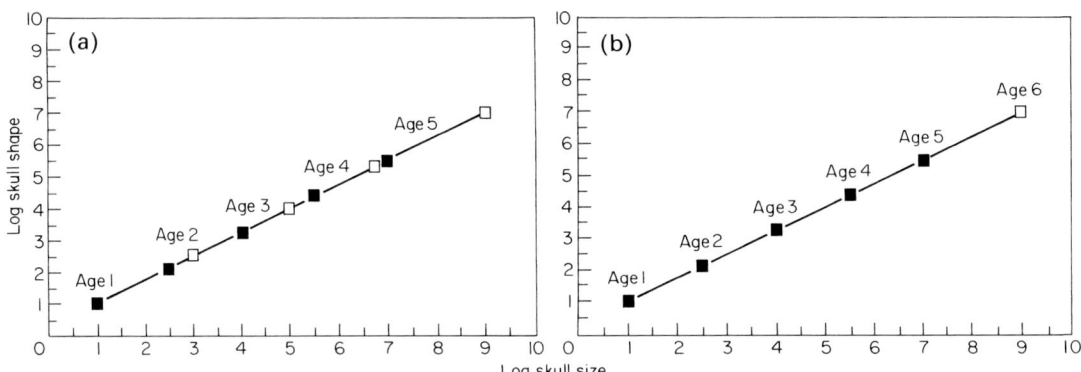

Figure 2. Ontogenetic scaling and rate and time hypermorphosis. A larger species (□, phyletic giant) and a smaller species (■, ancestral pattern) share similar growth trajectories for a measure of skull shape versus skull size. The larger species extends the smaller species ontogenetic scaling trend to attain larger adult sizes by (a) growing faster via rate hypermorphosis, i.e., individuals of the same age are dissimilar in size but grow for the same duration, and/or (b) growing longer via time hypermorphosis, i.e., individuals of the same age are similar in size but grow for different durations.

same chronological age (Figure 2a). Second, individuals of a species can grow for a longer duration via *time hypermorphosis*. That is, at the same non-adult age as the smaller species, members of the larger species have similarly-sized craniofacial dimensions but mature at a later chronological age (Figure 2b). As used in this context, faster (rate) and longer (time) comparisons are in relation to growth-in-time, not different coefficients of relative or allometric growth. Obviously, both processes can affect the evolution of species differences in overall size if developmental modifications occur first during ontogeny *and* become especially marked between adults.

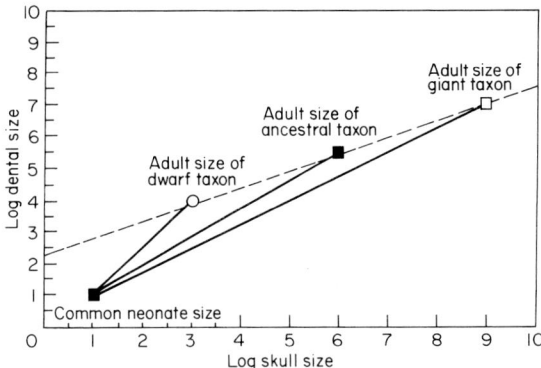

Figure 3. Ontogenetic scaling and dental scaling. A plot of dental size versus skull size for three sister taxa in which all other cranial proportions are ontogenetically scaled and all species happen to have the same neonatal size. One species maintains the ancestral or primitive pattern of relative growth (■). The other two species have the dissociated size/shape pattern predicted for a phyletic dwarf (○) and a phyletic giant (□). Within the dwarfed species, postcanine tooth size scales more positively relative to the ancestral pattern. However, within the phyletic giant, tooth size scales more negatively. Regardless of the direction of size change, tooth size will scale similarly across adults (– – –); moreover, this slope is lower than predicted in broad interspecific comparisons. As a result of selection on systemic growth processes, there is considerably more variation in skull size versus dental size.

When taxa demonstrate a consistent pattern of ontogenetic scaling of craniodental dimensions, brain size and postcanine tooth size are two components of the skull which may demonstrate a dissociation of shared patterns of allometric growth. This is because selection for rapid increases in overall body size apparently targets primarily postnatal systemic growth. Therefore, in comparisons between ontogenetically-scaled taxa, species differences in brain size—a structure which develops mostly prenatally—are predicted to be minimal as compared to differences in facial or somatic dimensions, which develop and enlarge primarily postnatally (Lande, 1979; Shea, 1983b,c, 1985, 1988; Riska & Atchley, 1985; Shea et al., 1987, 1990; Ravosa, 1991a; Ravosa & Ross, in press).

As noted above, dental scaling patterns tend to mirror those for brain size across and within ontogenetically-scaled taxa. For example, several workers suggest that species differences in postcanine tooth size should be relatively minor as opposed to differences in other regions of the facial skull, since genetic and epigenetic controls on dental morphogenesis appear to operate more independent of systemic effects on facial and somatic growth (Cochard, 1985, 1987; Shea, 1983c, 1986, 1988; Shea & Gomez, 1988; Shea et al., 1990; Ravosa, 1991a; Ravosa & Ross, in press). Therefore, in an ontogenetically-scaled series of close relatives, regardless of whether selection has acted to increase or decrease overall size inter- or intraspecifically, postcanine tooth size is expected to scale with negative allometry *across* adults such that larger individuals or taxa should have relatively smaller teeth (e.g., Gould, 1975; Cochard, 1985, 1987). However, during ontogeny *within* a species having undergone phyletic size change, tooth size should scale differently depending on whether selection has evinced an increase or a decrease in overall size (Figure 3).† That is, during growth in a phyletically

---

†While Figure 3 depicts all three species as having the same neonatal size, this condition is not necessarily a prerequisite for the ontogenetic model regarding patterns of dental scaling in phyletic giants and phyletic dwarves. Again, the important fact is that absolute differences in neonatal body size between species will be miniscule as compared to interspecific differences in adult body size.

dwarfed species, post-canine tooth size would be expected to show a higher (and presumably positive) allometric coefficient relative to the ancestral condition of its sister taxon, whereas phyletic giants would be expected to show a lower (and presumably negative) allometric coefficient for postcanine tooth size relative to the ancestral condition of its sister taxon (Figure 3). Again, as a result of selection primarily on postnatal facial growth, in ontogenetically-scaled taxa there will be a greater discrepancy in aspects of skull size relative to dental size.

Alternatively, postcanine tooth size might scale with positive allometry across adults of phyletic size-series, such that larger individuals or taxa have relatively larger teeth. In this case, apart from selection for increased body size, selection may specifically target tooth size since, within a monophyletic group, larger species tend to be more folivorous and therefore require corresponding increases in relative postcanine tooth size (Gould, 1975; Kay, 1975, 1978; Shea, 1983c). Lastly, it is important to stress that predictions of negative or positive allometry of postcanine tooth dimensions differ from empirical data for interspecific series of anthropoids (Kay, 1975) and other mammals (Creighton, 1980; Legendre & Roth, 1988), indicating that postcanine tooth size often scales with isometry versus body mass.

Until very recently, extensive interspecific examinations of allometry and heterochrony in primates have been limited mainly to studies of anthropoids (e.g., Shea, 1983a–c, 1984, 1986, 1988; Ravosa, 1991a; Ravosa & Ross, in press). In particular, most previous studies have focused on intraspecific investigations of the ontogeny of sexual dimorphism (e.g., Cochard, 1985; Shea, 1985, 1986; Cheverud & Richtsmeier, 1986; Leutenegger & Masterson, 1989a,b; Masterson & Leutenegger, 1990; Leigh & Cheverud, 1991; Ravosa, 1991a,b; Leigh, 1992). On the other hand, little information exists regarding the ontogenetic bases of interspecific differences in morphology among prosimians [but see Gomez (1992a,b) on lorises], despite that many prosimian congeners differ considerably in body size (Albrecht et al., 1990; Kappeler, 1990). Recent studies of the behavioral ecology and functional morphology of Malagasy primates now make it possible to offer some specific predictions about patterns of allometry and heterochrony among closely related taxa.

In this study, allometric comparisons are made among three closely related groups of size-differentiated Malagasy lemurs. The first group consists of *Propithecus verreauxi* (3780 g) and *P. diadema* (6500 g), respectively the western and diademed sifakas, with limited data for *P. tattersalli* (3300 g), the golden-crowned sifaka also considered here. The second group consists of *Hapalemur griseus* (880 g), the lesser or gentle bamboo lemur, and *H. simus* (2365 g), the greater or broad-nosed bamboo lemur; no data for the medium-sized *H. aureus* (1560 g), the golden bamboo lemur, presently exist. The third group consists of *Varecia variegata* (3800 g), the ruffed lemur, and the larger-bodied subfossil *Pachylemur insignis* (no body weight estimates exist). The purpose of this study is three-fold: (1) to evaluate whether differences in skull form between close relatives of differing body sizes develop via ontogenetic scaling; (2) to examine the effects of phyletic size change on patterns of facial, dental and neurocranial morphology (and alternatively, to provide an ontogenetic criterion of subtraction for a functional analysis of non-allometric differences in the form of the masticatory apparatus—Shea, 1983c; Ravosa, 1991a); and (3) to address, given a pervasive pattern of ontogenetic scaling of cranial morphology among sifakas, whether differences in skull size between diademed and western sifakas develop via rate hypermorphosis (*P. diadema* grows faster) and/or time hypermorphosis (*P. diadema* grows for a longer duration). As such, comparisons within these three groups of Malagasy primates will serve to highlight the developmental under-pinnings of ecogeographic size variation, and provide additional data regarding the ontogenetic bases of phyletic size change in a broader variety of mammals. Moreover, apart from some work on early hominids (i.e., Pilbeam & Gould, 1974; Shea, 1988), heterochronic analyses have been rarely extended to an assessment of extinct primates.

Materials and Methods

Samples
The sifaka ontogenetic samples consist of 70 crania of the western sifaka, *P. verreauxi* (42 adults and 28 non-adults), and 36 crania of the diademed sifaka, *P. diadema* (21 adults and 15 non-adults); all subspecies are represented in the adult samples of both species. The only cranial specimen of the newly found golden-crowned sifaka, *P. tattersalli* (Simons, 1988), that of the adult holotype, is also included in bivariate comparisons. For between-species comparisons of sifaka craniofacial development, the data are grouped into six dental age classes (no. 1 = infant; no. 6 = adult). In *P. verreauxi*, there are at least five skulls per non-adult dental age classes nos 1–5. The *P. diadema* sample has at least four specimens per dental ages nos 1, 4 and 5, whereas dental ages nos 2 and 3 have only one case apiece.

The ontogenetic sample of lesser bamboo lemurs, *H. griseus*, numbers 59 crania (38 adults and 21 non-adults); the adult sample of *H. griseus* is represented by members of all three subspecies. In *H. griseus*, there are at least five specimens per non-adult dental age classes nos 1–4. The sample of greater bamboo lemurs, *H. simus*, consists of 17 adult crania (12 subfossil and 5 recent specimens) as well as one subadult (age no. 4) and one juvenile (age no. 3) (without a lower jaw). The recently-described golden bamboo lemur, *H. aureus* (Meier *et al.*, 1987; Wright *et al.*, 1987), is not considered here since there are no known cranial specimens.

The ontogenetic sample of ruffed lemurs, *V. variegata*, numbers 45 crania (26 adults and 19 non-adults); both subspecies of *Varecia* are represented in the adult sample. In *Varecia*, there are no less than four skulls per non-adult dental age classes nos 1–4. There are 20 adult crania of subfossil *P. insignis*, which is sometimes referred to as *V. insignis*. This study follows the subspecific classification of the two size variants of subfossil *Pachylemur-P. "jullyi"* and *P. "insignis"* (Tattersall, 1982), both of which are included in this study.

Measurements
Craniometric data were recorded with digital calipers accurate to 0.1 mm. A total of 19 linear dimensions were taken on each specimen (Appendix 1). Measures from bilateral structures (e.g., ramus height) were taken on the right side of the skull. Body weight, postcanine toothrow length and neural volume were obtained only among adults; bigonial breadth, temporalis lever arm length and body weight were not available for *Pachylemur*. Information on adult body weight was taken from Jungers (1985), Simons (1988), Glander *et al.* (1989), the Duke University Primate Center records, and individual museum records [see also Kappeler (1990) for a review]. Neurocranial volume was determined by filling the braincase with barley until a level surface was reached at the foramen magnum and then pouring the barley into a cylinder graduated in milliliters; these adult values correspond closely to previously published data (see Ravosa, 1989).

Statistical analyses
Within each of the four ontogenetic series, least squares bivariate regression ($P \leqslant 0.05$) was applied to log-transformed craniometric data to describe allometric growth trajectories. In each case the data were grouped and averaged by dental age and subspecies so as to reduce

the effects of disproportionately large numbers of adults and/or large numbers of one subspecies on the slope of the regression line. For instance, in the former case this may result in a lower regression coefficient for statistical reasons, i.e., greater numbers of cases (adults) at one end of the size range, and not for biological reasons (Cheverud, 1982). Basicranial length was used as the independent variable in all bivariate comparisons. Analysis of covariance (ANCOVA, $P \leqslant 0.05$) was used to test for differences in patterns of relative growth between regression lines derived for diademed and western sifakas, i.e., whether both species are ontogenetically scaled. Since adequate ontogenetic series are unavailable for *P. insignis* and *H. simus* (and *P. tattersalli*), an alternative method was used to assess whether these taxa are scaled-up versions of their respective sister taxa *V. variegata* and *H. griseus*. As a first approximation, if the cranial ontogenetic trajectory for the smaller taxon is extended to a larger size *and* intersects the adult scatter for the larger taxon, then it is inferred that these forms are ontogenetically scaled. However, if the adult data scatter fall entirely above or below the regression line, then it is inferred that morphological differences between adults of each sister taxon are not due to the differential extension of common patterns of relative growth.

Analysis of variance (ANOVA, $P \leqslant 0.05$) was used to test for species differences in the size of cranial measures at common dental ages. Given a pervasive pattern of ontogenetic scaling among sifakas, for instance, if at common dental ages non-adult means for sifaka cranial measures are significantly different, then this would indicate that species differences in adult skull form develop via rate hypermorphosis, i.e., diademed sifakas grow at a faster rate, but not longer (Shea, 1983*a*, 1986, 1988; Ravosa, 1991*a*). However, if species differences are noted only among adult sifakas, then this would indicate that morphological differences are attained via time hypermorphosis, i.e., diademed sifakas grow for a longer duration, but not faster (Shea, 1983*a*, 1986, 1988; Ravosa, 1991*a*). Again, if size differences are noted between sifakas during the early stages of ontogeny and are particularly marked between adults, then this would suggest that overall size differentiation evolved via rate *and* time hypermorphosis. Also, as noted previously, faster (rate) and longer (time) characterizations refer to growth-in-time, not different coefficients (regression slopes) of relative growth.

In all three sets of species comparisons, ratios were calculated for the adult means of each cranial measure, i.e., larger species mean/smaller species mean; a value of 1.00 indicates that the species are on average the same size for that measure. Subsequently, these ratios were ranked in descending order to investigate the effect of phyletic size increases and dietary adaptation on differences in relative postcanine tooth size. Assuming an isometric condition, higher ranks for postcanine tooth size would suggest that apart from selection for systemic body size increases in the larger species, selection has targeted tooth size as well (due to a more obdurate diet). In such cases, dimensions of the mandibular corpus and symphysis might also be expected to have correspondingly higher rank values, since mandibular robusticity in prosimians is asociated with a structurally resistant diet (Ravosa, 1991*c*). On the other hand, a low rank would support the prediction that tooth size scales negatively across sister taxa in which selection has acted solely to increase overall body size; likewise, a low rank is predicted for brain volume. In this study, tooth size ratios are expected: (1) to be lower among sifakas, since both species ingest roughly similar amounts of fibrous leaves [compare 46% for *P. verreauxi* (Richard, 1985) to 53% for *P. diadema* (Wright, 1987; Wright *et al.*, 1987)]; (2) to be higher among bamboo lemurs, since *H. griseus* eats primarily bamboo leaf bases and vine pith whereas *H. simus* eats the tougher, fibrous culm pith of the giant bamboo (Wright *et al.*, 1987; Glander *et al.*, 1989; Wright & Randriamanantena, 1989); and (3) to be higher among

Table 1 *Propithecus diadema (P.d.)* **and** *P. verreauxi (P.v.)* **regression analyses (sexes pooled)**

vs. Basicranial length[1,2]	Y-intercept P.d./P.v.	Slope P.d./P.v.	95% C.I. P.d./P.v.	r P.d./P.v.
Outer biorbital breadth (NS)	0.084	0.971	±0.064	0.965
	−0.900	1.130	±0.051	0.971
Interorbital breadth (**)	−1.973	1.130	±0.080	0.960
	−2.886	1.287	±0.102	0.915
Upper palate breadth (*)	2.843	0.552	±0.059	0.882
	1.898	0.592	±0.061	0.867
Palate length (NS)	0.785	0.784	±0.079	0.924
	−0.199	0.949	±0.043	0.970
Lower skull length (NS)	−0.146	1.050	±0.048	0.983
	−0.121	1.048	±0.029	0.989
Anterior face length (NS)	−0.873	1.028	±0.113	0.911
	−2.019	1.221	±0.103	0.905
Bizygomatic breadth (NS)	−0.609	1.095	±0.070	0.967
	−1.913	1.304	±0.041	0.985
Bigonial breadth (NS)	0.475	0.823	±0.184	0.855
	2.027	0.557	±0.110	0.726
Bicondylar breadth (NS)	−0.197	0.994	±0.070	0.926
	0.201	0.931	±0.057	0.956
Bicoronoid breadth (NS)	−1.507	1.213	±0.105	0.948
	−1.112	1.146	±0.057	0.970
Ramus height (**)	−7.019	2.011	±0.112	0.976
	−7.438	2.110	±0.090	0.973
M2 Bite point length (NS)	−2.342	1.304	±0.094	0.959
	−4.550	1.658	±0.131	0.916
Masseter lever arm length (NS)	−2.088	1.252	±0.074	0.972
	−3.362	1.464	±0.080	0.957
Temporalis lever arm length (NS)	−1.992	1.134	±0.128	0.912
	−2.830	1.268	±0.136	0.959
Symphysis height (NS)	−0.959	1.005	±0.090	0.941
	−1.191	1.053	±0.061	0.952
Symphysis width (NS)	0.248	0.617	±0.210	0.862
	−0.540	0.745	±0.166	0.913
M2 Corpus height (NS)	−4.300	1.424	±0.126	0.943
	−4.488	1.475	±0.068	0.969
M2 Corpus width (NS)	2.874	0.296	±0.147	0.608
	2.403	0.349	±0.081	0.838

[1] For each variable, *P. diadema* regression analyses are indicated on the first line.
[2] NS = ANCOVA between species is not significant, $P > 0.05$. * = *P. diadema* line is transposed above that for *P. verreauxi*, ANCOVA, $P \leqslant 0.05$; ** = *P. verreauxi* line is transposed above that for *P. diadema*; ANCOVA, $P \leqslant 0.05$.

V. variegata and *P. insignis*, since studies of dental (Szalay & Delson, 1979) and mandibular (Ravosa, 1991c) morphology suggest that *Pachylemur* probably ate tougher or more fibrous foods, rather than the primarily frugivorous diet of *Varecia* (Richard, 1985).

Results

Sifaka comparisons

In *P. diadema* and *P. verreauxi*, all 18 bivariate regressions versus basicranial length are highly significant (Table 1). The regression coefficients for both species typically indicate slight negative allometry to slight positive allometry for most craniomandibular measures, though

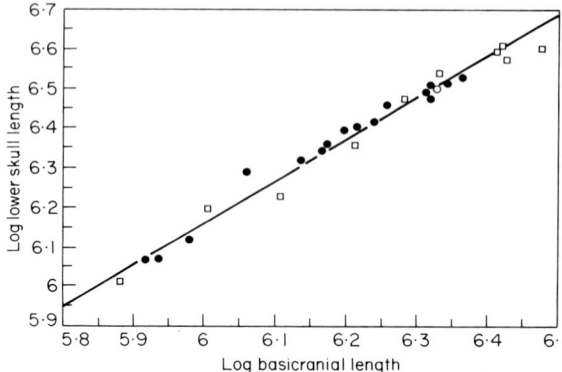

Figure 4. A plot of lower skull length versus basicranial length for *Propithecus*. Note that all three sifaka species lie along the same regression line, with differences between adults due to the differential extension of shared growth allometries; this scaling pattern typifies all but three sifaka comparisons. $\log_e$ of 1×10^{-1} mm. (●) *P. verreauxi*; (□) *P. diadema*; (○) *P. tattersalli*.

there are some exceptions (e.g., upper palate breadth and mandibular corpus width scale with strong negative allometry whereas ramus height scales with strong positive allometry). In 15 of 18 cases, between-species comparisons (ANCOVAs) of cranial growth trajectories are not significantly different (Table 1; e.g., Figure 4); only three measures show allometric transpositions between sifakas—upper palate breadth, interorbital breadth and ramus height. Generally, however, the results indicate that most morphological differences between adults are due to the differential extension of common patterns of relative growth; that is, cranial form among diademed and western sifakas is ontogenetically scaled. Since the single available specimen of *P. tattersalli*, an adult male, lies within the data scatter for adult *P. verreauxi* and on the regression line for both diademed and western sifakas, ontogenetic scaling of cranial proportions can be inferred for the golden-crowned sifaka as well (Figure 4).

ANOVAs between cranial measures for adult *P. diadema* and adult *P. verreauxi* are significant in 20 of 22 comparisons (Table 2); diademed sifakas are consistently larger in size. This pattern is demonstrated to a lesser extent at dental age no. 5 (subadult), where nine of 19 cases are significantly different. However, at dental age no. 4, only one ANOVA is significant. Likewise, at dental age no. 1 (infant), no significant differences are noted between cranial measures for each species; in fact, data on *P. verreauxi* at the Duke Primate Center and *P. diadema* in the wild (Ravosa *et al.*, in press) indicate that neonates of both species weigh about 150 g. While similar comparisons were not possible at dental ages nos 2 and 3, these results suggest that *P. diadema* and *P. verreauxi* have similar neonatal sizes and, therefore, *P. diadema* attains larger adult body and skull size mainly by growing at a higher rate. If the smaller overall adult size of the western sifaka is considered primitive for *Propithecus*, the growth pattern for *P. diadema* follows the prediction for rate hypermorphosis in that significant differences in the mean size of cranial measures become expressed with greater frequency in the later stages of ontogeny. Two factors argue against the presence of time hypermorphosis in the development of skull and body size differences. First and most convincing, data on wild populations of *P. diadema* (Ravosa *et al.*, in press) and *P. verreauxi* (Richard *et al.*, 1991) indicate that both species attain sexual maturity at about the age of 5 years. Second, not all differences between adult cranial measures are significant. Moreover,

Table 2   ***Propithecus* between-species ANOVAs at each dental age**[1]

Variable	P.d. vs. P.v.
Bizygomatic breadth	1, **4, 5, 6**
Outer biorbital breadth	1, 4, **5, 6**
Upper palate breadth	1, 4, **5, 6**
Lower skull length	1, 4, **5, 6**
M2 Bite point length	1, 4, **5, 6**
Symphysis height	1, 4, **5, 6**
Symphysis width	1, 4, **5, 6**
M2 Corpus width	1, 4, **5, 6**
Basicranial length	1, 4, **5, 6**
Interorbital breadth	1, 4, 5, **6**
Palate length	1, 4, 5, **6**
Anterior face length	1, 4, 5, **6**
Bigonial breadth	1, 4, 5, **6**
Bicondylar breadth	1, 4, 5, **6**
Bicoronoid breadth	1, 4, 5, **6**
Masseter lever arm length	1, 4, 5, **6**
Temporalis lever arm length	1, 4, 5, **6**
Body weight	**6**
Postcanine toothrow length	**6**
Neurocranial volume	**6**
Ramus height	1, 4, 5, 6
M2 Corpus height	1, 4, 5, 6

[1]Dental ages nos 1–6: cases where *P. diadema* measures are significantly larger than *P. verreauxi* are noted in **bold** ($P \leqslant 0.05$), otherwise ANOVAs are not significant. Not enough *P. diadema* data are available for ANOVAs at dental ages nos 2 and 3. Variables are grouped in ascending order by the dental age at which species differences (**bold**) develop.

absolute differences in adult sizes (dental age no. 6) of each species are not especially marked as compared to species differences at dental age no. 5 (subadult).

Ranked ratios of 22 cranial dimensions indicate, as predicted, that neurocranial volume has one of the lowest ratios (20th of 22); in contrast, body weight has the highest ratio (1st of 22) (Table 3). Postcanine toothrow length has a fairly low value (16th of 22), much as predicted based on the lack of substantial dietary differences between western and diademed sifakas. The ranked values for measures of the mandibular symphysis (height: 18th; width: 15th) and corpus (height: 19th; width: 7th) are similarly low or intermediate (for corpus width).

Bamboo lemur comparisons

In *H. griseus*, all 18 bivariate regressions versus basicranial length are highly significant (Table 4). The regression coefficients indicate a range of slight negative allometry to slight positive allometry for most craniomandibular measures in the lesser bamboo lemur, though there are several exceptions (e.g., mandibular corpus width scales with strong negative allometry and ramus height scales with strong positive allometry). In 13 of 18 cases, the *H. griseus* scaling trajectory intersects the data scatter for *H. simus* (Table 4; e.g., Figure 5). Therefore, it can be inferred that between-species comparisons of allometric growth trajectories are mostly not significant, except with regards to dimensions of the symphysis and corpus.

Table 3 *Propithecus* **ranked species ratios for adults**[1]

Variable	P.d./P.v.
Body weight	1·663
Bigonial breadth	1·269
Bizygomatic breadth	1·151
Bicoronoid breadth	1·144
Upper palate breadth	1·129
M2 Bite point length	1·125
M2 Corpus width	1·121
Temporalis lever arm length	1·117
Bicondylar breadth	1·114
Basicranial length	1·110
Lower skull length	1·108
Masseter lever arm length	1·097
Outer biorbital breadth	1·092
Interorbital breadth	1·061
Symphysis width	1·060
Postcanine toothrow length	1·054
Anterior face length	1·053
Symphysis height	1·040
M2 Corpus height	1·033
Neurocranial volume	1·032
Palate length	1·031
Ramus height	1·012

[1] Ranked ratios of adult *P. diadema* mean/adult *P. verreauxi* mean.

Thus, primarily in the facial skull, but not in the mandible, the greater bamboo lemur is a scaled-up or ontogenetically-scaled version of the lesser bamboo lemur. Unfortunately at present, it is impossible to detail the heterochronic process(es) by which this occurs, since neither life-history data nor adequate numbers of juvenile cranial specimens are available for *H. simus*.

Of those five comparisons where ontogenetic scaling cannot be inferred, four relate to dimensions of the mandibular corpus and symphysis. Moreover, in all five cases, the *H. simus* data scatter lies entirely above the *H. griseus* allometric trajectory (e.g., Figure 6). Previous interspecific analyses of the scaling of mandibular corpus and symphysis dimensions in prosimian primates similarly show that greater bamboo lemurs have more robust jaws than lesser bamboo lemurs (Ravosa, 1991*c*). Thus, differences in the pattern of relative growth of mandibular proportions correspond to the interspecific results, and suggest differences in the loading regime of the lower jaws of bamboo lemurs due to variation in the mechanical properties of their respective diets.

ANOVAs between cranial dimensions for adult *H. griseus* and adult *H. simus* are significant in all 22 comparisons (Table 5); greater bamboo lemurs are always larger in size. Ranked ratios of 22 cranial measures indicate, as predicted, that neurocranial volume has a very low ratio (21st of 22), and also as with sifakas, body weight has the highest ratio (1st of 22) (Table 5). In fact, both of these patterns are demonstrated in sexually dimorphic anthropoids as well (Ravosa, 1991*a*). On the other hand, postcanine toothrow length has a moderately high value (12th of 22), much as predicted based on known dietary differences between greater and lesser bamboo lemurs. As might be expected based on the higher postcanine toothrow ratio (reflecting a dietary difference), the ontogenetic comparisons and a previous

Table 4 *Hapalemur griseus* **regression analyses (sexes pooled)**

vs. Basicranial length[1,2]	Y-intercept	Slope	95% C.I.	r
Outer biorbital breadth (NS)	0·340	0·906	±0·076	0·955
Interorbital breadth (NS)	−0·808	0·854	±0·179	0·787
Upper palate breadth (NS)	1·181	0·786	±0·092	0·894
Palate length (NS)	−1·813	1·215	±0·054	0·986
Lower skull length (NS)	−1·368	1·255	±0·055	0·987
Anterior face length (NS)	−2·189	1·255	±0·099	0·959
Bizygomatic breadth (NS)	−0·743	1·120	±0·056	0·983
Bigonial breadth (NS)	0·419	0·843	±0·205	0·765
Bicondylar breadth (*)	0·304	0·915	±0·069	0·926
Bicoronoid breadth (NS)	0·513	0·876	±0·077	0·949
Ramus height (NS)	−6·998	2·035	±0·113	0·979
M2 Bite point length (NS)	−5·157	1·775	±0·123	0·968
Masseter lever arm length (NS)	−3·533	1·505	±0·078	0·982
Temporalis lever arm length (NS)	−4·326	1·537	±0·210	0·891
Symphysis height (*)	−1·073	0·933	±0·175	0·818
Symphysis width (*)	−1·279	0·880	±0·113	0·893
M2 Corpus height (*)	−2·769	1·174	±0·161	0·898
M2 Corpus width (*)	0·029	0·639	±0·157	0·735

[1] Lower jaw dimensions for the juvenile specimen of *H. simus* are not available, therefore, scaling inferences in such cases are based on adult data and one subadult specimen.

[2] NS = not significant; the *H. griseus* trajectory intersects the *H. simus* scatter; * = the *H. simus* scatter is transposed above the *H. griseus* trajectory.

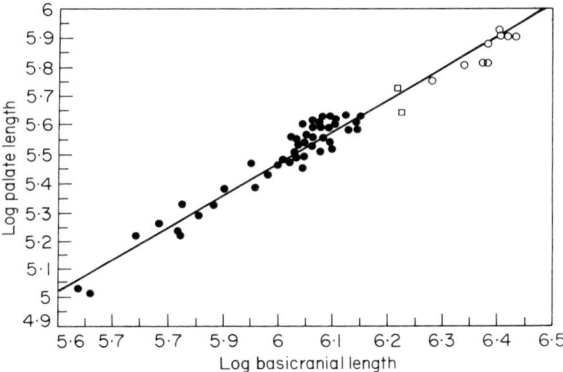

Figure 5. A plot of palate length versus basicranial length for *Hapalemur*. As in most comparisons, the ontogenetic scaling trajectory for *H. griseus* (●) intersects the primarily adult scatter for *H. simus* (□). Note also that the only two adult specimens of *H. g. alaotrensis* (○) lie along the common *Hapalemur* trajectory between the scatter for *H. simus* and somewhat apart from the remaining *H. griseus* scatter. Apparently, this is a case of subspecific differentiation via ontogenetic scaling. $\log_e$ of 1×10^{-1} mm.

interspecific analysis of mandibular size and scaling (Ravosa, 1991*c*), ratios for dimensions of the mandibular corpus (height: 4th; width: 8th) and mandibular symphysis (height: 7th; width: 5th) are quite high. In sum, the correspondingly higher ranks for postcanine tooth size

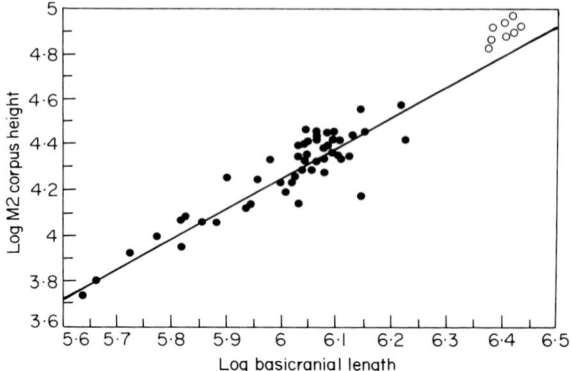

Figure 6. A plot of mandibular corpus height versus basicranial length for *Hapalemur*. Note that the ontogenetic scaling trajectory for *H. griseus* (●) does not intersect the adult scatter for *H. simus* (○). Instead, the adult scatter for greater bamboo lemurs are transposed above the regression line for lesser bamboo lemurs. This probably reflects differences in the loading regime of the mandible due to differences in the mechanical properties of the diet. $\log_e$ of 1×10^{-1} mm.

Table 5   ***Hapalemur* ranked species ratios for adults**[1,2]

Variable	H.s./H.g.
Body weight	2·688
Interorbital breadth	2·012
Bigonial breadth	1·732
M2 Corpus height	1·704
Symphysis width	1·619
Ramus height	1·606
Symphysis height	1·573
M2 Corpus width	1·538
Bicondylar breadth	1·538
Upper palate breadth	1·421
Masseter lever arm length	1·407
Postcanine toothrow length	1·405
Temporalis lever arm length	1·387
Bizygomatic breadth	1·383
Basicranial length	1·376
Palate length	1·357
Outer biorbital breadth	1·345
Lower skull length	1·328
Bicoronoid breadth	1·317
M2 Bite point length	1·312
Neurocranial volume	1·272
Anterior face length	1·230

[1] Ranked ratios of adult *H. simus* mean/adult *H. griseus* mean.
[2] Note that all ANOVAs between adult *H. simus* and adult *H. griseus* cranial measures are significant, $P \leqslant 0.001$.

and other features of the masticatory apparatus appear related to greater dietary variation between the bamboo lemurs, much as implied by the correlation of lower postcanine tooth size and lower mandibular size ranks with negligible dietary variation between sifakas.

Table 6 *Varecia variegata* **regression analyses (sexes pooled)**

vs. Basicranial length[1]	Y-intercept	Slope	95% C.I.	r
Outer biorbital breadth (NS)	0·280	0·927	±0·082	0·907
Interorbital breadth (NS)	−0·292	0·827	±0·192	0·744
Upper palate breadth (*)	1·417	0·677	±0·089	0·893
Palate length (**)	−1·553	1·197	±0·056	0·993
Lower skull length (NS)	−1·259	1·253	±0·034	0·997
Anterior face length (**)	−2·895	1·386	±0·103	0·969
Bizygomatic breadth (*)	−0·110	1·001	±0·079	0·930
Bigonial breadth (NA)	1·339	0·689	±0·089	0·894
Bicondylar breadth (NS)	0·024	0·958	±0·090	0·970
Bicoronoid breadth (NS)	0·344	0·909	±0·050	0·989
Ramus height (*)	−4·969	1·594	±0·218	0·940
M2 Bite point length (NS)	−4·316	1·608	±0·181	0·936
Masseter lever arm length (*)	−1·998	1·223	±0·051	0·994
Temporalis lever arm length (NA)	−3·483	1·341	±0·205	0·861
Symphysis height (*)	−1·869	0·902	±0·133	0·893
Symphysis width (*)	1·180	0·442	±0·139	0·648
M2 Corpus height (*)	−1·190	0·975	±0·101	0·920
M2 Corpus width (*)	−2·084	0·283	±0·117	0·543

[1]NS = not significant, the *Varecia* trajectory intersects the adult *Pachylemur* scatter; * = the adult *Pachylemur* scatter is transposed above the *Varecia* trajectory; ** = the adult *Pachylemur* scatter is transposed below the *Varecia* trajectory; NA = not applicable, data are not available for *Pachylemur*.

Ruffed lemur and Pachylemur *comparisons*

In *V. variegata*, all 18 bivariate regressions versus basicranial length are highly significant (Table 6). The regression coefficients typically indicate moderate negative allometry to moderate positive allometry for craniomandibular measures in the ruffed lemur, though there are some notable bivariate comparisons (e.g., mandibular corpus width scales with strong negative allometry whereas M2 bite point length and ramus height scale with strong positive allometry). In only six of 16 possible comparisons, the *V. variegata* allometric trajectory intersects the data scatter for adult *P. insignis* (Table 6). In the other 10 comparisons, the adult *Pachylemur* data are transposed above the *Varecia* regression line eight times and transposed below the regression line twice. Therefore, the majority of between-species comparisons suggest that subfossil *Pachylemur* is *not* simply a scaled-up version of its sister taxon the ruffed lemur (e.g., Figure 7).

ANOVAs between cranial dimensions for adult *V. variegata* and adult *P. insignis* are significant in all 19 comparisons (Table 7); subfossil *Pachylemur* is consistently larger in size. Ranked ratios of 19 cranial measures indicate, as predicted, that neurocranial volume has a very low ratio value (16th of 19) (Table 7). In contrast, postcanine toothrow length has a more moderate value (10th of 19), again much as predicted based on inferred dietary differences between these species. As might be expected based on the somewhat higher postcanine toothrow length ratio (reflecting a dietary difference), the ontogenetic comparisons and previous interspecific analyses of mandibular robusticity (Ravosa, 1991c); ratios for dimensions of the mandibular corpus (height: 1st; width: 5th) and mandibular symphysis (height: 2nd; width: 3rd) are similarly very high.

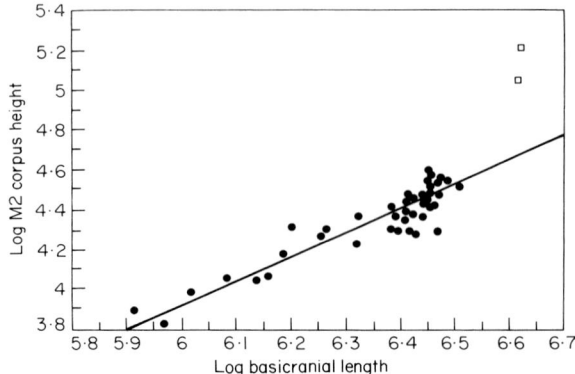

Figure 7. A plot of mandibular corpus height versus basicranial length for *Varecia* and *Pachylemur*. Note that the ontogenetic scaling trajectory for *V. variegata* (●) does not intersect the adult scatter for *P. insignis* (□). The scatter for *Pachylemur* fall above the regression line for ruffed lemurs, which is the typical pattern for most comparisons between these genera. This case probably reflects differences in the loading regime of the mandible. Log_e of 1×10^{-1} mm.

Table 7   ***Varecia* and *Pachylemur* ranked species ratios for adults**[1-3]

Variable	P.i./V.v.
M2 Corpus height	1·931
Symphysis height	1·567
Symphysis width	1·494
Ramus height	1·477
M2 Corpus width	1·397
Masseter lever arm length	1·323
Bizygomatic breadth	1·288
Interorbital breadth	1·258
M2 Bite point length	1·223
Postcanine toothrow length	1·184
Upper palate breadth	1·171
Bicondylar breadth	1·165
Lower skull length	1·135
Basicranial length	1·132
Anterior face length	1·123
Neurocranial volume	1·109
Outer biorbital breadth	1·073
Bicoronoid breadth	1·065
Palate length	1·063

[1] Ranked ratios of adult *Pachylemur* mean/adult *Varecia* mean.
[2] Note that all ANOVAs between adult *Pachylemur* and adult *Varecia* cranial measures are significant, $P \leqslant 0.001$.
[3] Note that *Pachylemur* bigonial breadth, temporalis lever arm length and body weight estimates are not available for ANOVAs and ranked ratios.

Discussion

The results of comparisons among the three groups of Malagasy lemurs highlight several important benefits of allometric and heterochronic analyses to studies of primate evolutionary biology. As noted previously, comparisons of cranial growth trajectories for diademed and western sifakas demonstrate that these species are ontogenetically scaled (Figure 4). Recently, Albrecht et al. (1990) argued that variation in adult body size among sifakas and other lemurs is correlated with the occupation of differing ecogeographic regions in Madagascar. This study serves to outline the ontogenetic mechanisms by which these differences (and those between bamboo lemurs) likely evolved (see also Ravosa et al., in press).

Heterochronic analysis suggests that adult diedemed sifakas attain larger overall size primarily via rate hypermorphosis; that is, *P. diadema* grows at a faster rate, but probably not for a longer duration, than *P. verreauxi*. Unfortunately, without adequate data on more subspecies of sifakas in the wild, it is presently difficult to address all of the behavioral and ecological factors underlying rate/timing differences in these taxa, and phyletic size increases in other Malagasy lemurs. Such data are essential to establishing a link between variation in socioecological factors and variation in adult morphology vis-à-vis the modification of developmental processes.

Another important aspect of the sifaka analyses relates to the ontogenetic bases of sexual dimorphism. A previous comparative study of sexual dimorphism in prosimian primates indicates that adult female diademed sifakas are significantly larger in body mass than adult males (Kappeler, 1990). Unpublished analyses of the cranial data used in this study likewise confirm that diademed sifakas are sexually dimorphic, i.e., females are significantly larger than males, whereas western sifakas are monomorphic. While not examined directly, the fact that the majority of the sifaka comparisons are ontogenetically scaled suggests that the growth trajectories for males and females are also coincidental. As compared to cases of sex dimorphism in anthropoid primates, where males are larger than females, it would be especially interesting to investigate further the ontogenetic patterns of size differentiation between female and male diademed sifakas, and subspecies for that matter.

Comparisons of cranial growth trajectories for the lesser bamboo lemur with data for greater bamboo lemurs suggest that, in all but features of the mandible, these species are ontogenetically scaled. Among bamboo lemurs, apparent differences in the relative growth of mandibular corpus and symphysis morphology strongly suggest the existence of dissimilar loading regimes in the lower jaws (Figure 6). Like the ontogenetic results, interspecific allometric comparisons among prosimians show that greater bamboo lemurs have relatively more robust mandibles than lesser bamboo lemurs (Ravosa, 1991c). This is probably correlated with differences in diet, since the greater bamboo lemur eats the much more obdurate, fibrous culm pith of the giant bamboo (Wright et al., 1987; Glander et al., 1989; Wright & Randriamanantena, 1989). Whether these morphological differences become increasingly manifested during the later stages of ontogeny and/or are already present in the neonates of each species remains indeterminable. Taken a step further, perhaps the extent of ecogeographic size variation among bamboo lemurs (almost a three-fold difference in body mass) in turn requires nonsize-related shape changes in specific functional components of the skull, such as those of the masticatory apparatus, which cannot be accommodated by the simple extrapolation of common patterns of relative growth (sensu Gould, 1975; Shea, 1983c).

However, as stated earlier, it is virtually impossible to detail the heterochronic bases of size differentiation, since appropriate life-history data (e.g., neonate size, age to maturity) are unavailable for *H. simus*. Thus, it is difficult to ascertain the extent to which there is a rate versus a timing component underlying the development of body size differences in bamboo lemurs.

While not a major focus of this study, it is interesting that the two adult specimens of *H. g. alaotrensis*, the largest subspecies of the lesser bamboo lemur, lie along the common *Hapalemur* trajectories, apart from the remaining *H. griseus* scatter towards the adult scatter for *H. simus* (Figure 5). Apparently, this is a case of subspecific morphological variation via ontogenetic scaling. Similar subspecific patterns of size differentiation are demonstrated in the slow loris, *Nycticebus coucang*, and the potto, *Perodicticus potto* (Ravosa, in prep.).

Comparisons between *Varecia* and its extinct sister taxon, *Pachylemur*, suggest that these taxa typically do not share common patterns of cranial growth. Four of the 10 comparisons where ontogenetic scaling cannot be inferred relate to dimensions of the mandibular corpus and symphysis. In all four cases the data for adult *Pachylemur* fall above the *Varecia* regression lines (Figure 7). As noted for the bamboo lemurs, apparent differences in the pattern of relative growth for mandibular proportions suggests species differences in the loading regime of the lower jaws. These differences in the loading regime of the mandible are presumably associated with substantial variation in the structural consistency of the diets of these two species (Szalay & Delson, 1979; Ravosa, 1991c). Two other cases where the *Pachylemur* fall above the *Varecia* regression line relate to bizygomatic breadth and masseter lever arm length. Relatively broad bizygomatic dimensions may indicate that *Pachylemur* had larger temporalis muscles as an adaptation to a more obdurate diet. Similarly, perhaps a relatively longer masseter lever arm indicates that *Pachylemur* required less masseter muscle force to generate a given bite force.

The only two comparisons where the *Pachylemur* scatter fall below the *Varecia* regression lines—palate length and anterior face length—may highlight additional morphological adaptations of the masticatory apparatus to an obdurate or fibrous diet. That is, a relatively shorter facial skull in *Pachylemur* would enable bite forces to be generated more efficiently than in a longer-faced species like *Varecia*. In fact, this mirrors the results of similar ontogenetic comparisons between a shorter-faced folivorous colobine monkey, *Nasalis larvatus*, and a longer-faced frugivorous cercopithecine monkey, *Macaca fascicularis* (Ravosa, 1991a).

The ruffed lemur and *Pachylemur* results also have important implications regarding the utility of ontogenetic analyses for phylogenetic inference and classification. Since differences in overall size have figured into arguments for a generic designation of *Lemur* and *Varecia*, Albrecht *et al.* (1990) argued recently that similar criteria should be applied to the generic separation of *Pachylemur* vis-à-vis *Varecia*. An alternative viewpoint is that *Pachylemur* should be only a subgeneric designation, i.e., *Varecia* (*Pachylemur*) *insignis*. In this paper, the generic status accorded to *Pachylemur* seems warranted, especially given the lack of ontogenetic scaling of cranial proportions inferred for *Pachylemur* and *Varecia*. Conversely, the apparent similarity of most cranial growth allometries for both bamboo lemur species lends support to the claim of Vuillaume-Randriamanantena *et al.* (1985) that *H. simus* and *H. griseus* should be separated only at the species level. Though it is clear that ontogenetic analyses enable a more comprehensive and informed evaluation of the homology of morphological features and purported phylogenetic affinities among taxa (Fink, 1982; Bonde, 1984; Gomez, 1992b; Yoder, 1992), unfortunately to date such approaches have as yet received only scant attention by primate systematists.

Craniodental scaling and phyletic size change
The dental comparisons performed here, as well as those from other studies, are of great relevance to predictions regarding expected patterns of tooth size variability among phyletic size series. Essentially, it appears that the degree to which dental size may become dissociated from shared ancestral patterns of relative growth may be a function of whether selection has acted to decrease overall size (phyletic dwarfing) or increase overall size (phyletic gigantism),

and/or whether selection has acted to influence inter- or intra-specific variation in overall size. Depending on which of these factors is operative, prior studies show varied results.

Work regarding the influence of phyletic size decreases on relative dental size among dwarfed taxa is somewhat equivocal. As predicted by Gould (1975) for mammals in general, dwarf marsupials (Marshall & Corruccini, 1978), dwarf elephants (Maglio, 1973) and human pygmies (Shea & Gomez, 1988), all have relatively larger postcanine teeth than their larger sister taxa or conspecifics. However, to the contrary, dwarf rhinoceroses (Prothero & Sereno, 1982), pygmy chimpanzees (Shea, 1984), callitrichid monkeys (Plavcan & Gomez, 1992) and pygmy slow lorises (Ravosa, in prep.) have proportionally-sized or relatively *smaller* teeth than their larger sister taxa. Thus, taken as a whole, studies to date indicate that the adults of the dwarfed forms, at best, only sometimes have relatively larger postcanine teeth than their larger-bodied sister taxa.

Investigations into the effect of phyletic size increases on relative dental size among the two species of bamboo lemurs, the ruffed lemur and *Pachylemur*, as well as comparisons between the common chimpanzee and gorilla (Shea, 1983c), indicate that there appears to have been increases in postcanine tooth size related to a more obdurate or tougher diet in the larger species. This is not the case for the sifakas. The fact that all sifakas are folivorous and have roughly similar percentages of fibrous leaves in their diets [compare 46% for *P. verreauxi* (Richard, 1985) to 53% for *P. diadema* (Wright, 1987; Wright *et al.*, 1987)], probably explains why diademed sifakas have absolutely, but not relatively, larger teeth than western sifakas. Perhaps as a rule, taxa which undergo substantial phyletic size increases tend to transcend adaptive thresholds, and thus are more likely to experience strong selective pressures for increased relative tooth size due to a dietary shift to more obdurate foods (Gould, 1975; Shea, 1983c), and larger mandibular corpus and symphysis dimensions related to increased masticatory stresses (e.g., bamboo lemurs). Therefore, from a functional perspective, it may be alright to be a phyletic dwarf with relatively larger teeth since this may not be necessarily maladaptive, whereas phyletic giants presumably *require* relatively larger postcanine teeth to perform adequately.

A remaining issue to consider is whether selection for overall size changes which target intraspecific variation (i.e., sexual dimorphism; clinal or ecogeographic size variation among subspecies) will result in the corresponding dissociation of dental scaling patterns among ontogenetically-scaled taxa. To date, ontogenetic analyses among sexually dimorphic anthropoids indicate that the larger-bodied adult males are scaled-up versions of, and have relatively smaller postcanine teeth than, smaller-bodied adult females (Cochard, 1985, 1987; Ravosa, 1991a; Ravosa & Ross, in press; see also Kay, 1978). In general, females of sexually dimorphic species typically, but not always, retain the ancestral body size (Shea, 1986; Leigh & Cheverud, 1991; Ravosa, 1991a). As compared to individuals of average stature, ontogenetically-scaled human "pygmies" have relatively larger postcanine teeth (Shea & Gomez, 1988). When ratios of 22 adult cranial measures are ranked for two subspecies of the lesser bamboo lemur [largest subspecies, *H. g. alaotrensis* (1520 g)/smallest subspecies, *H. g. griseus* (880 g)],‡ postcanine toothrow length has a very low ratio (18th of 22). Assuming that the size of *H. g. alaotrensis* represents the derived condition for the lesser bamboo lemur, then this subspecies likely represents a "giant" form with relatively smaller postcanine dentition as predicted by the ontogenetic scaling model. Similar comparisons among normal-sized and

‡The adult body weight of 1520 g for *Hapalemur griseus alaotrensis*, the lake Alaotro subspecies of lesser bamboo lemur, is based on the mean body weight for an adult female at the Duke Primate Center taken over three successive years. Note that this is similar to the body size of *H. aureus*, the golden bamboo lemur.

scaled-up, larger "transgenic" mice of the same species indicate that the transgenic forms have relatively smaller teeth (Shea *et al.*, 1990). In sum, while there are a lesser number of individual examples, the within-species analyses show consistently strong support for the prediction that within an ontogenetically-scaled taxon, as compared to adults of the ancestral or average body size, adults of larger body size will have relatively smaller teeth and adults of smaller body size will have relatively larger teeth.

Why might the inter- and intra-specific analyses of dental scaling differ in the extent of support for the ontogenetic scaling argument of Gould (1975) and others? One explanation is that there might be a difference between selection for ecogeographic or clinal size variation among subspecies, or dimorphism among the sexes, versus selection for body size increases between species. In the former two cases, selection acts necessarily to increase the entire range of intraspecific variation. For example, in a species having become more dimorphic over time for whatever reasons, selection would increase the overall mean body size of the species primarily by increasing the maximum body size; this would not necessarily require a similar alteration in the minimum body size (of the smaller sex or smallest subspecies). Conversely, selection for body size increases between species via character displacement likely acts primarily to alter the overall mean differences—both the maxima and minima for each species—but not necessarily the overall range of variation within each species. Perhaps this results in a closer correspondence between patterns of somatic, cranial and dental variation. Obviously, selection for ecogeographic size variation among subspecies may eventually result in speciation *per se*, since both maximum and minimum body sizes may be affected. In fact, such continued or prolonged pressures may coincide with selection for increased postcanine tooth size. An alternative explanation for why the interspecific and intraspecific results differ may be that perhaps it is rare that body size differences between adults of one species ever differ in magnitude to the extent observed among closely-related species. Thus, the heterochronic processes affecting intra- versus inter-specific patterns of dental variability may be qualitatively different as suggested here, although this argument clearly requires further comparative study.

Summary and conclusions

In general, most differences in skull form among sifakas and, to a lesser extent, bamboo lemurs result from the differential extension of common patterns of relative growth. This study details the developmental or heterochronic process by which ecogeographic size differentiation has apparently occurred among sifakas, namely rate hypermorphosis. On the other hand, few differences in cranial morphology between *Varecia* and *Pachylemur* appear to result from ontogenetic scaling; this pattern supports claims for the generic separation of these taxa. The allometric analyses considered here also impact upon previous interspecific studies of cranial morphology among Malagasy primates specifically by demonstrating the utility of an ontogenetic criterion of subtraction regarding mandibular and dental form and function. For instance, allometric comparisons among bamboo lemurs indicate a close correspondence between ontogenetic and interspecific differences in lower jaw form and function.

While this investigation aims to synthesize available information on allometry and heterochrony in Malagasy lemurs and other mammals, considerably more data on primate behavioral ecology and life history variation are essential in order to characterize better the developmental bases of morphological and evolutionary patterns of phyletic size change. In addition, further comparative analyses would clearly benefit our understanding of the effects of phyletic size change on dental form. Put another way, what we still do not know is whether among sexes or subspecies that differ in body size to the extent observed in gorillas and baboons are there corresponding constraints on variation in the degree of niche differentiation, since the larger individuals tend to have relatively smaller teeth? To this end, it is reassuring that some of these issues are being currently addressed.

Acknowledgements

Dr B. Shea, A. Gomez and three anonymous reviewers provided helpful comments on an earlier version of this paper. For access to primate collections and fossil materials, my sincerest appreciation is extended to the following museum curators and staff: M. Rutzmoser (Harvard Museum of Comparative Zoology); Dr I. Tattersall, Dr E. Delson, W. Fuchs, Dr G. Musser, Dr S. Anderson (American Museum of Natural History); Dr B. Patterson, Dr J. Kerbis (Field Museum of Natural History); Dr R. Thorington, L. Gordon, L. Coley (National Museum of Natural History—Smithsonian); Dr P. Andrews, Dr C. Stringer, P. Jenkins, R. Kruszynski, M. Sheridan, M. Sheldrick (British Museum of Natural History); Dr J. Roche, Dr M. Tranier, Dr D. Goujet, F. Petter (Museum National d'Histoire Naturelle); Dr E. Simons, P. Chatrath (Duke University Primate Center); Dr C. Smeenk, Dr M. Hoogmoed, D. Reider (Rijksmuseum van Natuurlijke Historie); Dr R. Angermann (Museum für Naturkunde–Humboldt Universität); Dr B. Latimer, L. Jellema, L. Linden (Cleveland Museum of Natural History); F. Sibley, M. Turner (Yale Peabody Museum of Natural History); T. Daeschler (Academy of Natural Sciences of Philadelphia); Dr M. Coombs (Pratt Museum of Natural History); and Dr C. Grigson (Odontological Museum–Royal College of Surgeons). Dr B. Rakotosamimanana of the Service de Paléontologie, Université de Madagascar, and Dr E. Simons are thanked for permission to examine subfossil *Hapalemur simus* curated at the Duke University Primate Center. Dr W. Jungers kindly measured specimens of *Pachylemur* located at the Université de Madagascar. B. Fox is thanked for assistance with various phases of this project. Dr S. Stack is thanked for continued support and encouragement. I also gladly acknowledge my network of friends that luckily for me reside in cities with museums. Financial support for this research was provided in part by the NIH (DE-05595), the NSF (BNS-8813220), Northwestern University (0100-510-110Y), the American Museum of Natural History, the American Philosophical Society, and the Department of Biological Anthropology and Anatomy, Duke University Medical Center. Subfossil *Hapalemur simus* examined in this study were collected with the support of NSF grant BNS-8911315 to Drs E. Simons, W. Jungers and L. Godfrey. This is Duke University Primate Center publication no. 519.

References

Albrecht, G. H., Jenkins, P. D. & Godfrey, L. R. (1990). Ecogeographic size variation among the living and subfossil prosimians of Madagascar. *Am. J. Primat.* **22,** 1–50.
Bonde, N. (1984). Primitive features and ontogeny in phylogenetic reconstructions. *Vidensk. Meddr dansk naturh. Foren.* **145,** 219–236.
Cheverud, J. M. (1982). Relationships among ontogenetic, static, and evolutionary allometry. *Am. J. phys. Anthrop.* **59,** 139–149.
Cheverud, J. M. & Richtsmeier, J. T. (1986). Finite-element scaling applied to sexual dimorphism in rhesus macaque (*Macaca mulatta*) facial growth. *Syst. Zool.* **35,** 381–399.
Cochard, L. R. (1985). Ontogenetic allometry of the skull and dentition of the rhesus monkey (*Macaca mulatta*). In (W. L. Jungers, Ed.) *Size and Scaling in Primate Biology*, pp. 231–256. New York: Plenum Press.

Cochard, L. R. (1987). Postcanine tooth size in female primates. *Am. J. phys. Anthrop.* **74,** 47–54.
Creighton, G. K. (1980). Static allometry of mammalian teeth and the correlation of tooth size and body size in contemporary mammals. *J. Zool.* **191,** 435–443.
Fink, W. L. (1982). The conceptual relationship between ontogeny and phylogeny. *Paleobiol.* **8,** 254–264.
Glander, K. E., Wright, P. C., Seigler, D. S., Randrianasolo, V. & Randrianasolo, B. (1989). Consumption of cyanogenic bamboo by a newly discovered species of bamboo lemur. *Am. J. Primat.* **19,** 119–124.
Gomez, A. M. (1992a). Ontogenetic scaling of the limbs and trunk of the Asian Lorisidae: *Nycticebus coucang, Nycticebus pygmaeus,* and *Loris tardigradus. Am. J. phys. Anthrop.* (in press).
Gomez, A. M. (1992b). Primitive and derived patterns of relative growth among the species of Lorisidae. *J. hum. Evol.* **23,** in press.
Gould, S. J. (1975). On the scaling of tooth size in mammals. *Am. Zool.* **15,** 351–362.
Jungers, W. L. (1985). Body size and scaling of limb proportions in primates. In (W. L. Jungers, Ed.), *Size and Scaling in Primate Biology,* pp. 345–382. New York: Plenum Press.
Kappeler, P. M. (1990). The evolution of sexual size dimorphism in prosimian primates. *Am. J. Primat.* **21,** 201–214.
Kay, R. F. (1975). Functional adaptations of primate molar teeth. *Am. J. phys. Anthrop.* **43,** 195–216.
Kay, R. F. (1978). Molar structure and diet in extant Cercopithecidae. In (K. Joysey & P. Butler, Eds) *Development, Function and Evolution of Teeth,* pp. 309–339. New York: Academic Press.
Lande, R. (1979). Quantitative genetic analysis of multivariate evolution, applied to brain:body size allometry. *Evol.* **33,** 402–416.
Legendre, S. & Roth, C. (1988). Correlation of carnassial tooth size and body weight in recent carnivores (Mammalia). *Hist. Biol.* **1,** 85–98.
Leigh, S. R. (1992). Patterns of variation in the ontogeny of primate body size dimorphism. *J. hum. Evol.* **23,** 27–50.
Leigh, S. R. & Cheverud, J. M. (1991). Sexual dimorphism in the baboon facial skeleton. *Am. J. phys. Anthrop.* **84,** 193–208.
Leutenegger, W. & Masterson, T. J. (1989a). The ontogeny of sexual dimorphism in the cranium of Bornean orangutans (*Pongo pygmaeus pygmaeus*): I. univariate analyses. *Z. Morph. Anthrop.* **78,** 1–14.
Leutenegger, W. & Masterson, T. J. (1989b). The ontogeny of sexual dimorphism in the cranium of Bornean orangutans (*Pongo pygmaeus pygmaeus*): II. allometry and heterochrony. *Z. Morph. Anthrop.* **78,** 15–24.
Maglio, V. J. (1973). Origin and evolution of the Elephantidae. *Trans. Am. phil. Soc.* **63,** 1–149.
Marshall, L. G. & Corruccini, R. S. (1978). Variability, evolutionary rates, and allometry in dwarfing lineages. *Paleobiol.* **4,** 101–119.
Masterson, T. J. & Leutenegger, W. (1990). The ontogeny of sexual dimorphism in the cranium of Bornean orangutans (*Pongo pygmaeus pygmaeus*) as detected by principal components analysis. *Int. J. Primatol.* **11,** 517–539.
Meier, B., Albignac, R., Peyrieras, A., Rumpler, Y. & Wright, P. (1987). A new species of *Hapalemur* (Primates) from South East Madagascar. *Folia primat.* **48,** 211–215.
Pilbeam, D. R. & Gould, S. J. (1974). Size and scaling in human evolution. *Science* **186,** 892–901.
Plavcan, J. M. & Gomez, A. M. (1992). Dental scaling in the Callitrichidae. *Int. J. Primat.* **13,** in press.
Prothero, D. R. & Sereno, P. C. (1982). Allometry and paleoecology of medial Miocene rhinoceroses from the Texas gulf coastal plain. *Paleobiol.* **8,** 16–30.
Ravosa, M. J. (1989). An evaluation of biomechanical and non-mechanical models of primate circumorbital morphology. Ph.D. Dissertation, Department of Anthropology, Northwestern University.
Ravosa, M. J. (1991a). The ontogeny of cranial sexual dimorphism in two Old World monkeys: *Macaca fascicularis* (Cercopithecinae) and *Nasalis larvatus* (Colobinae). *Int. J. Primatol.* **12,** 403–426.
Ravosa, M. J. (1991b). Ontogenetic perspective on mechanical and nonmechanical models of primate circumorbital morphology. *Am. J. phys. Anthrop.* **85,** 95–112.
Ravosa, M. J. (1991c). Structural allometry of the prosimian mandibular corpus and symphysis. *J. hum. Evol.* **20,** 3–20.
Ravosa, M. J. & Ross, C. F. (in press). Craniodental allometry and heterochrony in two howler monkeys: *Alovatta seniculus* and *A. Palliata. Am. J. Primat.*
Ravosa, M. J., Meyers, D. M. & Glander, K. E. (in press). Relative growth of the limbs and trunk in sifakas: heterochronic, ecological and functional considerations. *Am. J. phys. Anthrop.*
Richard, A. F. (1985). *Primates in Nature.* New York: Freeman.
Richard, A. F., Rakotomanga, P. & Schwartz, M. (1991). Demography of *Propithecus verreauxi* at Beza Mahafaly, Madagascar: sex ratio, survival, and fertility, 1984–1988. *Am. J. phys. Anthrop.* **84,** 307–322.
Riska, B. & Atchley, W. R. (1985). Genetics of growth predicts patterns of brain-size evolution. *Science* **229,** 668–671.
Shea, B. T. (1983a). Allometry and heterochrony in the African apes. *Am. J. phys. Anthrop.* **62,** 275–289.
Shea, B. T. (1983b). Phyletic size change and brain/body allometry: a consideration based on the African pongids and other primates. *Int. J. Primatol.* **4,** 33–62.
Shea, B. T. (1983c). Size and diet in the evolution of African ape craniodental form. *Folia primat.* **40,** 32–68.
Shea, B. T. (1984). An allometric perspective on the morphological and evolutionary relationships between pygmy (*Pan paniscus*) and common (*Pan troglodytes*) chimpanzees. In (R. L. Susman, Ed.) *The Pygmy Chimpanzee: Evolutionary Biology and Behavior,* pp. 89–130. New York: Plenum Press.

Shea, B. T. (1985). The ontogeny of sexual dimorphism in the African apes. *Am. J. Primat.* **8,** 183–188.
Shea, B. T. (1986). Ontogenetic approaches to sexual dimorphism in anthropoids. *Hum. Evol.* **1,** 97–110.
Shea, B. T. (1988). Heterochrony in primates. In (M. L. McKinney, Ed.) *Heterochrony in Evolution*, pp. 237–266. New York: Plenum Press.
Shea, B. T. & Gomez, A. M. (1988). Tooth scaling and evolutionary dwarfism: an investigation of allometry in human pygmies. *Am. J. phys. Anthrop.* **77,** 117–132.
Shea, B. T., Hammer, R. E. & Brinster, R. L. (1987). Growth allometry of the organs in giant transgenic mice. *Endocrinol.* **121,** 1924–1930.
Shea, B. T., Hammer, R. E., Brinster, R. L. & Ravosa, M. J. (1990). Relative growth of the skull and postcranium in giant transgenic mice. *Genet. Res.* **56,** 21–34.
Simons, E. L. (1988). A new species of *Propithecus* (Primates) from Northeast Madagascar. *Folia primat.* **50,** 143–151.
Szalay, F. S. & Delson, E. (1979). *Evolutionary History of the Primates*. New York: Academic Press.
Tattersall, I. (1982). *The Primates of Madagascar*. New York: Columbia University Press.
Vuillaume-Randriamanantena, M., Godfrey, L. R. & Sutherland, M. (1985). Revision of *Hapalemur* (*Prohapalemur*) *gallieni* (Standing, 1905). *Folia primat.* **45,** 89–116.
Wright, P. C. (1987). Diet and ranging patterns in *Propithecus diadema edwardsi* in Madagascar. *Am. J. phys. Anthrop.* **72,** 271.
Wright, P. C., Daniels, P. S., Meyers, D. M., Overdorff, D. J. & Rabesoa, J. (1987). A census and study of *Hapalemur* and *Propithecus* in Southeastern Madagascar. *Primate Conserv.* **8,** 84–88.
Wright, P. C. & Randriamanantena, M. (1989). Comparative ecology of three sympatric bamboo lemurs in Madagascar. *Am. J. phys. Anthrop.* **78,** 327.
Yoder, A. D. (1992). The applications and limitations of ontogenetic comparisons for phylogeny reconstruction: the case of the strepsirhine internal carotid artery. *J. hum. Evol.* **23,** 183–195.

Appendix 1

Craniodental measurements[1,2]
Basicranial length: nasion to basion
Lower skull length: prosthion to basion
Palate length: prosthion to posterior nasal spine
Anterior face length: prosthion to nasion
Outer biorbital breadth: frontomalare temporale to frontomalare temporale
Upper palate breadth: between upper, outer alveolar M1 junctions
Interorbital breadth: maxillofrontale to maxillofrontale
Bizygomatic breadth: between the outer lateral surfaces of the zygomatic arches
Symphysis height: infradentale to the inferior border of the symphysis
Symphysis width: perpendicular to symphysis height, labiolingually
Mandibular corpus height: alveolus at M2 to inferior border of the lower jaw
Mandibular corpus width: perpendicular to M2 corpus height, buccolingually
Ramus height: superior condylar surface to gonial angle
Bigonial breadth: gonial angle to gonial angle
Bicondylar breadth: between the outer, lateral condylar surfaces
Bicoronoid breadth: between the lateral aspects of the coronoid processes' tips
Masseter lever arm length: inner superior surface of the external auditory meatus to the anterior root of the zygomatic arch
Temporalis lever arm length: with jaws in occlusion, inner superior surface of the external auditory meatus to the anterior surface of the coronoid process tip
M2 Bite point length: inner superior surface of the external auditory meatus to the anterior, mesial alveolar surface of M2
Postcanine toothrow length: mesial alveolar junction of anteriormost premolar to the distal alveolar junction of the posteriormost molar
Neurocranial volume: filled to the foramen magnum with barley to the nearest cm^3

[1] Postcanine toothrow length and neurocranial volume were obtained only for adults.
[2] Bigonial breadth, temporalis lever arm length and body weight estimates were unavailable for *Pachylemur insignis*.

Anne M. Gomez

Department of Biological Anthropology and Anatomy, Duke University Medical Center, Box 3170, Durham, NC 27710, U.S.A.

Received 1 July 1991
Revision received 1 April 1992
and accepted 1 May 1992

Keywords: lorises, phylogeny, postcranial allometry, ontogeny.

Primitive and derived patterns of relative growth among species of lorisidae

Evolutionary lineages may be described as series of successively modified ontogenies (Gould, 1977). Because of this important link between evolutionary change and ontogenetic modification, insight into phylogenetic relationships may be gained through comparing patterns of growth among species. Here, patterns of relative growth among the species of Lorisidae (*Nycticebus coucang, Nycticebus pygmaeus, Loris tardigradus, Perodicticus potto* and *Arctocebus calabarensis*) are described and compared. External postcranial measurements were taken on captive *N. coucang, L. tardigradus, N. pygmaeus* and *P. potto* throughout growth, and comparable skeletal measurements were obtained for adult *A. calabaraensis*. Scaling trends were analysed with standard bivariate regression techniques and principal components analysis. *Nycticebus coucang, N. pygmaeus* and *P. potto* generally have coinciding growth trajectories, while those for *L. tardigradus* are consistently steeper than those for the other lorisids. Ontogenetic data for *N. coucang, N. pygmaeus* and *P. potto*, as well as adult data for *A. calabarensis*, were clearly separated from data for *L. tardigradus* along the second component in the principal components analysis. Postcranial growth patterns for *L. tardigradus* are most likely derived among the species of Lorisidae, while *Nycticebus, Perodicticus*, and perhaps *Arctocebus* have mainly retained primitive patterns of relative growth.

Journal of Human Evolution (1992) **23**, 219–233

Introduction

The importance of the study of the ontogeny of individuals as it relates to evolutionary trends is well established (Haeckel, 1866; de Beer, 1930; Rensch, 1959; Hennig, 1966; Gould, 1977). Ontogenetic studies are especially important for understanding the evolutionary morphology of closely-related species, as slight modifications in growth parameters may underlie many small-scale evolutionary differences between species (Rensch, 1959; Valentine & Campbell, 1975; Gould, 1977).

Some researchers have attempted to glean phylogenetic information from the study of individual ontogenies (Fink, 1982, 1988). Nelson (1978), for example, asserted that an "ontogenetic criterion" can be used to determine the polarity of characters in phylogenetic analyses. A few other researchers have utilised ontogenetic characters *per se* in phylogenetic analyses. Creighton & Strauss (1986) used growth curves as characters in analysing the relationships of species of cricetine rodents. They discovered that the phylogeny that was generated was similar to phylogenies generated using other types of data. Jungers & Hartman (1988) analysed great ape postcranial growth data both cladistically and phenetically. They found that neither analysis yielded results consistent with the general body of evidence for the phylogenetic relationships within this group. Instead, they concluded that the results of their analyses were "dominated by the impact of functional specialization in appendicular proportions" (p. 357).

Clearly, comparison of ontogenetic trends among species is one way of investigating the relation between ontogeny and phylogeny. The purpose of this research is to investigate the evolution of size and shape among the members of Lorisidae through an examination of postcranial scaling patterns during ontogeny. Specifically, focusing on the importance of truncation and extrapolation of shared patterns of relative growth and of the dissociation of these patterns, in the evolution of differences in postcranial proportions among lorisid

0047–2484/92/090219 + 15 $08.00/0

© 1992 Academic Press Limited

species. Finally, patterns of relative growth among lorisids are compared with a phylogenetic outlook, in order to determine the primitive versus derived status of various ontogenetic scaling patterns.

Two main hypotheses of lorisid relationships have been proposed. The first hypothesis is the traditional one, first espoused in the 1950s and 1960s by Hill (1953), Ellis & Montagna (1963) and later by Groves (1971), Goodman *et al.* (1974) and Rumpler *et al.* (1987). They proposed that *Nycticebus* and *Loris*, and *Perodicticus* and *Arctocebus* are sister groups. A second hypothesis was recently proposed by Yoder (1989, 1992), which places *P. potto* as the sister group of a clade made up of *Arctocebus*, *Loris* and *Nycticebus*; within this clade, *Arctocebus* is the sister taxon of a clade made up of *Nycticebus* and *Loris*. Her analysis was based on a variety of cranial and postcranial skeletal characters.

Materials and methods

Sample

The Lorisidae is comprised of five species: *Nycticebus coucang* (the slow loris), *Nycticebus pygmaeus* (the pygmy slow loris) (Groves, 1971), *Loris tardigradus* (the slender loris), *Perodicticus potto* (the potto) and *Arctocebus calabarensis* (the golden potto or angwantibo) (Fleagle, 1988). The species span a wide range of body weights, from about 200 g for adult *L. tardigradus* to about 1200 g for adult *N. coucang*. *Nycticebus coucang* is distributed throughout southeast Asia, while *N. pygmaeus* is confined to Vietnam and Laos (Dao, 1960; Groves, 1971; Eudey, 1987). *Loris tardigradus* inhabits India and Sri Lanka. *Perodicticus potto* and *A. calabarensis* are African; their distributions overlap in western Africa (Charles-Dominique, 1977). All lorisid species exhibit the slow climbing mode of locomotion (Charles-Dominique, 1977; Gebo, 1987; Demes *et al.*, 1990). Animals used in this analysis were housed at Duke Primate Center (DPC).

Loris tardigradus has the smallest body weight among lorisids (190 g, $n=24$; from DPC records), although some wild-caught specimens of *L. t. nordicus* and other subspecies have had considerably higher weights (Nieschalk & Meier, 1984). *Nycticebus pygmaeus* and *A. calabarensis* are also fairly small-bodied, with *N. pygmaeus* weighing about 440 g (DPC records, $n=10$) and *Arctocebus* about 210 g (Charles-Dominique, 1977; wild-caught specimens, $n=30$). *Nycticebus coucang* weighs about 1200 g (DPC records, $n=24$), although some weights for *N. c. bengalensis* reported by MacPhee & Jacobs (1986) range up to 2000 g. *Perodicticus potto* weighs about 1100 g (Charles-Dominique, 1977; wild-caught specimens, $n=33$), with some variation among subspecies (MacPhee & Jacobs, 1986).

Body weights and a series of postcranial measurements (trunk height, thigh length, leg length, arm length and forearm length; see Appendix 1) were taken on four *N. coucang*, four *L. tardigradus*, two *N. pygmaeus* and two *P. potto* throughout the growth period, at approximately 2 week intervals. Measurements were obtained from unanaesthetized animals with the help of a DPC technician. Sliding calipers were used for linear measurements, which were taken to the nearest 1 mm. Body weights were accurate to the nearest gram. Most animals were measured beginning at 2 weeks to 1 month of age until growth in linear measurements ceased (about 6 months for *N. coucang*, 5 months for *L. tardigradus* and *P. potto* and 4 months for *N. pygmaeus*). Data were included that were incomplete if, for example, the animal died before reaching adult size. Ten sets of adult measurements from captive DPC *N. pygmaeus* were added to the ontogenetic analysis of *N. pygmaeus*. Measurements were also included from an infant cadaver. Data from an infant cadaver were also included in the analyses of *N. coucang* (both cadavers are from DPC).

Captive adults of each species from DPC were also measured (*N. coucang*, $n=21$; *L. tardigradus*, $n=13$; *N. pygmaeus*, $n=10$; *P. potto*, $n=1$). Adult cadaver measurements from one adult *N. coucang* (DPC) and four *L. tardigradus* (from the Field Museum of Natural History) were also included in the adult samples. For animal and specimen numbers, see Appendix 1.

Comparable skeletal measurements, but no body weights, on 19 adult *Arctocebus calabarensis* were supplied by Dr William Jungers. All other linear measurements in this study were external measurements as defined in the Appendix 1 and were taken on live animals or cadavers. Points of measurement for the *Arctocebus* skeletal measurements were similar to those used on live animals and cadavers.

Statistical analysis

Adult differences. Two sets of ratios were compared: (1) limb measurements divided by trunk height (Jungers, 1979) for all lorisid species and (2) limb measurements and trunk height divided by the cube root of body weight (Sneath & Sokal, 1973) for *N. coucang*, *N. pygmaeus*, *L. tardigradus* and *P. potto*. Analysis of variance (ANOVA) was used to detect differences between species for these ratios ($P<0.05$). Planned comparisons (*F*-tests) were used to test for statistical significance (adult measurements of *P. potto* were not compared statistically because of the small sample size).

Ontogenetic allometry. Bivariate growth allometries were computed utilizing Huxley's (1932) power function, transformed using natural logs: $\ln y = \ln a + b(\ln x)$. "*x*" and "*y*" are the variables being investigated, "*b*" is the slope of the regression of ln *y* on ln *x*, and "*a*" is the *y*-intercept of this line. Data were treated as mixed cross-sectional (Cock, 1966) and regressions were computed with both the least-squares method and the reduced major axis method (for both, $P<0.05$). Three sets of regressions were performed (for *N. coucang*, *L. tardigradus*, *N. pygmaeus* and *P. potto*): (1) limb and trunk measurements versus body weight; (2) limb measurements versus trunk height; (3) limb measurements versus each other. Least squares slopes and intercepts were compared with analysis of covariance (ANCOVA, $P<0.05$), in order to identify shared patterns of growth versus those that differ in slope, elevation (*y*-intercept), or both. Reduced major axis slopes were compared with a computer program developed by Tim Cole of the State University of New York at Stonybrook, based on Clarke (1980).

A principal components analysis was performed on the ontogenetic data for lorisid species, with and without the adult data for *A. calabarensis* included. Variables were log-transformed. Forelimb and hindlimb lengths were not included, as these would be redundant with other limb measures, e.g., arm and forearm lengths.

A principal components analysis involves the geometric rotation of axes describing the variables, resulting in new axes (principal components) whereby the first principal component explains the highest amount of variation. Subsequent principal components are orthogonal and account for progressively less of the variation. Principal component loadings, or eigenvectors, describe how much each variable contributes to or is "loaded on" each component, while eigenvalues express the percentage of total variation accounted for by each of the components (Albrecht, 1978; Blackith *et al.*, 1984; Shea, 1985).

Table 1 Means (in mm), standard deviations and results of ANOVAs for four limb measurements and trunk height, all divided by the cube root of body weight for adult lorisids

	N. coucang (n=20)		Loris (n=13)		N. pygmaeus (n=10)		P. potto[1] (n=1)		Nc/Lt	Nc/Np	Np/Lt
	Mean	S.D.	Mean	S.D.	Mean	S.D.	Mean	S.D.			
Trunk	19.018	1.461	23.474	1.813	19.231	1.717	17.540	—	***	ns	***
Thigh	7.653	0.467	11.682	0.676	7.759	0.540	7.494	—	***	ns	***
Leg	7.283	0.558	11.587	0.936	7.923	0.350	7.084	—	***	**	***
Arm	7.205	0.468	10.075	0.807	7.302	0.473	7.334	—	***	ns	***
Forearm	6.939	0.515	11.444	0.678	7.513	0.420	7.050	—	***	**	***

[1]*Perodicticus potto* was not included in the ANOVA because of small sample size.
$*P<0.05$, $**P<0.01$, $***P<0.001$, ns = not significant.

Results

Interspecific adult comparisons
The results of comparisons between ratios of variables for adult lorises are given in Tables 1 and 2. Relative to measures of body size, *L. tardigradus* has markedly long limbs compared to the other three species. Values of these ratios for *N. coucang*, *N. pygmaeus*, *P. potto* and *A. calabarensis* are in many cases not significantly different from each other [for example, comparisons between *N. coucang* and *N. pygmaeus* (Tables 1 & 2) and *N. coucang* and *A. calabarensis* (Table 2)]. The ratios for the one *P. potto* measured are also closer to the ratios of *Nycticebus* and *A. calabarensis* than they are to *L. tardigradus*. Some elongation of the limbs of *N. pygmaeus* is indicated by several significant differences with ratios for *N. coucang* (Tables 1 & 2) and *A. calabarensis* (Table 2).

Interspecific ontogentic comparisons
Comparison of least-squares slopes and intercepts and reduced major axis slopes are given in Tables 3 and 4. For regressions of linear measures and trunk height versus body weight and limb measures versus trunk height, trajectories for *L. tardigradus* differ in slope in many cases from trajectories for the other species, especially those for *N. coucang* (Figures 1 & 2). Some least squares slopes are not significantly different between *L. tardigradus* and the other species (trunk height versus weight or leg length versus weight between *L. tardigradus* and *N. pygmaeus*). In these instances, y-intercept values differ. In nearly all of these cases, reduced major axis slopes are significantly different between the species compared.

For regressions of limb and limb segment measures versus each other, all lorisid species show many similarities in growth patterns (Tables 3 & 4; Figure 3). While most slope values are not significantly different, a few comparisons (such as arm length versus thigh length and forearm length versus arm length, for *N. coucang* and *L. tardigradus*) demonstrate y-intercept differences.

Results of the principal components analyses run on the five log-transformed variables for lorisid species are given in Table 5 and Figure 4. Results of the analyses run with and without data for adult *A. calabarensis* are very similar, so only the results including the *A. calabarensis* data will be discussed here.

Individuals are arranged along the first principal component, which accounts for over 94% of the variation, mostly according to size, with the youngest (and smallest) individuals

Table 2 Means (in mm), standard deviations and results of ANOVAs for four limb measurements divided by trunk height ($\times 100$) in adult lorisids

	N. coucang (n=22)		Loris (n=17)		N. pygmaeus (n=10)		P. potto[1] (n=1)		Arctocebus (n=19)		Nc/Lt	Nc/Np	Nc/Ac	Np/Lt	Np/Ac	Lt/Ac
	Mean	S.D.	Mean	S.D.	Mean	S.D.	Mean	S.D.	Mean	S.D.						
Thigh	40·166	3·489	50·229	3·961	40·585	3·921	42·727	—	41·294	1·700	***	ns	ns	***	ns	***
Leg	38·436	3·665	49·350	4·655	41·420	3·114	40·390	—	37·637	1·423	***	*	ns	***	***	***
Arm	37·671	3·781	43·004	4·002	38·265	4·440	41·818	—	34·751	1·236	***	ns	**	**	**	***
Forearm	36·388	3·437	48·868	3·849	39·293	3·457	40·195	—	35·531	1·594	***	*	ns	***	***	***

[1] *P. potto* was not included in the ANOVA because of small sample size.
Symbols are as in Table 1.

Table 3 Results of regressions of ontogenetic data for species of Lorisidae. Variables are log-transformed

y	x	N. coucang (n=32)				L. lardigradus (n=36)				N. pygmaeus (n=24)				P. potto (n=12)			
		LS	SEE	RMA	r	LS	SEE	RMA	r	LS	SEE	RMA	r	LS	SEE	RMA	r
Trunk height	Body weight	0·397	0·114	0·440	0·902	0·554	0·087	0·578	0·959	0·483	0·076	0·495	0·976	0·415	0·772	0·430	0·964
Thigh length	Body weight	0·329	0·094	0·364	0·904	0·530	0·081	0·552	0·960	0·389	0·568	0·398	0·975	0·257	0·906	0·290	0·887
Leg length	Body weight	0·321	0·073	0·343	0·936	0·470	0·091	0·500	0·939	0·418	0·046	0·423	0·988	0·280	0·104	0·319	0·877
Arm length	Body weight	0·310	0·091	0·345	0·898	0·473	0·073	0·492	0·960	0·352	0·066	0·364	0·966	0·336	0·538	0·345	0·973
Forearm length	Body weight	0·258	0·104	0·310	0·831	0·517	0·091	0·545	0·949	0·378	0·059	0·379	0·976	0·272	0·061	0·286	0·950
Thigh length	Trunk height	0·646	0·125	0·777	0·831	0·880	0·112	0·943	0·933	0·714	0·071	0·743	0·961	0·614	0·081	0·674	0·911
Leg length	Trunk height	0·717	0·080	0·768	0·933	0·774	0·112	0·845	0·913	0·836	0·060	0·854	0·979	0·632	0·113	0·743	0·851
Arm length	Trunk height	0·656	0·097	0·891	0·736	0·811	0·089	0·855	0·949	0·697	0·081	0·734	0·949	0·766	0·069	0·801	0·956
Forearm length	Trunk height	0·636	0·838	0·911	0·698	0·861	0·109	0·877	0·943	0·759	0·645	0·782	0·971	0·622	0·069	0·725	0·934
Arm length	Thigh length	0·864	0·872	0·946	0·913	0·870	0·081	0·908	0·958	0·957	0·064	0·989	0·968	1·081	0·097	1·188	0·910
Forearm length	Leg length	0·818	0·088	0·909	0·900	1·042	0·095	1·094	0·950	0·904	0·040	0·914	0·989	0·836	0·070	0·897	0·932
Leg length	Thigh length	0·907	0·089	0·989	0·917	0·863	0·074	0·895	0·964	1·131	0·052	1·153	0·968	0·993	0·937	1·102	0·901
Forearm length	Arm length	0·866	0·831	0·950	0·912	1·031	0·095	1·082	0·950	1·002	0·090	1·063	0·944	0·798	0·546	0·831	0·960
Forelimb length	Hindlimb length	0·866	0·072	0·925	0·936	0·970	0·058	0·990	0·980	0·934	0·031	0·941	0·933	1·007	0·061	1·052	0·957

For regressions of linear measures versus body weight, isometry is indicated by a scaling coefficient of 0·333. For regressions of linear measures versus each other, isometry is 1·00.
LS = Least squares regression coefficient, SEE = standard error of the estimate, RMA = reduced major axis regression coefficient, r = correlation coefficient.

Table 4 Comparisons between least squares slopes and intercepts, and reduced major axis slopes for regressions of log-transformed ontogenetic data among species of Lorisidae

	Nc vs. Lt			Nc vs. Np			Nc vs. Pp			Lt vs. Np			Lt vs. Pp			Np vs. Pp		
	LS	INT	RMA	LS	INT	RMA	LS	INT	RMA	LS	INT	RMA	LS	INT	RMA	LS	INT	RMA
TH vs. BW	**	—	***	ns	ns	ns	ns	***	ns	ns	***	**	*	—	**	ns	ns	ns
TL vs. BW	***	—	***	ns	ns	ns	ns	ns	ns	***	—	***	***	—	***	**	—	ns
LL vs. BW	***	—	***	ns	ns	ns	ns	ns	ns	***	—	***	***	—	***	ns	ns	ns
AL vs. BW	***	—	***	**	—	ns	ns	ns	ns	***	—	***	***	—	***	**	—	*
FL vs. BW	***	—	***	ns	—	ns	ns	ns	ns	***	—	***	***	—	***	**	—	ns
TL vs. TH	*	—	*	ns	ns	ns	ns	ns	ns	*	—	**	*	—	*	ns	ns	ns
LL vs. TH	ns	*	ns	ns	ns	ns	ns	ns	ns	ns	***	ns	ns	***	ns	ns	ns	ns
AL vs. TH	*	—	*	ns	ns	ns	ns	ns	ns	ns	ns	ns	ns	*	ns	ns	ns	ns
FL vs. TH	**	—	*	ns	ns	ns	ns	ns	ns	ns	***	*	ns	***	*	ns	ns	ns
AL vs. TL	ns	***	ns	ns	ns	ns	ns	ns	ns	ns	***	ns	ns	**	ns	ns	ns	ns
FL vs. LL	*	—	ns	*	—	ns	ns	ns	ns	ns	ns	**	ns	ns	ns	ns	ns	ns
LL vs. TL	ns	ns	ns	ns	ns	ns	ns	ns	ns	***	—	***	ns	ns	ns	ns	ns	ns
FL vs. AL	ns	***	ns	ns	ns	ns	ns	ns	ns	ns	***	ns	ns	***	ns	ns	ns	ns
FO vs. HL	ns	ns	ns	ns	ns	ns	ns	ns	ns	ns	***	ns	ns	ns	ns	ns	ns	ns

Symbols are as in Tables 1 and 3; INT = least-squares y-intercept.

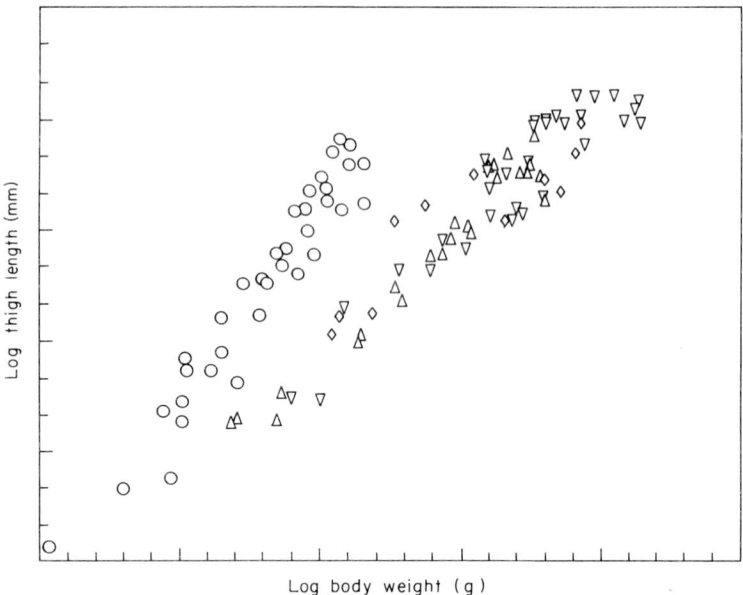

Figure 1. A bivariate plot of thigh length (in mm) versus body weight (in g) for four lorisid species. In this and subsequent plots, all variables are ln-transformed. The growth trajectory for *L. tardigradus* (○) clearly diverges from those of the other species (▽, *N. coucang*; △, *N. pygmaeus*; ◇, *P. potto*).

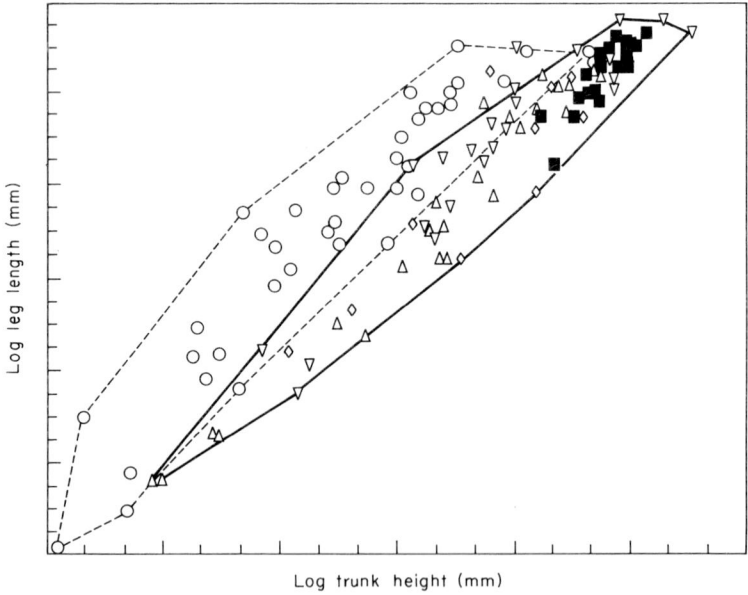

Figure 2. A plot of log leg length versus log trunk height (both in mm) for five lorisid species. The upper envelope (——) contains *L. tardigradus* data points, while the lower envelope (- - -) contains data points for the other four species of lorisids. *Arctocebus calabarensis* points (■) are adult data only. Symbols as in Figure 1.

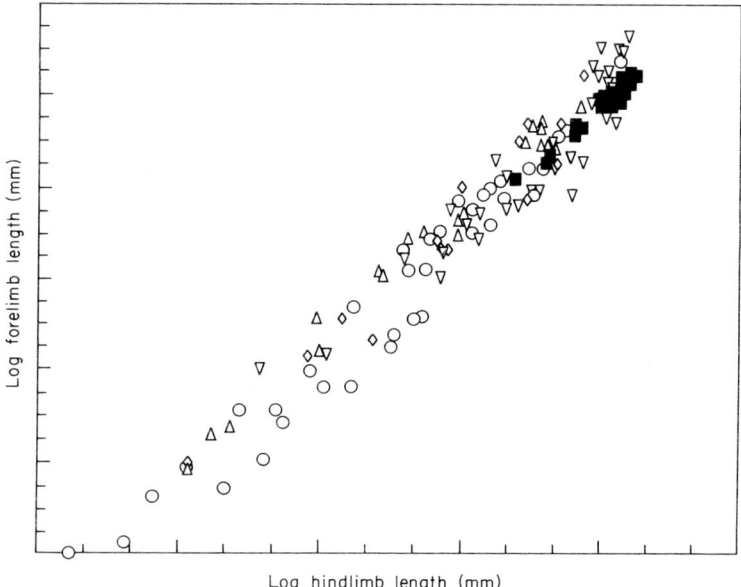

Figure 3. A plot of log forelimb length versus log hindlimb length (both in mm) for five lorisid species. In regressions of limb lengths against each other, the trajectories for *L. tardigradus* sometimes coincide with those of the other species, in contrast to regressions of limb lengths versus body weight or trunk height in which the *L. tardigradus* trajectories deviate from those of the other species. Symbols as in Figure 2.

Table 5   **Results of principal components analyses on log-transformed ontogenetic data for *N. coucang*, *N. pygmaeus* and *P. potto*, with adult data for *A. calabarensis* included in analysis 1 only**

	Variable	Component 1	Component 2
Analysis 1	Trunk height	0·359	−0·084
	Thigh length	0·274	0·040
	Leg length	0·256	0·042
	Arm length	0·266	−0·014
	Forearm length	0·246	0·050
	Total % (eigenvalue)	94·493	3·083
Analysis 2	Trunk height	0·349	−0·089
	Thigh length	0·267	0·046
	Leg length	0·259	0·042
	Arm length	0·268	−0·019
	Forearm length	0·248	0·052
	Total % (eigenvalue)	93·802	3·528

Variable loadings on the first two components are shown, as well as the total variation accounted for by these components. The third and subsequent components account for less than 1% of total variation each.

having lower values of this component. The second component generally separates *L. tardigradus* from the other lorisids, although a small amount of overlap exists. The variable most responsible for the separation (the one with the most extreme loading) is trunk height.

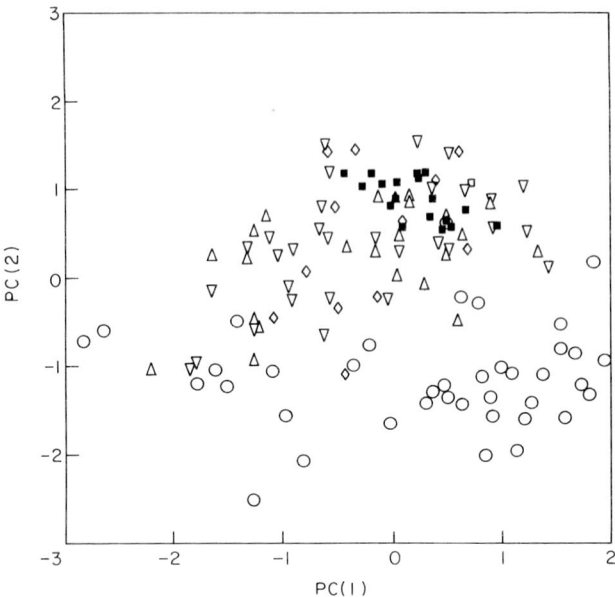

Figure 4. Results of the principal components analysis of ontogenetic data for *N. coucang*, *N. pygmaeus* and *P. potto* and adult data for *A. calabarensis* (i.e., Analysis 1). Principal component (PC) 2, which separates *L. tardigradus* from the other species of Lorisidae, is plotted against PC 1, on which individuals are arranged mostly according to size. Symbols as in Figure 2.

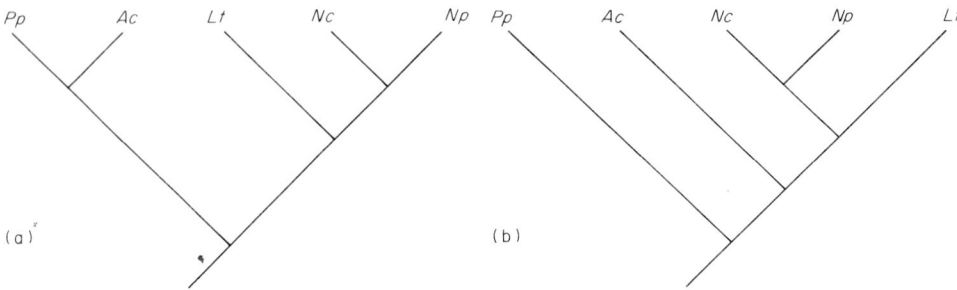

Figure 5. Two hypotheses of relationships among lorisid species: (a) the traditional hypothesis of relationships among lorisids; and (b) the recent hypothesis proposed by Yoder (1989, 1992).

Thigh length, leg length and forearm length have similar positive loadings, while arm length has a small negative loading. Subsequent components each account for less than one percent of the variation.

In summary, patterns of relative growth for *L. tardigradus* differ markedly from those of *N. coucang*, *P. potto* and *N. pygmaeus* for bivariate regressions of limb and trunk measures versus body weight and limb measures versus trunk height. These results are summarized by the principal components analysis, in which *L. tardigradus* is separated clearly from the other species (including *A. calabarensis*) along the second principal component. For bivariate regressions of limb measurements versus each other, patterns of relative growth for *L.*

tardigradus do not differ as markedly from those for the other species as they do for limb versus body size regressions.

Discussion

Adaptation and growth

This investigation focused on the importance of shared growth patterns versus dissociated growth patterns for the postcranium of lorisid species. While *N. coucang*, *P. potto*, *N. pygmaeus* and presumably *A. calabarensis* tend to follow shared patterns of growth, postcranial growth patterns for *L. tardigradus* are largely dissociated from those of the other species.

This dichotomy between shared versus dissociated patterns of growth has been recognized as important in understanding the selective pressures behind evolutionary size and shape change.

Some researchers have asserted that evolutionary size increase or decrease is most "easily" accomplished through maintaining ancestral growth patterns and merely truncating or extrapolating these patterns to larger or smaller size ranges via ontogenetic scaling (Gould, 1966, 1977; Marshall & Corruccini, 1978; Riska & Atchley, 1985; Shea, 1985). Thus, evolutionary body size change that occurs through the extrapolation or truncation of ancestral growth trajectories may be especially important for small-scale, rapid evolutionary change. Evolution on a larger scale, perhaps brought about through strong selection for morphological changes, may necessarily involve dissociations of ancestral patterns of relative growth. Therefore, selection on size alone or on length of ontogeny may bring about important adaptive changes in morphology (Gould, 1977; Rowell, 1977; Shea, 1983); however, other evolutionary shape modifications may involve selection directly on morphology. Several examples of these two scenarios may be found through comparisons of postcranial patterns of relative growth among the members of Lorisidae.

There is pervasive coincidence of ontogenetic patterns for many aspects of limb and trunk growth among *N. coucang*, *N. pygmaeus* and *P. potto* (and presumably *A. calabarensis*) (Tables 3 & 4; Figures 1 & 2). Thus, selection may have been targeting body size or length of ontogeny in one or more of these species in relation to an ancestral condition (Gomez, 1990; 1992). However, there are deviations from ontogenetic scaling among these species that may be functionally important. For example, there are several instances of *y*-intercept differences or vertical shifts in growth trajectories, but no slope differences, between *L. tardigradus* and the other species for regressions of limb lengths versus trunk height. The proportional relationship between trunk height and limb lengths may be more canalized among lorisid species than the scaling relationship between limb lengths and body weight, in which slope differences are pervasive between *L. tardigradus*' trajectories and those of the other species. For example, the ontogenetic trajectory for forearm length (relative to growth in body weight) is steeper for *N. pygmaeus* than it is for *N. coucang*. This may be an example of a size-required shape change: small-bodied species of lorises may need relatively longer forelimbs than larger-bodied species to catch insects (McArdle, 1981) or for specific locomotor and positional behaviors.

Cartmill (1974) & Jungers (1979) have proposed that large-bodied primates generally require relatively longer limbs than do smaller primates in order to maintain vertical climbing ability. However, Jungers (1984, 1985) notes that lorisids do not follow this pattern because the larger species have lower centers of gravity in order to improve stabilization in quadrupedal climbing. Furthermore, Jungers (1984, 1985) notes that lorisids deviate from the general primate pattern of a positive relationship between body size and intermembral index; larger species of primates generally have longer forelimbs relative to hindlimbs than do smaller primates, also related to maintaining vertical climbing ability. The relatively elongated forelimbs of the smaller lorisid species are explicable in terms of the more "suspensory" nature of their locomotion among the lianes, foliage, and fine branches of the forest (Jungers, 1979).

Loris tardigradus deviates from the patterns generally shared by *Nycticebus* and *P. potto*. This is especially evident for bivariate comparisons of limb measures versus weight and in the principal components analysis (Tables 3 & 4; Figures 1, 2 & 4). This deviation results in relatively longer limbs in adults of *L. tardigradus* as compared to adults of other lorisids (Tables 1 & 2). Again, relatively longer limbs may be necessary in the small-bodied *L. tardigradus* for locomotor reasons. Demes *et al.* (1990) have proposed that relatively long limbs in *L. tardigradus* may result in its high walking speed compared with other lorisids. It has also been proposed that the relatively long limbs of *L. tardigradus* are related to the fact that this species may engage in more bridging behavior than does *N. coucang* (Jungers, 1979; see also Gomez, in press). While *N. coucang* appears to inhabit the upper canopy and locomote on thicker, more continuous substrates, *L. tardigradus* is found more often in the lower canopy and undergrowth of the forest, where more bridging between supports would be necessary (Petter & Hladik, 1970).

Evolution of lorisid size and shape
The results of this study indicate that one lorisid species, *L. tardigradus*, has a unique pattern of postcranial growth among loris species. There are three possibilities for the evolution of this pattern. First, the *Nycticebus/Perodicticus/Arctocebus* pattern is primitive and the *Loris* pattern is derived. Second, the *L. tardigradus* pattern is primitive and has been lost in all other species. Third, the *L. tardigradus* pattern is derived for a small clade within Lorisidae (for example, *L. tardigradus*, *N. coucang* and *N. pygmaeus*) and has been lost in all taxa except *L. tardigradus*. If either proposed phylogeny of lorisid relationships is correct (Figure 5), then the second and third possibilities would require parallel evolution of the *Nycticebus/Perodicticus/Arctocebus* pattern, or a reversal to this pattern. It seems most likely that the general pattern of growth displayed by *L. tardigradus* is uniquely derived among Lorisidae.

Although *L. tardigradus*' scaling trends are generally unique among Lorisidae, two of the bivariate ontogenetic trajectories (leg length versus weight and forearm length versus weight) are shared between *N. pygmaeus* and *L. tardigradus* (this is true only for comparisons of the least squares slopes for these regressions; Table 4). These two patterns of relative growth are both related to elongation of the distal limb segments and may have arisen in parallel in these two small-bodied species (see McArdle, 1981; Demes *et al.*, 1990).

The ancestral body size of Lorisidae cannot be reliably determined. Outgroup data are not very useful in this regard. Galago species range in weight from 61 g for *Galagoides demidovii* (Charles-Dominique, 1977; $n=66$) to 1000–2000 g for *Otolemur crassicaudatus* (Napier & Napier, 1967; no sample sizes given). Cheirogaleid species, which are generally small-bodied [from 80 g for *Microcebus murinus* to 400 g for *Phaner furcifer* and *Cheirogaleus medius* (Harvey *et al.*, 1986; no sample sizes given)] have also been held to be closely-related to Lorisidae. However, recent research by Yoder (1992) has demonstrated that the cheirogaleids are closely-related to Lemuridae and that lorises and galagos form a monophyletic group.

Lorisid fossil material is extremely rare. There is a diversity of lorisoid material from East Africa, representing species which range in size from that of a living *G. demidovii* to *O.*

crassicaudatus (Walker, 1970). The most complete Asian skeleton is of a Miocene *Nycticebus*-like form from Pakistan, *Nycticeboides simpsoni*, which has an estimated body weight of 500 g (MacPhee & Jacobs, 1986).

If the ancestral loris is *N. coucang*- or *P. potto*-sized (Gebo, 1989), then *A. calabarensis*, *N. pygmaeus* and *L. tardigradus* must have gone through presumably independent evolutionary reductions in body size. *Nycticebus pygmaeus* and *A. calabarensis* (from the results presented here) reduced in body size mainly by truncating ancestral lorisid postcranial growth patterns (see Gomez, 1990; 1992). In contrast, postcranial growth patterns for *L. tardigradus* are dissociated from ancestral patterns.

Small body size may be adaptive to these three small species for several reasons. All three species are highly insectivorous compared to *N. coucang* and *P. potto* (Charles-Dominique, 1977; Rasmussen & Izard, 1986; K. Weisenseel, pers. comm.). As discussed above, there may be an association between small-body size, postcranial proportions and habitat and substrate usage (Jungers, 1979). Reduction in body size may also be related to selection on duration of ontogeny (Levitch, 1986; Rowell, 1977), especially in *N. pygmaeus* (Gomez, 1990; 1992). Because two patterns of body size reduction have occurred during lorisid evolution, this may suggest differing selective pressures on *N. pygmaeus* and *A. calabarensis* as opposed to *L. tardigradus*.

Summary and conclusions

This study has demonstrated that postcranial growth patterns are very similar for four lorisid species: *N. coucang*, *N. pygmaeus*, *P. potto* and probably *A. calabarensis*. Growth trajectories for *L. tardigradus* consistently have higher slopes than those of the other species, which results in relatively elongated limbs in adult *L. tardigradus*.

When these results are viewed in a phylogenetic framework, it is demonstrated that many postcranial growth patterns for *L. tardigradus* are probably derived from a primitive condition similar to that displayed by *N. coucang*, *N. pygmaeus*, *P. potto* and *A. calabarensis*. Other patterns, shared between *N. pygmaeus* and *L. tardigradus*, may have arisen in parallel in these small-bodied lorisids.

If the ancestral lorisid was large-bodied, then *N. pygmaeus* and *A. calabarensis* underwent body size reduction from this ancestral condition by truncating ancestral growth trajectories, while *L. tardigradus* dissociated postcranial growth patterns from ancestral patterns. Body size reduction in these species may be related to dietary or locomotor adaptations, or selection on duration of ontogeny.

Acknowledgements

Dr Matt Ravosa and three reviewers provided helpful comments on drafts of this manuscript. I wish to thank Dr Elwyn Simons for his permission to carry out this project, as well as Dr Kay Izard and the staff of the Duke Primate Center (especially Karen Weisenseel and Vicki Steirhoff) for their assistance with this research. I also thank Prithijit Chatrath of the Duke Primate Center and Dr Julian Kerbis and Dr Bruce Patterson of the Field Museum of Natural History in Chicago for allowing me access to specimens in their care. Callum Ross and Dr Matt Ravosa provided assistance with drafting the figures. This research was funded in part by a Sigma Xi Grant-in-Aid of Research and was conducted while under the tenures of a Ford Foundation Predoctoral Fellowship and a Duke Endowment Fellowship. This is Duke Primate Center Publication no. 513.

References

Albrecht, G. H. (1978). The craniofacial morphology of the Sulawesi macaques. Multivariate approaches to biological problems. *Contr. Primat.* **13,** 1–151.
Biegert, J. & Maurer, R. (1972). Rumpfskelettlange, Allometrian und Korporproportionen bei catarrhinen Primaten. *Folia primatol.* **17,** 142–156.
Blackith, R. A., Reyment, R. E. & Campbell, N. A. (1984). In *Multivariate Morphometrics*, 2nd edn. New York: Academic Press.
Cartmill, M. (1974). Pads and claws in arboreal locomotion. In (F. A. Jenkins, Ed.) *Primate Locomotion*, pp. 45–83. New York: Academic Press.
Charles-Dominique, P. (1977). *Ecology and Behavior of Nocturnal Primates*. New York: Columbia University Press.
Clarke, M. R. B. (1980). The reduced major axis of a bivariate sample. *Biometrika* **67,** 441–446.
Cock, A. G. (1966). Genetical aspects of form and growth in animals. *Q. Rev. Biol.* **41,** 131–190.
Creighton, G. K. & Strauss, R. E. (1986). Comparative patterns of growth and development in cricetine rodents and the evolution of ontogeny. *Evolution* **40,** 94–106.
Dao, V. T. (1960). Sur une nouvelle espece de *Nycticebus* au Vietnam. *Zool. Anz.* **164,** 240–243.
de Beer, G. R. (1930). *Embryology and Evolution*. Oxford: Clarendon Press.
Demes, B., Jungers, W. L. & Nieschalk, U. (1990). Size- and speed-related aspects of quadrupedal walking in slender and slow lorises. In (F. K. Jouffroy, M. H. Stack & C. Niemitz, Eds) *Gravity, Posture, and Locomotion in Primates*, pp. 175–197. Firenze: Il Sedicesimo.
Ellis, R. A. & Montagna, W. (1963). In (J. Beuttner-Janusch, Ed.) *Evolutionary and Genetic Biology of Primates*, pp. 197–228. New York: Academic Press.
Eudey, A. A. (1987). *IUCN/SSC Primate Specialist Group Action Plan for Asian Primate Conservation: 1987–91*. Lochhaven, Pennsylvania: Consolidated Business Forms.
Fink, W. L. (1982). The conceptual relation between ontogeny and phylogeny. *Paleobiol.* **8,** 254–264.
Fink, W. L. (1988). Phylogenetic analysis and the detection of ontogenetic patterns. In (M. L. McKinney, Ed.) *Heterochrony in Evolution*, pp. 71–91. New York: Plenum Press.
Fleagle, J. G. (1988). *Primate Adaptation and Evolution*. New York: Academic Press.
Gebo, D. L. (1987). Locomotor diversity in prosimian primates. *Am. J. Primat.* **13,** 271–281.
Gebo, D. L. (1989). Postcranial adaptation and evolution in Lorisidae. *Primates* **30,** 347–367.
Gomez, A. M. (1990). Heterochrony and the evolution of the lorises. *Am. J. phys. Anthrop.* **81,** 230.
Gomez, A. M. (1992). Ontogenetic scaling trends among African and Asian Lorisidae: functional, heterochronic and phylogenetic considerations. Master's Thesis, Duke University.
Gomez, A. M. (1993). Ontogenetic scaling of the limbs and trunk of the Asian Lorisidae: *Nyticebus coucang*, *Nycticebus pygmaeus*, and *Loris tardigradus*. *Am. J. phys. Anthrop.*
Goodman, M., Farris Jr., W., Moore, W., Prychodko, W., Poulik, E. & Sorenson, M. (1974). Immunodiffusion systematics of the primates II: findings on *Tarsius*, Lorisidae, and Tupaiidae. In (G. A. Doyle, R. D. Martin & A. C. Walker, Eds) *Prosimian Biology*, pp. 881–890. Pittsburgh: University of Pittsburgh Press.
Gould, S. J. (1966). Allometry and size in ontogeny and phylogeny. *Biol. Rev.* **41,** 587–640.
Gould, S. J. (1977). *Ontogeny and Phylogeny*. Cambridge: Harvard University Press.
Groves, C. P. (1971). Systematics of the genus *Nycticebus*. *Proc. 3rd Int. Congr. Primat., Zurich 1970.* **1,** pp. 44–53.
Haeckel, E. (1866). *General Morphologie der Organismen*. Berlin: Georg Reimer.
Harvey, P. H., Martin, R. D. & Clutton-Brock, T. H. (1986). Life histories in comparative perspective. In (B. B. Smuts, D. L. Cheney, R. M. Seyfarth, R. W. Wrangham & T. H. Struhsaker, Eds) *Primate Societies*, pp. 181–196. Chicago: University of Chicago Press.
Hennig, W. (1966). *Phylogenetic Systematics*. Urbana: University of Illinois Press.
Hill, W. C. O. (1953). *Primates: Comparative Anatomy and Taxonomy. I. Strepsirhini*. Edinburgh: Edinburgh University Press.
Huxley, J. S. (1932). *Problems of Relative Growth*. London: Methuen.
Jungers, W. L. (1977). Hindlimb and pelvic adaptations to vertical climbing and clinging in *Megaladapis*, a giant subfossil prosimian from Madagascar. *Yearb. phys. Anthrop.* **20,** 508–524.
Jungers, W. L. (1979). Locomotion, limb proportions, and skeletal allometry in lemurs and lorises. *Folia primatol.* **32,** 8–28.
Jungers, W. L. (1984). Aspects of size and scaling in primate biology with special reference to the locomotor skeleton. *Yearb. phys. Anthrop.* **27,** 73–97.
Jungers, W. L. (1985). Body size and scaling of limb proportions in primates. In (W. L. Jungers, Ed.) *Size and Scaling in Primate Biology*, pp. 345–381. New York: Plenum Press.
Jungers, W. L. & Hartman, S. E. (1988). Relative growth of the locomotor skeleton in orangutans and other large-bodied hominoids. In (J. H. Schwartz, Ed.) *Orang-utan Biology*, pp. 347–359. New York: Oxford University Press.

Jungers, W. L. & Susman, R. L. (1984). Body size and skeletal allometry in African apes. In (R. L. Susman, Ed.) *The Pygmy chimpanzee: Evolutionary Biology and Behavior*, pp. 131–177. New York: Plenum Press.
Levitch, L. (1986). Ontogenetic allometry of the postcranial skeleton in platyrrhines, with special emphasis on its relationship to the evolution of small body size in the callitrichidae. Ph.D. Dissertation, University of Washington.
MacPhee, R. D. E. & Jacobs, L. L. (1986). *Nycticeboides simpsoni* and the morphology, adaptations, and relationships of Miocene Siwalik Lorisidae. In (K. M. Flanagan & J. A. Lillegraven, Eds) *Vertebrates, Phylogeny and Philosophy, Contributions to Geology, Special Paper 3*, pp. 131–161. Laramie: University of Wyoming.
Marshall, L. G. & Corruccini, R. S. (1978). Variability, evolutionary rates, and allometry in dwarfing lineages. *Paleobiol.* **1,** 101–119.
McArdle, J. E. (1981). Functional morphology of the hip and thigh of the Lorisiformes. *Contr. Primat.* **17**. Basel: Karger.
Napier, J. R. & Napier, P. H. (1967). *A Handbook of Living Primates*. London: Academic Press.
Nelson, G. J. (1978). Ontogeny, phylogeny, paleontology and the biogenetic law. *Syst. Zool.* **27,** 324–345.
Nieschalk, U. & Meier, B. (1984). Haltung and Zucht von Schlankloris. *Z. Kolner Zoo.* **27,** 95–100.
Petter, J. J. & Hladik, C. M. (1970). Observations sur la domaine vitale et la densité de population de *Loris tardigradus* dans les forêts de Ceylon. *Mammalia* **34,** 395–409.
Rasmussen, D. T. & Izard, M. K. (1986). Scaling of growth and life history traits relative to body size, brain size, and metabolic rate in lorises and galagos (Lorisidae, Primates). *Am. J. phys. Anthrop.* **75,** 357–367.
Rensch, B. (1959). *Evolution Above the Species Level*. New York: Columbia University Press.
Riska, B. & Atchley, W. R. (1985). Genetics of growth predict patterns of brain-size evolution. *Science* **229,** 668–671.
Rowell, T. E. (1977). Variation in age at puberty in monkeys. *Folia primatol.* **27,** 284–296.
Rumpler, Y., Warter, S., Meier, B., Preuschoft, H. & Dutrillaux, B. (1987). Chromosomal phylogeny of three Lorisidae: *Loris tardigradus*, *Nycticebus coucang*, and *Perodicticus potto*. *Folia primatol.* **48,** 216–220.
Schultz, A. H. (1929). The technique of measuring the outer body of human fetuses and of primates in general. *Contr. Embryology* **20** (117), 213–257.
Shea, B. T. (1983). Allometry and heterochrony in the African apes. *Am. J. phys. Anthrop.* **62,** 275–289.
Shea, B. T. (1985). Bivariate and multivariate growth allometry: statistical and biological considerations. *J. Zool., Lond.* **206,** 267–290.
Sneath, P. H. A. & Sokal, R. R. (1973). *Numerical Taxonomy*. San Francisco: W. H. Freeman.
Valentine, J. W. & Campbell, C. A. (1975). Genetic regulation and the fossil record. *Am. Sci.* **63,** 673–680.
Walker, A. (1970). Post-cranial remains of the Miocene Lorisidae of East Africa. *Am. J. phys. Anthrop.* **33,** 249–262.
Yoder, A. D. (1989). A phylogenetic systematic study of the lorises. *Am. J. phys. Anthrop.* **78,** 327.
Yoder, A. D. (1992). The phylogenetic affinities of the cheirogalidae: morphological and molecular analysis. Ph.D. Dissertation, Duke University.

Appendix 1

Measurements taken on live animals and cadavers (from Schultz, 1929)
Trunk height: suprasternale (upper edge of sternum) to symphysion (upper edge of pubic symphysis).
Thigh length: trochanterion summun (tip of the greater trochanter) to femorale (most distal point on the lateral condyle of the femur).
Leg length: tibiale (highest point on the medial condyle of the tibia) to sphyrion (most distal point on the medial malleolus of the tibia).
(Upper) arm length: Acromion (the most lateral point on the acromial process of the scapula) to radiale (most proximal point on the lateral side of the capitulum of the radius).
Forearm length: radiale to stylion (most distal point on the styloid process of the radius).
Forelimb length: arm length + forearm length.
Hindlimb length: thigh length + leg length.
Skeletal measurements (taken on Arctocebus calabarensis*) supplied by Dr William Jungers*
Trunk length (Biergert & Maurer, 1972): same as trunk height, above.
Maximum femur length, maximum tibia length, maximum humerus length, and maximum radius length (Junger, 1977; pers. comm.): intercondular distances.
Animal and specimen numbers (DPC = Duke Primate Center, FMNH = Field Museum of Natural History)
Nycticebus coucang: DPC 1941m, 1940f, 1946f, 1945m, 970m, 1924f, 961m, 1914f, 1936f, 1949m, 1941m, 982f, 979m, 1911f, 993m, 1903f, 980m, 1915f, 1940f, 981f, 1920m, 989f, 1937m, 966f, 1913m.
Loris tardigradus: DPC 1948m, 1943f, 1950m, 1947f, 1921f, 1918m, 976f, 984f, 994f, 975m, 1906m, 1902f, 977f, 978f, 985m, 991f, 990m; FMNH 60574, 60575, 60451, 60123.
Nycticebus pygmaeus: DPC 1951m, 1952m, 1927m, 1929m, 1931f, 1925m, 1926f, 1935f, 1933m, 1928f, 1932m, 1930f, 1942f.
Perodicticus potto: DPC 925f, 922m.

Timothy G. Bromage

Hard Tissue Research Unit, Department of Anthropology, Hunter College, CUNY, 695 Park Ave., New York, N.Y. 10021, U.S.A.

Received 1 August 1991
Revision received 1 March 1992
and accepted 27 May 1992

Keywords: craniofacial architecture, developmental constraint, chimpanzee, hominid.

The ontogeny of *Pan troglodytes* craniofacial architectural relationships and implications for early hominids

Developmental constraints characterize functional growth boundaries of the mammalian face. Enlow and colleagues identified boundary conditions and planes of the midface in humans and other mammal species that are defined on the basis of important growth sites and the developmental disposition of neural and pharyngeal matrices. With few exceptions a conservative mammalian architecture is said to be achieved by the adult stage. Three aspects of this architecture are investigated here for a cross-sectional ontogenetic series of *Pan troglodytes* crania: (1) a line passing from the maxillary tuberosity through the junction of middle and anterior cranial fossae is perpendicular to the neutral horizontal axis of the orbit (PM–NHA angle), (2) an average 45° angle, whose origin is the external auditory meatus, separates the maxillary tuberosity from the midpoint of the orbital opening (meatus angle) and (3) the base of the brain, maxillary tuberosity and prosthion are on or close to the same plane (anterior maxillary hypoplasia).

This investigation considers to what extent samples of chimpanzees and early hominids reflect Enlow's characterization of mammalian craniofacial architecture. Evidence derived from the chimpanzee sample bears out the perpendicularity of the PM–NHA angle and a near 45° meatus angle (though slightly higher overall), but does not conform to the absence of marked anterior maxillary hypoplasia seen in other mammals. *Pan* is instead characterized by considerable ontogenetic variation (noted by Enlow amongst anthropoids) influenced to some extent by sex. Preliminary early hominid data on the meatus angle indicates that *Australopithecus* specimens are similar to the widespread and probably primitive condition, while *Paranthropus* and early *Homo* specimens portray an uncharacteristic mammalian architectural relationship.

Journal of Human Evolution (1992) **23**, 235–251

Introduction

This study has two objectives: (1) to examine architectural features of the developing chimpanzee skull and (2) to examine these structural relationships among a select collection of Plio–Pleistocene hominid skulls. Enlow & Azuma (1975) observed a suite of architectural relationships, based on developmental criteria, for a variety of mature mammals. The *ontogeny* of these relationships has yet to be investigated for a given species. Should these relationships be maintained over the course of growth and development, then this may be a suitable demonstration of developmental constraints concerning hominoid facial growth. These data may then be employed for the understanding of fossil hominid taxa as well, particularly those for which we have some ontogenetic information (e.g., Bromage, 1989).

Background

Workers over the last half century have identified commonalities of mammalian craniofacial development and architecture that form the structural foundation of the mammalian skull. Thompson (1942) illustrated successive deformations of one form to another with his "Theory of Transformations" (a mathematical treatment of continuous transformations) using Decartes "Methods of Coordinates". His transformations of closely related species often illustrated divergent growth properties of their skulls, but Thompson was also able to accomplish global transformations of mammalian skulls which led him to state that "... there is something, an essential and indispensable something, which is common to them all, something which is the subject of all our transformations, and remains *invariant* ..." (1942:1085).

Heintz (1966), operating under another mensurational tradition, similarly noted divergent ontogenetic trajectories of certain craniofacial dimensions amongst primate families and genera. She found, nevertheless, that *Australopithecus*, the apes and humans followed common ontogenetic trajectories of basion–prosthion (BA–PR) against nasion–prosthion (NA–PR) length (her Graphique 55), NA–PR dimensions against upper facial length (BA–NA) (Graphique 57), as well as an angular measure of prognathism (Graphique 91), demonstrating a correspondence between increasing facial projection and height (Heintz, 1966).

Delattre & Fenart (1960) oriented skulls according to the horizontal vestibular axis and likewise showed that the ontogeny of the facial skeleton in the chimpanzee, gorilla and humans followed the same general pattern of down and forward growth. Swindler *et al.* (1973) illustrate isometry of the upper and lower facial elongation for both *Papio* and *Macaca*, and Sirianni & Newell-Morris (1980) also note a common fetal craniofacial growth pattern and ossification sequence in humans and macaques.

Between the 1960s and the 1980s, Enlow and co-workers explicitly combined histological interpretations of human facial bone growth remodeling with serial radiographic data in order to promote a new cephalometric system. A cephalometric system determined by *actual* sites of growth would, by definition, portray the underlying and developmentally constrained architectural features of the mammalian skull, as observed by the orderliness of skull ontogeny and the overwhelming similarity of facial growth across primates and other mammals. Enlow (1966) noted that "apparent" growth measured using conventional cephalometric methods was apparent simply because the comparison of serial cephalographs was based on registering the skull on fixed points. Although Enlow acknowledged the value of such analyses in providing data on overall growth pattern, he superimposed serial cephalographs in such a way that they would correspond to known growth and remodeling sites, permitting "actual" growth to be visualized as well (1966). Enlow (1968) and Enlow & Hunter (1968) championed this new system and outlined a series of anatomical parts and counterparts which could be evaluated on the basis of growth equivalents and growth compensations between them. Such an analysis, for instance, would assess the correspondence between the horizontal dimensions of an individuals' maxillary (part) and mandibular (counterpart) dental arches. Facial growth was characterized by Enlow and colleagues in cephalometric applications which demonstrated a correlation between horizontal equivalents and vertical equivalents responsible for the stability and balance of craniofacial relationships.

Enlarging on this approach, Enlow *et al.* (1969, 1971a,b), Enlow & Moyers (1971) and Enlow (1974) described a number of points and planes on lateral cephalographs corresponding to architectural units conforming to sites of growth, remodeling and displacement. A procedure was developed to explain *how* a pattern was produced as opposed to systems of cephalometric evaluation which explained *what* craniofacial pattern resulted due to growth. Bhat & Enlow (1985:270–272) subsequently noted that the part–counterpart procedure was more sensitive than traditional cephalometric methods in studies of cranial base relations to craniofacial variability because "an angular value such as basion-sella-nasion is based on midline points, none of which are involved in the actual articular fitting of basicranium, maxilla, and mandible to each other, or in the anatomic basis of bilateral positioning among the respective parts; nor do they represent growth sites directly participating in this three-part relationship".

In 1975, Enlow & Azuma investigated the prevalence of certain architectural relationships among a diversity of mammalian skulls. Their comparative sample included lateral radiographs of 116 human subjects of at least 10 years of age and 45 mammal species represented by rodents, lagomorphs, artiodactyls, carnivores and non-human primates. These authors defined several architectural relationships that depend on important growth sites and the developmental disposition of neural and pharyngeal matrices as follows (see Figure 1): (1) a line passing from the maxillary tuberosity (MT) through the junction of middle and anterior cranial fossae is close to perpendicular to the Neutral horizontal axis (NHA) of the orbit, (2) the base of the brain, maxillary tuberosity and prosthion (P) are on or close to the same plane and (3) an average 45° angle, whose origin is the external auditory meatus (EAM), separates the maxillary tuberosity from the midpoint of the orbital opening. These relationships were found to hold for mammals in general. Members of the Anthropoidea, however, were found by Enlow & Azuma (1975) to exhibit a characteristic vertical hypoplasia of the anterior maxilla, meaning that prosthion lay significantly above the inferior brain-to-maxilla plane (IBMP). Thus, while these structural features appear to be developmentally constrained for many mammals, anthropoids were found to depart from a broadly primitive mammalian condition in maxillary development.

An operational definition for developmental constraint is provided by Smith *et al.* (1985: 265) as ". . . a bias on the production of variant phenotypes or a limitation on phenotypic variability caused by the structure, character, composition, or dynamics of the developmental system". The reference planes and angles investigated by Enlow & Azuma (1975), and investigated here concerning the ontogeny of the chimpanzee craniofacial region, characterize craniofacial *relations*. These relations between the brain and facial growth vectors determine the size and disposition of the mammalian facial complex and the space available for the pharyngeal contents. Here, developmental constraints over the maintenance of these relations among various mammals are considered homologous—a concept in accordance with De Beer (1937), who emphasized that the issue of homology rested in the preserved morphological "relations" between parts of the skull.

Architectural relations
This study addresses architectural relations in an ontogenetic series, albeit a cross-sectional one. This makes the study particularly relevant to the question of developmental constraint and can add to the empirical foundation established previously on more mature individuals (i.e., Enlow & Azuma, 1975). The brief review (below) of three architectural relationships studied here should serve as a demonstration that concepts of craniofacial growth are well served by methods that include known functional–structural units and key sites of growth in the overall methodological approach. The following investigation of the chimpanzee is based on lateral cephalograms using the development-dependent cephalometric system described by Enlow & Azuma (1975). Figure 1 portrays the architectural relations (landmarks, planes and axes) investigated here.

(1) The posterior maxillary (PM) plane (Figure 1), or boundary, described by Enlow & Azuma (1975), is a natural anatomical interface between the neurocranium and the face and, hence, helps to establish facial growth vectors. (See also descriptions of landmarks, planes and axes below.) Along the top of this boundary, the ethmomaxillary complex and

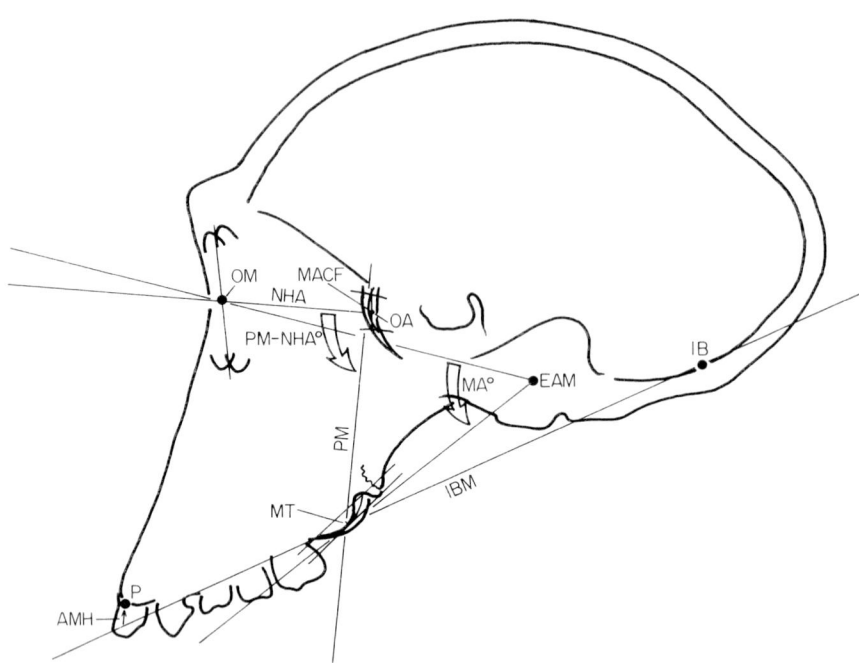

Figure 1. *Orbital midpoint (OM)* is between the superior and inferior orbital rims. *Orbital apex (OA)* is between the superior rims of the superior orbital fissure and the inferior rims of the optic canals. *Neutral horizontal axis (NHA)* is a plane formed through the OM and the posterior cone-shaped apex of the orbit (OM–OA). *Middle-anterior cranial fossae (MACF)* is the junction between the middle and anterior cranial fossae. *Maxillary tuberosity (MT)* is the posterior–inferior point on the MT. *Posterior maxillary (PM) plane* is formed through the junction between the MACF and the MT. *Posterior maxillary plane–neutral horizontal axis angle (PM–NHA)* is the angle constructed between the PM and NHA planes. *External auditory meatus (EAM)* is the central point within the auditory opening. *Meatus angle (MA)* is formed between the OM, to the EAM, to the MT. *Inferior brain (IB)* is the inferior-most extent of the brain. *Inferior brain-to-maxilla plane (IBMP)* is a plane constructed between the IB point and the MT. *Prosthion (P)* is the most anteroinferior point on alveolar bone between the central incisors. *Anterior maxillary hypoplasia (AMH)* is the vertical height difference between the IBMP and P.

anterior cranial fossa are separated from the pharyngeal space and the middle cranial fossa. Below, the posterior maxillary boundary intersects the maxillary tuberosity separating the jaws from the pharyngeal space. This boundary relates to important sites of growth and bony remodeling: the expanding neurocranium and anterior remodeling of the pterygoid–maxillary tuberosity region. Thus, the posterior maxillary plane translates anteriorly during growth and is structurally related to the neutral horizontal axis (NHA) of the orbit normally disposed at 90° to each other. This relationship is maintained throughout growth and developmentally constrained among mammals in general.

(2) The alignment of the orbits and the jaws with respect to the external auditory meatus is similarly constrained during growth. The angle originating from the external auditory meatus, to the orbital midpoint (OM) on the one hand, and the maxillary tuberosity on the other, approximates 45° (meatus angle, Figure 1). This suggests that ontogenetic trajectories of the orbit and jaws are isometric and patterned so that the feeding, sight and auditory functions are optimally positioned together. An alternative, or additional, explanation for this relationship may relate to head balance during locomotory behavior (cf. Bramble, 1990).

(3) Enlow & Azuma (1975) also noted that the base of the brain, maxillary tuberosity and prosthion are on or close to the same plane for most mammals (inferior brain-to-maxilla plane, Figure 1). The inferior extent of brain growth marks the inferior limit to which both prosthion and maxillary tuberosity descend during growth. It was noted, however, that anthropoids typically exhibit a vertical hypoplasia of the anterior maxilla so that prosthion is positioned above the inferior brain-to-maxilla plane. Whether this is due to a repositioning of maxillary tuberosity downward or is an actual result of diminished downward remodeling of prosthion, has yet to be determined.

Materials and methods

Chimpanzee specimens
A cross-sectional ontogenetic series of 45 wild-shot *Pan troglodytes* crania, housed at the American Museum of Natural History and the Cleveland Museum of Natural History, were investigated. Specimens were mounted in Frankfurt Horizontal on a Todd Craniostat which rested on a central stage calibrated in degrees. Each specimen was radiographed according to standard cephalometric convention for both lateral ($0°$) and posterior–anterior ($90°$) views.

These specimens were divided between three age groups (Table 1) on the basis of the eruption status of the permanent molars. This was done because of the significant contribution to palatal lengthening by the formation and eruption of permanent molars and the resulting translational and transformational movements of landmarks, planes and axes studied here. Group 1 consists of individuals prior to eruption of maxillary first permanent molars and includes as its youngest member an individual with maxillary second deciduous molars still in their crypts. Group 2 consists of juveniles prior to eruption of maxillary second permanent molars. Group 3 consists of older juveniles prior to eruption of maxillary third molars.

The architectural points relevant to the relationships identified by Enlow & Azuma (1975) and investigated here were traced onto acetate sheets and were measured to the nearest millimeter or degree (Figure 1). The superimposition of bilateral features, being less than precise on lateral cephalograms, meant tracing both features and rendering a central tracing between them, herein referred to as "averaged", from which measures were taken.

Cephalometric landmarks, planes and axes (Figure 1) Orbital midpoint (OM). Superior and inferior orbital rims were traced from the lateral radiograph, they were averaged and the maximum vertical height of the orbital opening was obtained and checked with the posterior–anterior view. This measurement was divided in half and the midpoint registered on the lateral tracing.

Orbital apex (OA). The superior rims of the superior orbital fissure and the inferior rims of the optic canals were traced from the lateral radiograph. The distance between the rims was measured along their posterior aspects confluent with the anterior bulge of middle cranial fossa (which is just near the apex), the measurement was halved and the midpoint was registered on the lateral tracing.

Neutral horizontal axis (NHA). This plane was formed through the orbital midpoint and the posterior cone-shaped apex of the orbit (OM–OA).

Table 1 *Pan* craniofacial architecture by dental age categories

	Group 1: dm2-M1 in crypts				Group 2: M2 in crypts				Group 3: M3 in crypts		
Specimen no.	PM-NHA angle	Meatus angle	AMH distance	Specimen no.	PM-NHA angle	Meatus angle	AMH distance	Specimen no.	PM-NHA angle	Meatus angle	AMH distance
1	85	44	2	10	85	44	0	26	80	49	30
2	85	45	1	11	85	47	10	27	81	44	20
3	86	43	−1	12	88	47	7	28F	81	50	30
4	89	45	−2	13	89	44	1	29	85	50	30
5M	90	38	9	14	89	46	5	30	87	48	14
6M	90	42	2	15F	89	47	15	31F	87	50	22
7	91	45	4	16	90	47	−3	32	88	46	19
8M	92	40	0	17	90	48	2	33	88	49	23
9	95	44	5	18	91	44	6	34M	88	63	5
Mean	89.22	42.99	2.22	19	91	48	1	35M	89	45	16
SD	3.38	2.47	3.38	20	92	46	6	36	89	46	21
t-test	0.69ns	2.56*	1.97ns	21	92	49	1	37	89	46	19
				22	93	45	−1	38F	89	54	40
				23	93	47	8	39	90	49	17
				24	95	52	8	40	90	50	23
				25	96	47	3	41	90	51	16
				Mean	90.50	46.75	4.31	42F	92	48	22
				SD	3.08	2.05	4.67	43	93	47	15
				t-test	0.65ns	3.42**	3.69**	44	93	49	1
								45	95	48	26
								Mean	88.20	48.60	20.45
								SD	3.99	2.52	8.73
								t-test	2.01ns	6.38**	10.47**
								Grand mean	89.25	46.72	10.73

Listed are specimen numbers, posterior maxillary–neutral horizontal axis (PM–NHA) angle, the meatus angle and the degree of anterior maxillary hypoplasia (AMH) (see text for details). Individual specimens are listed in ascending order of PM–NHA angles within dental age categories. M and F specimen numbers denote male and female individuals, respectively. Students *t*-tests were performed to test for significant differences from 90° (PM–NHA column), 45° (meatus column) and no hypoplasia (0·0 mm) (AMH column) within dental age categories.

*$P < 0.05$ significance level, **$P < 0.01$ significance level; ns, not significant.

Middle-anterior cranial fossae (MACF). The junction between the middle and anterior cranial fossae was defined as the anterior-most extent of the middle cranial fossa determined by the first contact made by a line pivoted on and swung from the maxillary tuberosity. This was basically the same as the anterior-most extent defined in relation to the skull held in the Frankfurt Horizontal.

Maxillary tuberosity (MT). The posterior–inferior point on the MT was defined as a tangent to the tuberosity parallel to a line drawn between the alveolar bone behind the last tooth to erupt and the contact between the tuberosity and the palatine bone.

Posterior maxillary (PM) plane. The PM plane was formed through the junction between the MACF and the MT.

Posterior maxillary plane–neutral horizontal axis angle (PM–NHA). This angle is constructed between the PM and NHA planes and enclosing the facial pocket.

External auditory meatus (EAM). Vertical and horizontal midpoints of the external auditory meati were averaged between left and right sides and the calculated point registered on the lateral tracing.

Meatus angle (MA). The MA was formed from the OM, to the EAM, to the MT.

Inferior brain (IB). The inferior-most extent of the brain was defined by the first contact with the cranial cavity by a line pivoted on and swung from the MT (through the occipital bone).

Inferior brain-to-maxilla plane (IBMP). This plane was defined as a result of defining the IB point from the MT.

Prosthion (P). Prosthion was defined as the most anteroinferior point on alveolar bone between the central incisors.

Anterior maxillary hypoplasia (AMH). The vertical height difference between the IBMP and P, taken parallel to the PM plane, was drawn and measured on the lateral tracing.

Early hominid specimens
Natural size camera lucida lateral tracings were made from casts of crania representing the following Plio–Pleistocene hominid species: the *Australopithecus afarensis* reconstruction by Kimbel & White (1988), *A. africanus* represented by Sts 5 and Sts 71, *A. aethiopicus* represented by KNM-WT 17000, *Paranthropus boisei* represented by KNM-ER 406, early *Homo* represented by KNM-ER 1470 and KNM-ER 1813 and African *Homo erectus* represented by KNM-ER 3733.

These specimens represent the most complete and undistorted adult Plio–Pleistocene hominid crania that preserve the relevant anatomy of interest. KNM-ER 3733 required only slight reconstruction of the MT behind the missing third molar. KNM-ER 1470 required positioning of the face to the braincase as outlined by Bromage (in prep.). The position of the KNM-ER 1470 MT was established according to the architectural relationships outlined by Enlow & Azuma (1975) and examined in this study (the resulting craniofacial architecture—

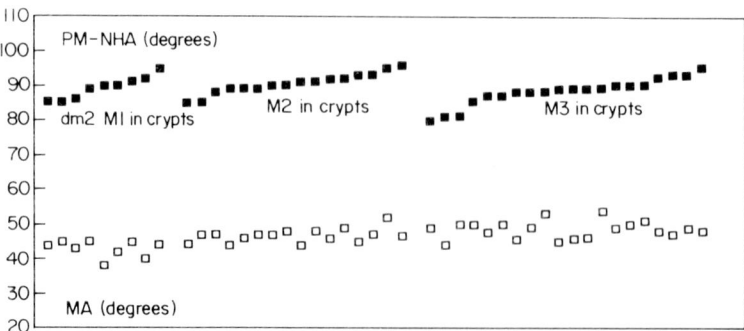

Figure 2. Individual specimens are listed in ascending order of PM–NHA angle within dental age categories (see Table 1). PM–NHA angles (top) are plotted with MAs. Meatus angles vary in each group, including both low and high values, although the mean MA increases with dental age categories.

as regards the hafting of the facial skeleton onto the braincase and the position of P—is similar to other prognathic early hominids). Other South African *Paranthropus* crania were not investigated at this time because of evident distortion. OH 5 is not included here since the specimen lacks any true bony contact between the face and cranium.

The MA was drawn onto the lateral tracing and measured in degrees. The IBMP was approximated and drawn onto the lateral tracing, but no measurement of AMH was attempted, leaving this to qualitative assessment. Determinations of PM–NHA angles were not formally attempted at this time except for KNM-ER 1470 (90°) where this was an integral part of the reconstruction procedure (Bromage, in prep.). (A first order attempt to provide PM–NHA angles for the other early hominid specimens, based upon observations of a large whole-skull comparative chimpanzee series housed at the American Museum of Natural History, resulted in values entirely within the range of values obtained from radiographs of Group 3 chimpanzees described below.)

Results

Three measurements, PM–NHA angle, MA and AMH distance were tabulated for the chimpanzee sample. Table 1 lists them according to the three major dental age groupings described above. It is evident from these data that mean PM–NHA angles are very near to 90° in all groups, though there are cases 5 or 6° above and below the mean within each group and 9 and 10° below in Group 3 (Table 1; Figure 2). Tests found no significant differences between PH–NHA group means from 90° (Group 1: mean = 89·22°, $t = 0·69$ N.S.; Group 2: mean = 90·50°, $t = 0·65$ N.S.; Group 3: mean = 88·20°, $t = 2·01$ N.S.; see Table 1). Seventy one percent of the study sample is within $\pm 3°$.

Mean meatus angles, however, do depart from 45° within dental age groups. The hypothesis that mean MAs do not depart significantly from 45° was rejected for all groups (Group 1: mean = 42·89°, $t = 2·56*$; Group 2: mean = 46·75°, $t = 3·42**$; Group 3: mean = 48·60°, $t = 6·38**$; see Table 1). Sixty seven percent of the study sample is within $\pm 3°$. While the Group 1 mean is less than 45°, Groups 2 and 3 are higher, resulting in a grand mean of 46·72°, not significantly different from 45° ($t = 1·5187$ N.S.). However, tests comparing Group means rejected the hypothesis of equal angles (Groups 1 & 2: $t = 4·21**$; Groups 1 & 3:

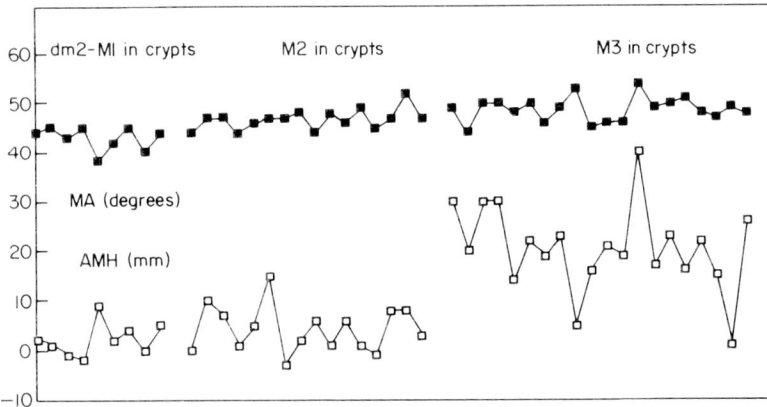

Figure 3. Individual specimens are listed in ascending order of PM–NHA angle within dental age categories (see Table 1). Meatus angles (top) are plotted with degree of AMH, illustrating generally higher AMH values in the later age group.

Table 2 **Correlation matrix between craniofacial measurements**

	PM–NHA angle	Meatus angle	AMH distance
PM–NHA angle	—		
Meatus angle	−0·01 (0·19)	—	
AMH distance	−0·32 (−0·36)*	0·50 (0·53*)	—

Simple correlations (above) are given between craniofacial measurements. Partial correlations are given parenthetically below, controlling for the third variable.
*Significant difference from zero at $P<0·05$.

$t=5·68**$; Groups 2 & 3: $t=2·37**$) suggesting that there is indeed a small increase in mean MA with age.

Figure 3 graphically portrays the increase in Group 3 AMH values, together with an increased range, above those of Groups 1 and 2 (see also Group 3 statistics, Table 1). Tests found no significant difference of AMH group means from 0·0 mm for Group 1 (Group 1: mean = 2·22 mm, $t=1·97$ N.S.) but rejected this hypothesis for Groups 2 and 3 (Group 2: mean = 4·31 mm, $t=3·69**$; Group 3: mean = 20·45 mm, $t=10·47**$; see Table 1).

For any one group of individuals with the same PM–NHA angle there is a range of MA values to the extent that a correlation coefficient of 0·0 exists between these two variables (Table 2). When AMH data are investigated, one also finds low and high values within PM–NHA angle groups, though when assessing the correlation between these variables as PM–NHA angle increases, there is a tendency (however unimpressive) for a reduction in AMH values (Table 2).

An examination of the AMH data in relation to MAs within PM–NHA angle groups reveals that in some cases as MAs rise so also do AMH values, while the reverse is true for other individual cases (see also Figure 3). Overall, the simple correlation between AMH and MA is positive (0·50) and improved slightly by partial correlation between AMH and MA, with the effect of PM–NHA being held constant (0·53) (Table 2).

Table 3 summarizes the architectural data by sex categories. There are 10 specimens in the study sample having field identifications of sex (see Table 1) and it transpires that four of five females have increasing AMH values ($\geqslant 15$ mm) with increasing MAs ($\geqslant 47°$) when compared to other individuals with the same PM–NHA angle, while four of five males have relatively low AMH values ($\leqslant 9$ mm) with variably low or high MAs (38–53°) when compared to other individuals with the same PM–NHA angle. The hypothesis that male and female mean PM–NHA angles depart from 90° was rejected (male mean = 89·80°, $t = 0·13$ N.S.; female mean = 87·25°, $t = 0·60$ N.S.) as was also the hypothesis that male and female mean MAs depart from 45° (male mean = 43·60°, $t = 0·24$ N.S.; female mean = 50·50°, $t = 2·18$ N.S.—the latter significant only at the 10% level). However, while AMH distance means do not depart significantly from 0·0 mm (male mean = 6·40 mm, $t = 1·01$ N.S.), female AMH distance means are significantly different (female mean = 28·50 mm, $t = 3·34*$). Test comparing male and female group means suggest that PM–NHA angles are not significantly different ($t = 1·30$ N.S.) but reject the hypothesis of equal MAs ($t = 2·38*$) and equal AMH distance ($t = 4·17**$) between the sexes.

Figure 4 portrays distinctive male and female ontogenetic paths for landmarks anterior to the EAM; there is a general ventral placement of female OA, OM, P and MT points relative to those of males. Vertical lines beneath P points on Figure 4 are the amounts of AMH in mm (Table 3).

The results for the early hominid sample—MA and IBMP—are provided together with the lateral tracings in Figure 5. All *Australopithecus* specimens except Sts 71 illustrate MAs well within the range, indeed close to the mean, for chimpanzees (Table 1), humans and other mammals (Enlow & Azuma, 1975). One can question the propriety of accepting a result based on a major reconstruction, as in the case of the *A. afarensis* cranium. However, given compelling morphological evidence for its primitiveness (Kimbel *et al.*, 1984), its architectural similarity to *A. africanus* specimen Sts 5 (Figure 5), as well as its architectural similarity (Figure 5) and numerous primitive features shared with *A. aethiopicus* specimen KNM-WT 17000 (Kimbel *et al.*, 1988), there is every indication that the reconstruction is accurate.

Australopithecus africanus specimen Sts 71, *Paranthropus* and early *Homo* specimens have MAs at the extreme high end of the chimpanzee range (Table 1) and beyond the range of modern humans and other mammals (Enlow & Azuma, 1975). *Homo erectus* specimen KNM-ER 3733 has an MA of 66°, which is beyond the known range of any mammal.

Discussion

The results of the present study on an ontogenetic series of chimpanzee crania are more or less consistent with those of Enlow & Azuma (1975) concerning humans and a variety of other mammals. Most individuals are within a few degrees (plus or minus) of the perpendicular for PM–NHA angle. Had this angle, for instance, illustrated a tendency to increase with age and, presumably, size, then one could question that this is a structural feature exhibiting developmental constraint. One finds, instead, a similar range of values in each juvenile age

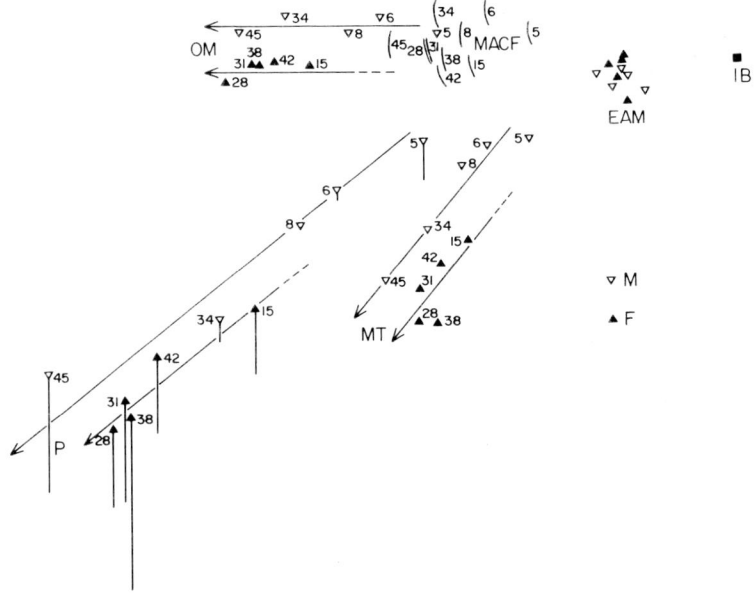

Figure 4. Architectural points for 10 individuals of known sex (see Table 3): male specimens (▽) 5, 6, 8, 34 and 35; female specimens (▲) 15, 28, 31, 38 and 42. IB point is base of brain for all individuals. Middle-anterior *cranial fossae* points are represented here as the vertical curvilinear distance of the orbital apex (OA) confluent with the superimposed anterior limit of the middle cranial fossa. The OA would be halfway along this curve and slightly behind. Vertical lines beneath P points are the amounts of AMH in mm. Note the inferiorward placement of female architectural points anterior to EAM (MACF, OM, P and MT points) while registering on base of brain and keeping the PM planes parallel.

Table 3 *Pan* **craniofacial architecture by sex categories**

Males	PM–NHA angle	Meatus angle	AMH distance	Females	PM–NHA angle	Meatus angle	AMH distance
5M	90	38	9	15F	89	47	15
6M	90	42	2	28F	81	50	30
8M	92	40	0	31F	87	50	22
34M	88	53	5	38F	89	54	40
35M	89	45	16	42F	92	48	22
Mean	89·80	43·60	6·40		87·25	50·50	28·50
SD	1·48	5·86	6·35		4·65	2·52	8·54
t-test	0·13ns	0·24ns	1·01ns		0·60ns	2·18ns	3·34*

Listed are specimen numbers, posterior maxillary–neutral horizontal axis (PM–NHA) angle, the meatus angle and the degree of anterior maxillary hypoplasia (AMH) (see text for details). Students *t*-tests were performed to test for significant differences from 90° (PM–NHA column), 45° (meatus column) and no hypoplasia (0·0 mm) (AMH column) within dental age categories.
*$P<0·05$ significance level; ns, not significant.

group (Table 1, Figure 2) which strongly suggests that individuals inherit a prescribed developmental disposition—and PM–NHA angle—that is maintained throughout growth in the midst of other developmental–structural adjustments in the growing skull. Furthermore, these data support the interpretation by Enlow & Azuma (1975) that the PM boundary

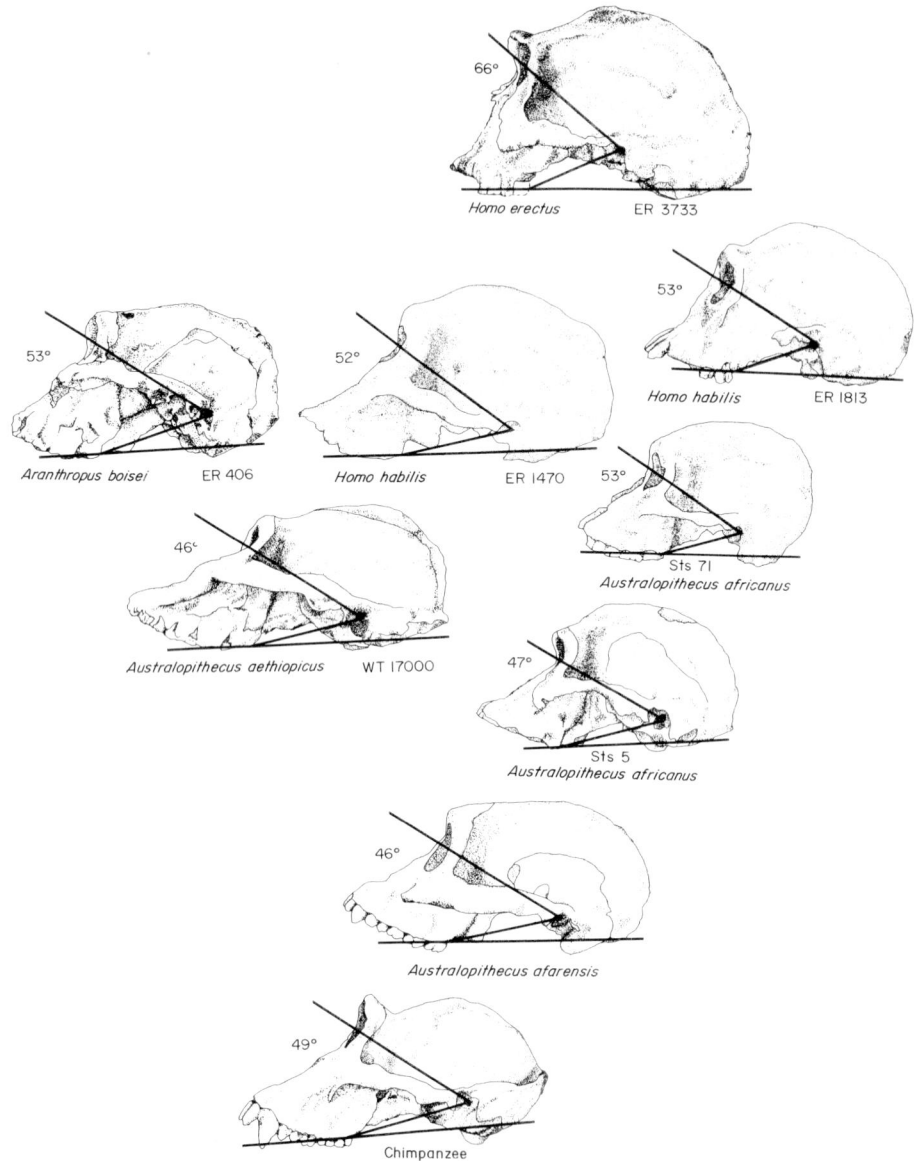

Figure 5. Early hominid specimens with calculated MAs (in degrees) and an estimated IBMP drawn so as to interpret relative AMH (see text for description). Alternative positions of the KNM-ER 1470 P and MT points have been indicated taking into account the range of chimpanzee PM–NHA angles (see text) in the less prognathic direction only. A chimpanzee at bottom (with a relatively high MA) has been provided for comparison.

is a developmental interface between the neurocranium and facial complex that moves uniformly forward throughout growth and development.

Meatus angles were also observed to be within a few degrees (plus or minus) from 45° for most individuals (Table 1, Figure 2) (the 95% confidence interval for the grand mean of

46·7° includes 45°). Furthermore, while there is extensive overlap in MAs between age groups, the mean angle does tend to increase with age. The data support the notion that while PM–NHA angles remain relatively stable during growth, increases in posterior facial height (the height of the PM boundary) and hence, the descent of MT, slightly outpace the forward horizontal repositioning of the tuberosity, resulting in greater MAs.

Enlow & Azuma (1975) noted that anthropoids exhibited a characteristic vertical hypoplasia of the anterior maxilla with P lying significantly above the IBMP. This may be due either to a vertical hyperplasia of posterior facial height (cf. Enlow et al., 1971b) or to a true underdevelopment of the anterior maxilla downward. Age Groups 1 and 2 illustrate a low but variable amount of vertical hypoplasia, while Group 3 exhibits an increase in the mean AMH value (Table 1, Figure 3). It was expected that AMH would increase with increasing MAs insofar as MA is a reflection of posterior facial height. Indeed, for most PM–NHA angle groups of 90° and less (except the 88° group), it appears that while MAs increase, AMH values also tend to increase. Indeed, a modest relationship is borne out with the simple correlation ($r=0.50$) between AMH and MA (Table 2) (the partial correlation between AMH and MA, controlling for PM–NHA, is 0·53). While the 92, 93 and 95 PM–NHA degree angle groups also suggest the same relationship, there exists a single individual in each group at the highest MA which has a low AMH value.

In an attempt to understand the sex differences in AMH described in the Results (Table 3), radiographs of specimens of known sex were superimposed to compare directly the amount of hypoplasia while simultaneously portraying the generalized down and forward facial growth vector of the face. Evidence from ontogenetic studies of the brain and cranial base indicate that the occipital–cerebellar portions of the cranial base complete more of their growth earlier than other parts of the brain (cf. Dean, 1988) and indeed, registering on IB while keeping PM boundaries of individual specimens parallel resulted in the portrayal of the anteroinferior facial growth vector (Figure 4). Minimal scatter of EAM points, with no apparent directionality, confirmed the precocity of this portion of the cranial base relative to any other registration. This registration had the added advantage of maintaining the parallel growth transgression of the PM boundary.

The most striking feature of Figure 4 is the separation of male and female ontogenetic growth trajectories which are evident prior to eruption of the second permanent molar. Indeed, it appears that these male and female trajectories are already established prior to completion of the deciduous dentition (the smallest inner circle of points representing a male with second deciduous molars still in their crypts: Figure 4, specimen 5). The impression from these lateral tracings is that males have a more steeply inclined posterior cranial base (this remains true even should the upward vertical displacement of all male points be due to a lower base of brain—the architectural point used for registration on Figure 4). This might otherwise constrict the horizontal dimension available for the pharyngeal space which would instead be recompensed by a further anteriorward remodeling shift of MT compared to females. Furthermore, it appears that the PM boundary is indeed longer in Group 3 females compared to their male counterparts (Figure 4), resulting in a net increase in posterior maxillary hyperplasia. This may account for the consistently high AMH values compared to males (represented as vertical lines beneath P points, Figure 4; Table 3). These patterns within sex groups suggest, at least, that some developmentally constrained architectural difference characterizes male and female chimpanzees from an early age.

A discussion of the early hominid material is complicated by a host of unforgiving sources of variation that include individual differences, differences due to sex (as described above), variation over time within and between lineages, geographic differences as well as the problem of incompleteness and deformation of the material. This is an admission, after all that has been said and observed for the early hominid sample (above), that some architectural constraints such as the MA are not as "constrained"—or constrained in the same way—as thought before. The *A. afarensis* reconstruction, *A. africanus* specimen Sts 5 and *A. aethiopicus* specimen KNM-WT 17000 conform to the architectural principles common to the chimpanzee: MAs in the 46–47° range and pronounced AMH. The *A. africanus* specimen Sts 71, *Paranthropus boisei* specimen KNM-ER 406 and early *Homo* specimens KNM-ER 1470 and KNM-ER 1813 have values at the extreme high end of the chimpanzee range: MAs in the 52–53° range and apparently less pronounced AMH (except perhaps for KNM-ER 1470). As for the African *H. erectus* specimen KNM-ER 3733, unless we are to posit an enormous range of variation or the likelihood of a craniofacial aberration, one is left to hypothesize the existence of increased MAs (e.g., 66° in KNM-ER 3733) in this taxon. Even allowing for some repositioning of the KNM-ER 3733 face and a requisite increased prognathism of the reconstruction, it is not likely that this MA would diminish by more than 5°. We are left with a craniofacial architecture not observed for any other early hominid, modern human or mammal.

Whatever the factors are that control MA variation—and we know that posterior facial height is one of them—it seems that they effected a definite increase for *Paranthropus*, early *Homo* specimens and African *H. erectus* (e.g., KNM-ER 3733) in the study sample (Figure 5). Because of the limited study sample it must be acknowledged that it remains to be seen whether a more comprehensive study and new discoveries of early hominid crania may sustain this accounting of the data. Meanwhile it may be hypothesized that something extraordinary happened with MA temporal variation, something related perhaps to a changing laryngeal and pharyngeal relationship. Presumably, evolutionary forces stabilized this new relationship and the MA returned to the mammalian condition as seen in modern *Homo*.

Should there be any significance to the uniformly high MA values for Sts 71, KNM-ER 406, KNM-ER 1470 and KNM-ER 1813, though 52–53° was close to the mean for their taxa, then it might be to unite features of *Paranthropus* and early *Homo* morphotypes (it is important here to note that Sts 71 is sometimes cited in respect to a *Paranthropus* morphotype—e.g., Clarke, 1988). A persistent theme in paleoanthropology has been the identification of many morphological (e.g., Dean & Wood, 1981, 1982; Kimbel *et al.*, 1984; Dean, 1986; Skelton *et al.*, 1986; Tobias, 1988) and ontogenetic (Bromage, 1989) similarities between *Paranthropus* and *Homo*—most of which have been interpreted as homoplasies. Here we may wish to add an architectural feature—the MA—for these Late Pliocene–Early Pleistocene fossils.

The list of so-called homoplasies that one may draw up from studies of *Paranthropus* and early *Homo* (e.g., Bromage, 1989; Dean, 1986; Dean & Wood, 1981, 1982; Kimbel *et al.*, 1984; Skelton *et al.*, 1986; Tobias, 1988) is long and is provoking reappraisal (e.g., Tobias, 1988). Given the preliminary architectural data presented here it may be relevant to hypothesize that the similarity in architectural relations between *Paranthropus* and early *Homo* is a departure from a primitive architectural program [i.e., like that observed for *A. afarensis* (reconstruction) and *A. africanus* (e.g., Sts 5)] and derived from a common ancestor.

The very high MA of KNM-ER 3733 suggests that the resumption of a near 45° MA in modern *Homo* is relatively recent. If one accepts that this individual was typical of African *H. erectus* and, furthermore, that this specimen is a representative of the lineage leading to modern *Homo*, then either the range of variation has diminished considerably to the present time or the modern human condition is secondarily derived, returning to the mammalian condition.

The structural relationships (as presented here) and the morphology of some specimens of South African *A. africanus* (here represented by Sts 71) are akin to the *Paranthropus* morphotype and should perhaps take another taxonomic nomen [Clarke's (1988) "Hominid B"]. The alternative to taxonomic splitting of Sterkfontein australopithecines is to include the variation between, for instance, Sts 5 and Sts 71 within one quite variable *A. africanus*. Both alternatives admittedly highlight quintessential problems in species recognition.

Another specimen from East Africa, contemporary with the South African australopithecines, *A. aethiopicus* (KNM-WT 17000), clearly has the stamp of early australopithecine structural relationships (*A. afarensis* reconstruction; Sts 5) and craniofacial similarities to both *A. afarensis* and *Paranthropus* (e.g., Walker *et al.*, 1986; Kimbel *et al.*, 1988). It may transpire that whatever we find, wherever we find it, that fossils of this age (c. 2·5 m.y.a.) may show a mixture of *Australopithecus* and *Paranthropus*—primitive and advanced—features. Remembering that *Paranthropus* illustrates some features that also characterize early *Homo*, we have a basis for interpreting or hypothesizing, at least in general, a common ancestral heritage for subsequently distinct *Paranthropus* and *Homo* grades of evolution.

Summary and conclusions

Data presented here for *Pan troglodytes* on developmental boundaries and relationships between the face and brain, and between the sensing organs and the jaws, are comparable to those data for humans and other mammals (Enlow & Azuma, 1975). However, the hypoplastic nature of the chimpanzee anterior maxilla clearly emphasizes Enlow & Azuma's (1975) earlier observation that anthropoids exhibit specific deviations from the mammalian architectural condition. While it is likely that with more data one would find much more overlap between the sexes, the data presented here do bear out fundamental architectural differences between the sexes that are evident prior to completion of the deciduous dentition. These deviations and sex differences make the study of the ontogeny and evolution of craniofacial architectural constraints all the more interesting and important for primate evolutionary studies.

Furthermore, the data on modern humans (Enlow & Azuma, 1975) and chimpanzees may not serve as appropriate models on which to assess variation in architectural relationships over human evolutionary time. While modern human and chimpanzee craniofacial architectural relationships are to some extent similar to other mammals studied (cf. Enlow & Azuma, 1975), one of the most interesting transformations of early hominid evolution may have been the increased MA of *Paranthropos* and early *Homo*, both departing from the primitive mammalian architecture.

While the early hominid data presented here are limited by relative incompleteness of the fossil record, their interpretation can be considered as new hypotheses to be tested with the recovery of more relatively complete early hominid craniofacial skeletons. For instance, one intriguing possibility is that the developmental pattern responsible for the architectural modification of *Paranthropus* and early *Homo* craniofacial complexes was shared in a common ancestor. *Australopithecus africanus* specimens, such as Sts 71, and *A. aethiopicus* specimen KNM-WT 17000, that portray a mixture of *Australopithecus* and *Paranthropus* characters, may provide us with glimpses of this heritage.

Acknowledgements

I thank Don Enlow, Fred Szalay, B. Holly Smith and two anonymous reviewers for the depth of their interest and their valuable comments. Don Johanson and Bill Kimbel (then of Cleveland Museum of Natural History) and Guy Musser (American Museum of Natural History) kindly provided the chimpanzee crania for study and Rolph Berhents made it possible to radiograph the specimens. The L.S.B. Leakey Foundation and Foundation for Research into the Origins of Man sponsored the data collection. James Panteleon rendered his artistic talents on the early hominid cranial specimens (Figure 5). I am grateful to Matt Ravosa and Anne Gomez for their invitation to present this paper on behalf of the symposium "Ontogenetic Perspectives on Primate Evolutionary Biology".

References

Bhat, M. & Enlow, D. H. (1985). Facial variations related to headform type. *Angle Orthod.* **55,** 269–280.
Bramble, D. M. (1990). Head stabilization and locomotor behavior in the Hominidae. *Am. J. phys. Anthrop.* **81,** 197–198.
Bromage, T. G. (1989). Ontogeny of the early hominid face. *J. hum. Evol.* **18,** 751–773.
Bromage, T. G. (in prep.). Architectural basis for the reconstruction of KNM-ER 1470 and implications for early hominid systematics.
Clarke, R. J. (1988). A new *Australopithecus* cranium from Sterkfontein and its bearing on the ancestry of *Paranthropus*. In (F. E. Grine, Ed.) *Evolutionary History of the "Robust" Australopithecines*, pp. 285–292. New York: Aldine de Gruyter.
De Beer, G. R. (1937). *The Development of the Vertebrate Skull*. Oxford: Clarendon Press.
Dean, M. C. (1986). *Homo* and *Paranthropus*: similarities in the cranial base and developing dentition. In (B. A. Wood, L. Martin & P. Andrews, Eds) *Major Topics in Primate and Human Evolution*, pp. 249–265. Cambridge: Cambridge University Press.
Dean, M. C. (1988). Growth processes in the cranial base of hominoids and their bearing on morphological similarities that exist in the cranial base of *Homo* and *Paranthropus*. In (F. E. Grine, Ed.) *Evolutionary History of the "Robust" Australopithecines*, pp. 107–112. New York: Aldine de Gruyter.
Dean, M. C. & Wood, B. A. (1981). Metrical analysis of the basicranium of extant hominoids and *Australopithecus*. *Am. J. phys. Anthrop.* **54,** 63–71.
Dean, M. C. & Wood, B. A. (1982). Basicranial anatomy of Plio–Pleistocene hominids from East and South Africa. *Am. J. phys. Anthrop.* **59,** 157–174.
Delattre, A. & Fenart, R. (1960). *L'Hominisation du Crâne Etudiée par la Methode Vestibulaire*. Paris: Editions du Centre national de la Recherche Scientifique.
Enlow, D. H. (1966). A morphogenetic analysis of facial growth. *Am. J. Orthod.* **52,** 283–299.
Enlow, D. H. (1968). *The Human Face: An Account of the Postnatal Growth and Development of the Craniofacial Skeleton*. New York: Harper and Row.
Enlow, D. H. (1974). Croissance et architecture de la face. *Pedod. Fr.* **6,** 122–144.
Enlow, D. H. & Azuma, M. (1975). Functional growth boundaries in the human and mammalian face. In (J. Langman, Ed.) *Morphogenesis and Malformations of the Face and Brain*, pp. 217–230. New York: The National Foundation.
Enlow, D. H. & Hunter, W. S. (1968). The growth of the face in relation to the cranial base. *Eur. Orthod. Soc., Congr. Rep.* **44,** 321–335.
Enlow, D. H., Kuroda, T. & Lewis, A. B. (1971*a*). The morphological and morphogenetic basis for craniofacial form and pattern. *Angle Orthod.* **41,** 161–188.
Enlow, D. H., Kuroda, T. & Lewis, A. B. (1971*b*). Intrinsic craniofacial compensations. *Angle Orthod.* **41,** 271–285.
Enlow, D. H. & Moyers, R. E. (1971). Growth and architecture of the face. *J. Am. Dent. Assoc.* **82,** 763–774.
Enlow, D. H., Moyers, R. E., Hunter, W. S. & McNamara, Jr, J. A. (1969). A procedure for the analysis of intrinsic facial form and growth. *Am. J. Orthod.* **56,** 6–23.
Heintz, N. (1966). Le crâne des anthropomorphes. Croissance relative, variabilité, évolution. Graphiques. *Mus. R. Afr. Cent.-Tervuren, Belgique, Ann. N.S. 4-Sci. Zool.* **6,** 1–94.
Kimbel, W. H. & White, T. D. (1988). A revised reconstruction of the adult skull of *Australopithecus afarensis*. *J. hum. Evol.* **17,** 545–550.
Kimbel, W. H., White, T. D. & Johanson, D. C. (1984). Cranial morphology of *Australopithecus afarensis*: a comparative study based on a composite reconstruction of the adult skull. *Am. J. phys. Anthrop.* **64,** 337–388.

Kimbel, W. H., White, T. D. & Johanson, D. C. (1988). Implications of KNM-WT 17000 for the evolution of "robust" *Australopithecus*. In (F. E. Grine, Ed.) *Evolutionary History of the "Robust" Australopithecines*, pp. 259–268. New York: Aldine de Gruyter.

Skelton, R. R., McHenry, H. M. & Drawhorn, G. M. (1986). Phylogenetic analysis of early hominids. *Curr. Anthrop.* **27,** 21–43.

Smith, J. M., Burian, R., Kauffman, S., Alberch, P., Campbell, J., Goodwin, B., Lande, R., Raup, D. & Wolpert, L. (1985). Developmental constraints and evolution. *Q. Rev. Biol.* **60,** 265–287.

Sirianni, J. E. & Newell-Morris, L. (1980). Craniofacial growth of fetal *Macaca nemestrina*: a cephalometric roentgenographic study. *Am. J. phys. Anthrop.* **53,** 407–421.

Swindler, D. R., Sirianni, J. E. & Tarrant, L. H. (1973). A longitudinal study of cephalofacial growth in *Papio cynocephalus* and *Macaca nemestrina* from three months to three years. In (M. R. Zingeser, Ed.) *Craniofacial Biology of Primates*, pp. 227–240. Basel: S. Karger.

Thompson, D'Arcy W. (1942). *On Growth and Form*. London: Cambridge University Press.

Tobias, P. V. (1988). Numerous apparently synapomorphic features in *Australopithecus robustus, Australopithecus boisei* and *Homo habilis*: support for the Skelton–McHenry–Drawhorn hypothesis. In (F. E. Grine, Ed.) *Evolutionary History of the "Robust" Australopithecines*, pp. 293–308. New York: Aldine de Gruyter.

Walker, A. C., Leakey, R. E. F., Harris, J. M. & Brown, F. H. (1986). 2·5-Myr *Australopithecus boisei* from West of Lake Turkana, Kenya. *Nature* **322,** 517–522.

Theodore M. Cole III

Doctoral Program in Anthropological Sciences, State University of New York at Stony Brook, Stony Brook, NY 11794, U.S.A.

Received 30 June 1991
Revision received 15 April 1992
and accepted 4 May 1992

Keywords: Cebus, mastication, heterochrony, growth and development, feeding ecology.

Postnatal heterochrony of the masticatory apparatus in *Cebus apella* and *Cebus albifrons*

Species of the genus *Cebus* have been characterized as omnivorous, although *Cebus apella* differs significantly from the other *Cebus* species in its ability to routinely consume hard fruits and nuts. Hard-object feeding behavior in *C. apella* is reflected in a suite of craniomandibular features that are related to the generation and dissipation of higher masticatory stresses. This study examines the biomechanical differences between *C. apella* and *Cebus albifrons*, a species that does not exploit hard foods, by considering the ontogenetic sources of anatomical differences among adults. When adult forms are compared, *C. apella* are found to have greater relative depth and thickness of the corpus at M_1 and greater relative thickness at the symphysis. In addition, the corporal and symphyseal cross-sections are absolutely larger. This overall pattern is consistent with the expectation of a need for increased resistance to parasagittal bending and twisting stresses in the corpus, increased resistance to wishboning stresses in the symphysis and greater resistance to shear stresses in both the corpus and symphysis. When patterns of postnatal growth are compared, it is found that there are few differences in growth trajectories, as indicated by similarity in regression line slopes. Instead, most of the biomechanical differences that distinguish adults are present early in postnatal life, as indicated by differences in regression line elevations. It is concluded that the modifications for hard-object feeding seen in *C. apella* adults are most probably the result of selection on prenatal growth processes.

Journal of Human Evolution (1992) **23,** 253–282

Introduction

Interspecific comparisons of adult primates are useful in identifying and describing relationships between anatomical form and function. However, such "static" comparisons do not necessarily provide explanations for how form differences originate in the course of evolution. That is, they cannot identify the evolutionary mechanisms or developmental processes of change in form. The incorporation of ontogenetic data in comparative studies of functional anatomy serves to add a more "dynamic" perspective, and thus provides such analyses with added explanatory power (Alberch *et al.*, 1979). Studies considering the explicit role of growth modification as a mechanism for evolutionary change deal with the processes of *heterochrony*. McKinney (1988a:17) defines heterochrony as "evolution via change in timing (and/or rate) of development" and states that

> "... virtually *all* evolution involves changes somewhere in the chain of developmental events. Whether size, shape, or behavior, phylogenetic change almost invariably springs from a change in rate or timing in the ontogeny of descendant individuals."

This study attempts to utilize ontogenetic data as a means of adding such a dynamic dimension to a comparative study of biomechanical function in the Neotropical primate *Cebus*. *Cebus* is generally characterized as omnivorous, consuming a variety of fruits, insects, small vertebrates, flowers and leaves (Thorington, 1967; Izawa, 1975, 1979; Moynihan, 1976; Defler, 1979a,b; Fleagle *et al.*, 1981; Freese & Oppenheimer, 1981; Terborgh, 1983). However, the tufted capuchin (*C. apella*) tends to differ from members of the untufted species group (*C. albifrons, C. capucinus, C. olivaceus*) by exploiting hard fruits and nuts as a regular part of its diet (Moynihan, 1976; Izawa, 1975, 1979; Struhsaker & Leland, 1977; Terborgh, 1983). The habitual consumption of hard food objects requires anatomical modifications for the production and subsequent dissipation of higher masticatory forces. *Cebus apella* is thus characterized by a suite of features that have been cited as correlates of hard-object feeding: (1) a large and robust face and mandible (Kinzey, 1974; Bouvier, 1986; Daegling, 1991); (2) teeth with relatively thick enamel (Rosenberger & Kinzey, 1976; Kay, 1981; Teaford, 1985); and (3) a variety of cranial correlates of an enlarged masticatory musculature: a larger infratemporal foramen, enclosed by widely flaring and robust zygomatic arches (Kinzey, 1974), sagittal cresting in large adult males (Hofer, 1973; Torres de Assumpcao, 1983) and larger, more flaring, pterygoid plates (Hershkovitz, 1949).

This study focuses specifically on differences in masticatory form between *C. apella* and *C. albifrons*. The analysis will pose and attempt to answer several questions:
(1) What are the expected biomechanical differences between a hard-object feeder and a closely-related animal that does not routinely exploit hard foods?
(2) How well do data from adult samples of *C. apella* and *C. albifrons*, respectively, conform to these expectations?
(3) How do variations in ontogeny act to produce adult differences in form?
(4) What information can a comparative consideration of ontogeny provide about the processes of dietary adaptation?

Masticatory apparatus biomechanics and shape

The effects that the proportions of the masticatory system have on function (particularly with respect to hard-object feeding) can be divided into two basic and related components: generation of external forces applied to food items and the dissipation of forces within the mandible.

External forces
The harder foods routinely consumed by *C. apella* require that greater bite forces be applied to food objects. If the ancestral form that gave rise to *C. apella* was not adapted for the habitual consumption of palm nuts and similar foods, there are at least five basic ways in which its form may have been altered for the production of greater forces.

First, there could be an increase in the absolute cross-sectional area of the masticatory muscles through an increase in overall body size. The amount of force that a muscle is capable of exerting (ignoring the complicating effects of muscle pinnation) is proportional to its physiological cross-sectional area (Schumacher, 1961). Selection for absolutely larger body size (for whatever reason) should result in absolutely larger muscles which, in turn, would produce the greater forces necessary for processing hard food objects. However, increasing body size is not a particularly efficient means of increasing jaw adducting force because muscle force scales to the two-thirds power of muscle mass (Kiltie, 1982; Cachel, 1984; Hylander, 1985).

Second, there could be an increase in the cross-sectional area of the muscles while the same overall body size was maintained. The absolute size of the masticatory muscles could be increased through an increased rate of muscle growth or an earlier onset of muscle growth, relative to the growth of the masticatory skeleton.

Third, there could be an improvement in the mechanical advantage of the masticatory apparatus. Changes in the shape of the mandible and upper face would produce greater efficiency in bit force production, without necessarily requiring an increase in muscle size.

The mechanical advantage of the mandible could be increased by lengthening the jaw adductor lever arm relative to the bite point load arm.

Fourth, changes in force could be brought about by modifications in muscle architecture. Increased pinnation of a muscle increases its effective physiological cross-section (Herring, 1980). Also, changes in the proportions of different muscle fiber types could increase the magnitudes of muscle forces (with perhaps no change in muscle size) and the periods of time over which the forces could be applied [see Dechow & Carlson (1990) for a recent review].

Finally, it is possible that increases in jaw adducting forces could be attained through alterations in muscle recruitment patterns (e.g., Ravosa & Hylander, n.d.).

Of these five factors, the last two are obviously not observable when studying dried skulls. The presence of the first three effects may be detected provided certain assumptions are made; these assumptions will be discussed further when defining the osteometric variables used in this study.

Forces within the mandible

Hard-object feeding not only requires the generation of higher bite forces, but must also provide a means of effectively dissipating the internal forces acting within the upper face and mandible.* *In vivo* experimental studies of macaques by Hylander (1979, 1981, 1984, 1985, 1986) provide a set of expectations for mandibular stress and strain environments in anthropoid primates. Together with theoretical expectations based on beam theory (Popov, 1976), these results provide a set of predictions of the sort of modifications in mandibular form that might be necessary for resisting the increased stresses that would result from adopting a hard-object feeding strategy.†

In unilateral mastication, the mandibular corpora experience a complex combination of stresses. On the working side, there is an accumulation of dorsoventral shearing stress directly beneath the bite point. At the same time, the transverse components of the forces exerted by the superficial masseter (pulling laterally) and by the food object (acting medially against the lower teeth) tend to twist the working side about its longitudinal axis. Of these two types of stress, Hylander (1988) considers the twisting stresses to be more important in influencing corporal design. On the balancing side, there is a tendency for parasagittal bending of the corpus to predominate, due to the vertical component of the balancing-side muscle force, such that the alveolus is placed in tension and the lower margin is compressed.

The symphysis is also loaded in a complex pattern during unilateral mastication. Perhaps the most predominant stresses affecting the symphysis are "wishboning" stresses (Hylander, 1984, 1985; Hylander *et al.*, 1987). During molar biting and chewing, at the end of the power stroke, the laterally directed component of the bite force on the working side and the prolonged activity of the balancing-side deep masseter have the effect of pulling the corpora away from each other. At the symphysis, this lateral transverse bending of the corpora places the labial aspect of the symphysis in compression, while the lingual aspect undergoes tension.

Of particular interest in this study are the stresses that the mandible experiences during incision. Field studies by Izawa (1975; Izawa & Mizuno, 1977; see also Terborgh, 1983) have

---

*It should be noted that higher peak forces are not necessarily the only important factor influencing skeletal design in hard-object feeding primates. Repetitive loading may also be a concern (i.e., forces may not only have greater magnitudes but may be experienced with greater frequency), so that skeletal design is altered for increased fatigue resistance (Hylander, 1979).

†Hylander (1979, 1984, 1985) and Daegling (1990) provide explicit discussions of the need to use *in vivo* experimental data to evaluate the validity of the assumptions and conclusions of comparative studies of the masticatory skeleton.

shown that the anterior teeth (incisors, canines and premolars) are used for the initial processing of hard food objects. Hylander (1979, 1984) predicted that incision will produce similar stress patterns in both corpora (because there is no distinction between working and balancing sides), which will be both bent parasagittally and twisted about their long axes. If both corpora are twisted, then the symphysis will experience vertical bending stress, with the superior aspect in compression and the inferior aspect in tension.

Are differences in mandibular form between *C. albifrons* and *C. apella* consistent with what might be expected on the basis of Hylander's experimental results? If we assume that the magnitudes of internal forces are simply greater (without the basic patterns of stress distribution changing), then the following differences might be expected: (1) to better resist parasagittal bending of the working side during unilateral mastication, the corpus is expected to be relatively deeper in *C. apella*. In addition, increased balancing-side muscle activity would be expected to increase dorsoventral shear in the symphysis, so that the symphyseal cross-section is expected to be absolutely larger in *C. apella*; (2) to increase the resistance to twisting about its long axis, the corpus of *C. apella* is expected to be relatively thicker; (3) increased twisting forces would result in higher vertical bending stresses at the symphysis; therefore, the symphysis of *C. apella* is expected to be relatively deeper; (4) higher "wishboning" stresses in *C. apella* are expected to be countered by increased symphyseal thickness; (5) because greater bite forces have the potential for producing correspondingly higher dorsoventral shearing loads, both the corpus and symphysis of *C. apella* are expected to have absolutely greater cross-sectional areas.

Heterochronic terminology and hypotheses

The heterochronic comparisons made in this study involve *ontogenetic scaling*.* This approach uses bivariate ("allometric") analyses of ontogenetic data that are designed to examine the proportional changes that organisms experience during growth. McKinney (1988a:24–25) has proposed the term *allometric heterochrony* to encompass studies that consider patterns of ontogenetic covariance among variables (i.e., proportional variation), but which do not require that individuals be of known age. Instead of plotting shape against age (see Alberch *et al.*, 1979), different measurements are plotted against each other, so that their relative changes with respect to size can be examined. When ontogenetic scaling analyses are applied to this type of mixed cross-sectional data (Cock, 1966), heterochronic processes cannot be distinguished with the same precision that would be possible if ages were known (Shea, 1983). However, McKinney (1988a) argues that scaling relationships can provide useful insights in detecting fundamental shifts in developmental patterns when the chronological ages of individuals are unavailable (as for museum collections of wild-shot primates).

Heterochronic comparisons of growth patterns are most often made in terms of "ancestors" and "descendants". Alberch *et al.* (1979) identify a basic set of perturbations in the growth patterns of an ancestral species that can result in alterations in the adult form of a descendant (see also Gould, 1977; Alberch, 1980; Shea, 1983; McKinney, 1984, 1986, 1988a,b; McNamara, 1986). The perturbations fall into three categories: (1) change in the duration or rate of growth (size change); (2) change in the rate of shape change as size

---

*For the purposes of this study, the term "ontogenetic scaling" refers to relationship between size and shape in an ontogenetic series of organisms. Shea (1983, 1985), on the other hand, uses the term more specifically to describe a situation in which two or more samples have similar ontogenetic trajectories.

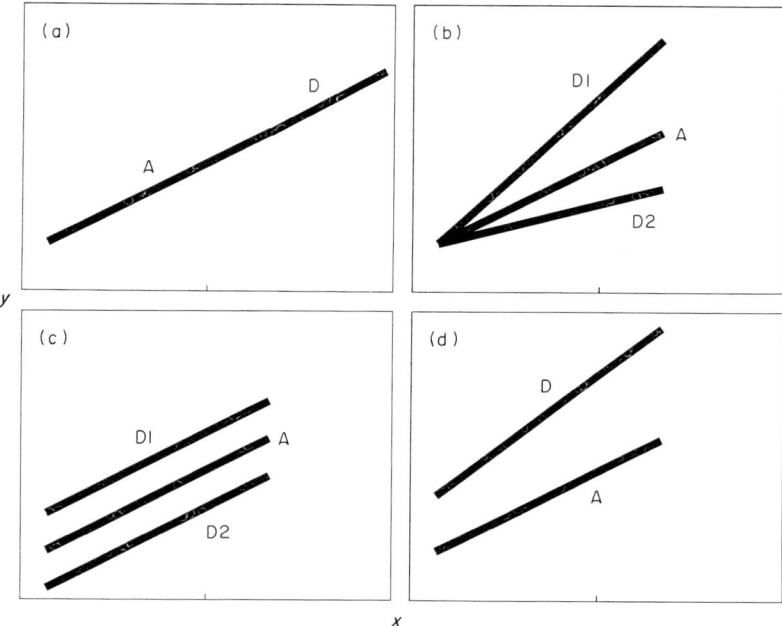

Figure 1. Perturbations of ancestral growth trajectories that will lead to significant differences in shape between ancestral and descendant species. (a) Hypermorphosis: if shape changes during growth for the ancestral species (A), the descendant species (D) will have a different adult shape if it grows to a different adult size along the same trajectory; (b) changes in rate of shape change: a difference in adult shape that is attained by the descendant species (D1) because the rate of change in y relative to the rate of change in x is greater than in the ancestral species (A) is termed *acceleration*. The opposite condition (D2) is termed *neoteny*. In both cases, adult sizes may be similar; (c) changes in growth onset: adult shape differences may be present from the beginning of postnatal growth. *Pre-displacement* occurs if a descendant species (D1) begins growth with a relative greater value of y than the ancestor (A). *Post-displacement* occurs when the descendant (D2) begins with a relatively smaller value of y; (d) combinations of patterns: the example shown in one in which differences in adult shape are produced through both pre-displacement and acceleration of the descendant growth trajectory (D), relative to that of the ancestral species (A).

changes (allometry); and (3) changes in the timing of growth onset. The analytical protocol used here has been constructed to compare patterns of postnatal development in terms of these categories.

Change in the duration or rate of growth
Shape differences in descendant species may be brought about by growing along the ancestral growth trajectory to a different adult size. If size and shape are correlated in such a case (that is, if growth is allometric), then the shape differences observed between species will be attributable to interspecific size differences among adults. *Hypermorphosis* (Figure 1a) is the term describing the condition in which shape changes occur in a descendant species that has "extended" the growth trajectory of its ancestor (McNamara, 1986). *Progenesis* is the opposite of hypermorphosis; the descendant reaches a smaller adult size via truncation of the ancestral trajectory. In a case of progenesis, adult descendants resemble immature ancestors in both size and shape. Hypermorphosis and progenesis can occur through alterations in either the rate of growth or the duration of growth. For example, Shea (1983) makes a distinction between *rate hypermorphosis* (where the rate of size increase is greater in the descendant,

although the duration of growth is the same) versus *time hypermorphosis* (where the descendant grows at the same rate as its ancestor, but for a longer duration).

Change in the rate of shape change as size changes
The second class of growth perturbations involves the rate of *shape* change during growth, where the ancestral and descendant growth trajectories become *dissociated*. These patterns are distinguished from the previous category which considered rates of change in *size* alone. If the rate of shape change during growth is more positively allometric in the descendant species than in the ancestral species, *acceleration* is said to occur (Figure 1b). The opposite condition, in which the rate of shape change is more negatively allometric in the descendant, is termed *neoteny* (Figure 1b). In either of these instances, the adults of the ancestral and descendant species need not differ in size or age at the cessation of growth (McNamara, 1986; McKinney, 1988a).

Changes in the timing of growth onset
The third class of growth perturbations considers shape differences that are already apparent at the observed onset of growth (Figure 1c). In some cases, the descendant species may exhibit a higher value y for any given value of x than ancestral individuals. This pattern is called *pre-displacement*. The opposite condition is *post-displacement*, where the descendant exhibits a lower value of y for any given x. In contrast to acceleration or neoteny, there are no modifications in the ancestral rates of shape change with growth, so that the postnatal growth trajectories remain parallel. However, as for changes in the rate of shape change, the ancestral and descendant patterns are said to be dissociated.

These different classes of perturbations are not necessarily mutually exclusive and may occur in combination. Figure 1d provides a hypothetical example in which the descendant species (D) displays a more peramorphic (i.e., "adult-like") shape at growth onset than the ancestral species (A). During growth, the relatively peramorphic form of D is exaggerated by an accelerated rate of shape change. Divergence in the adult form of D could be increased further if species D were to grow faster or for a longer amount of time.

Analyses of shape variation will be carried out using the following research design. For species comparisons, *C. albifrons* will be considered the "ancestral" species, while *C. apella* will be considered its "descendant". The phylogenetic relationships between *Cebus* species have yet to be fully resolved (but see Dickinson *et al.*, 1990); however, the generalized form of the masticatory apparatus in the untufted species group (compared to the specialized *C. apella* adaptations for hard-object feeding) suggests that *C. albifrons* is more likely to retain a primitive pattern of growth, as least as far as this particular functional complex is concerned. The null model is that the two species grow along the same ontogenetic trajectory. Differences in rate of shape change (acceleration or neoteny) or in shape at growth onset (pre-displacement or post-displacement) will result in a rejection of the null hypothesis. Because of marked sexual size and shape dimorphism in the crania of both species, sexes will be analysed separately. Within species, sex differences in growth will be analysed by arbitrarily assigning "ancestral" status to females. As with the sex-specific species comparisons, the null hypothesis is that the ontogenetic trajectories of the sexes will be the same. Once again, following the definition of "allometric heterochrony" proposed by McKinney (1988a), there are no assumptions of chronological age being made here; the null hypotheses have been constructed as tests for similar relationships between size increase during growth and size-related shape change.

Table 1 Age and sex composition of the study sample, by species

Species	Sex	Ontogenetic	Adult
Cebus apella	Males	90	49
	Females	48	39
	Unknown	3	0
Cebus albifrons	Males	69	31
	Females	41	37
	Unknown	2	0

Materials and methods

The study sample consists of 409 *Cebus* specimens (229 *C. apella* and 180 *C. albifrons*) drawn from the collections of the American Museum of Natural History (Department of Mammalogy) and the United States National Museum of Natural History (Department of Vertebrate Zoology). A summary of the sex and age composition of the sample is presented in Table 1. A specimen was considered to be adult if the sphenobasilar synchondrosis had undergone partial or complete fusion. All measurements were taken with digital sliding calipers and rounded to the nearest 0·1 mm.

Definition of variables

In this study, the form of the masticatory apparatus is described by the use of "shape variables" derived from linear measurements of the cranium and mandible. When a complex biomechanical system is studied using only its skeletal elements, simplifying assumptions are unavoidable (Weijs, 1980). Therefore, the shape variables described here should be thought of as "first approximations" of the biomechanical properties of the masticatory system. For each of the eight shape variables defined below, the maximum length of the mandible (denoted as LENGTH) will serve as a general "size" variable against which other measurements will be compared. In some cases, LENGTH actually represents a biomechanical lever arm; in other cases (particularly where a lever arm cannot be approximated using skeletal material), it serves as an overall "size surrogate" (Daegling, 1990). LENGTH is defined as the distance from the posterior border of the ramus to infradentale, taken parallel to the occlusal plane.

The area of the infratemporal foramen approximates the size of the jaw adducting musculature. It is expected to be significantly correlated with the cross-sectional size of the temporalis muscle, as well as the length of the masseter origin (Corruccini, 1980; Demes *et al.*, 1986; Demes & Creel, 1988). Infratemporal foramen size (IFS) is estimated using the formula for the area of the ellipse and then converting this value to a linear scale by taking its square root:

$$\text{IFS} = [\pi(\text{Foramen length}/2)(\text{Foramen width}/2)]^{1/2} \quad (1)$$

Foramen length is the greatest inside dimension of the infratemporal foramen. Foramen width is the greatest inside dimension that is perpendicular to the length. Relative infratemporal foramen size (RIFS) is the ratio of IFS to length. RIFS approximates the relative sizes of the muscular and skeletal components of the masticatory apparatus.

The variable LEVER approximates the position of the vertical component of the resultant force of the jaw adducting muscles (temporalis, masseter and medial pterygoid). It is the distance (taken parallel to the occlusal plane) between the posterior and anterior borders of the ramus. The anterior endpoint is defined as the point at which, in lateral view, the ramus border appears to intersect the alveolar margin. The ratio of LEVER to LENGTH defines the mechanical advantage in incision (MAI). In this case, LENGTH serves as an estimate of the bite-point load arm. An anterior bite point is chosen because field studies suggest that the incisors, canines and premolars are the teeth used by *C. apella* for breaking hard food objects (Izawa, 1975; Izawa & Mizuno, 1977; Terborgh, 1983).

The corpus depth at M_1 (CD) is the depth of the corpus at the midpoint of the M_1 crown or, in infants, at the estimated midpoint of the M_1 crypt. The first permanent molar (as opposed to the second or third) was chosen as the point for recording corpus depth to maximize the age range across which the measurement could be taken. Corpus thickness at M_1 (CT) is the transverse diameter of the corpus, taken at the same point along its length as CD. The relative corpus depth (RCD = CD/LENGTH) reflects the resistance of the corpus to parasagittal bending. The lever arms of the muscular and bite forces that tend to twist the corpus cannot be precisely estimated from the skeleton (Hylander, 1988; Daegling, 1990). Therefore, relative corpus thickness (RCT = CT/LENGTH) should be considered as indicative of twisting resistance (where higher values indicate greater resistance) in only a general sense. Corpus size is a linear estimate of corporal cross-sectional area, calculated as an ellipse:

$$CS = [\pi(CD/2)(CT/2)]^{1/2} \qquad (2)$$

The absolute cross-sectional area of a beam is proportional to its ability to withstand shearing stresses that act perpendicularly to the longitudinal axis. Relative corpus size (RCS = CS/LENGTH) does not have an explicit biomechanical interpretation, but serves as an overall measure of the robusticity of the corpus. Symphysis depth (SD) is the distance between infradentale and the most distant point in the midline of the posteroinferior aspect of the symphysis. Relative symphysis depth (RSD = SD/LENGTH) approximates the resistance of the symphysis to bending stresses that occur during corporal twisting. Symphysis thickness (ST) is the minimum diameter of the symphysis. Relative symphysis thickness (RST = ST/LENGTH) describes the ability of the symphysis to withstand wishboning stresses, where LENGTH approximates the lever arm of the lateral component of the balancing-side jaw muscles (Daegling, 1990). Symphysis size (SS) is a linear estimate of symphyseal area:

$$SS = [\pi(SD/2)(ST/2)]^{1/2} \qquad (3)$$

In absolute terms, it describes the resistance of the symphysis to dorsoventral shear stresses, which occur during unilateral biting and mastication (Hylander, 1979). Relative symphysis size (RSS = SS/LENGTH) is a general indication of symphyseal robusticity.

Shape comparisons

The analysis begins with an assessment of shape variation in the adult sample. The "shape" of an individual may be defined explicitly as the ratio of a variable *y* to LENGTH in each of the cases defined above (Mosimann, 1970; Mosimann & James, 1979; Alberch *et al.*, 1979):

$$\text{"Shape"} = y/\text{LENGTH} \qquad (4)$$

The end results (i.e., adult shapes) of ontogenetic trajectories are compared using analysis of variance (ANOVA) on species- and sex-specific shape means. If overall heterogeneity of means is found, then Tukey's (HSD) multiple comparisons test (Sokal & Rohlf, 1981; Zar,

1984) is employed to determine *a posteriori* which pairs of means are significantly different. The significance tests associated with ANOVA require the assumptions of normality and homogeneity of variances. Mosimann (1970; Mosimann & James, 1979) has suggested that shape ratios are often log-normally distributed; therefore, the shape variables were (natural) logarithmically transformed. Prior analysis of the log-transformed shape data showed that the assumptions of normality and homogeneity of variances required by ANOVA are met. All calculations for this phase of the analysis are performed using the GLM procedure of the SAS (1985) mainframe statistical package.

Assessing the rate of shape change during growth
Growth-related proportional change (produced by differing rates of growth among the body's different parts) is quantified here through the use of the bivariate "power function" (Huxley, 1924, 1932; Nomura, 1926):

$$y = bx^k \tag{5}$$

where y and x are the variables of interest and b is a scale factor. The coefficient k is the "ratio of the specific growth rates" of y and x (Shea, 1985a:369). Equation (5) is usually logarithmically transformed to produce a first-order linear equation:

$$\ln(y) = \ln(b) + k[\ln(x)] \tag{6}$$

where $\ln(b)$ is the y-intercept and k is the slope. In logarithmic space (and with ontogenetic data), k describes the rate of size change in $\ln(y)$ relative to the rate of size change in $\ln(x)$. It should be noted that some authors (e.g., Gelvin & Albrecht, 1987; Albrecht & Gelvin, 1987) have advocated the use of a "full model", in which an intercept term is added to Equation (5), such that

$$y = a + bx^k \tag{7}$$

where the regression statistics are calculated in the original (i.e., untransformed) data space. This adjustment may be potentially important when empirical coefficients are compared to a specific theoretical model (e.g., metabolic scaling) in raw space; however, when size-related *proportionality* is studied, the log-transformation of the original function (Equation 5) is sufficient (Smith, 1984; Jungers, 1990, pers. comm.).*

Reduced major axis (Model II) regression is used to describe and compare species- and sex-specific patterns of ontogenetic shape change in the non-adult portion of the sample. Descriptions of the computational details of reduced major axis (RMA) regression are provided by numerous sources (e.g., Imbrie, 1956; Jolicoeur & Mosimann, 1968; Kuhry & Marcus, 1977; Clarke, 1980; Sokal & Rohlf, 1981; Ricker, 1984; Rayner, 1985; Plotnick, 1989). Slopes are judged to be allometric (indicating growth-related shape change) if the 95% confidence intervals of the slope estimate (calculated following Jolicoeur & Mosimann, 1968) do not contain the predicted isometric slope of $k = 1.0$. Species and sex slopes are compared using Clarke's (1980) T statistic. Because a multiple comparison of slopes is involved, the alpha levels for testing the significance of slope differences must be adjusted to reduce the probability of a Type I error (where the null hypothesis of equal slopes is

---

*Arguments over which model ("full" or "reduced") is more appropriate for characterizing growth-related shape change are ultimately not of crucial importance, as Mosimann (1970) and Mosimann & James (1979) have demonstrated that questions of "allometry" do not necessarily require the application of regression analysis at all. Regression is used here simply as a convenient means of comparing rates of specific growth for multiple samples.

wrongly rejected). The experimental alpha level (in this case, $\alpha = 0.0083$) is determined by a Bonferroni adjustment, dividing the single-comparison level ($\alpha = 0.05$) by the six comparisons made (all possible comparisons of the four sex- and species-specific samples).

Calculations of regression equations, confidence intervals and Clarke's T statistics were performed on an IBM compatible personal computer, using a program written with Turbo Pascal (Borland International, Inc). The program is based on a modification of the BASIC program published by Plotnick (1989).

Assessing "shape" variation at the onset of postnatal growth
The most direct and accurate method of assessing shape differences at postnatal growth onset (birth) is to compare samples of newborn individuals (e.g., Falsetti & Cole, 1992). Alternative approaches must be used in samples where chronological age is unknown and where very young individuals may not be adequately represented. Using regression, proportional differences at growth onset might be assessed approximately by comparing y-intercepts, which indicate the relative "elevations" of the regression lines. However, the value of $\ln(x) = 0$ (equal to one millimeter in raw variable space) is well below the size range of the data; statistical inferences made by extrapolating so far outside that range would be ill advised. In addition, if what is desired is an estimate of whether species or sexes are different at birth (onset of postnatal growth), the extrapolations into prenatal size ranges do not serve any clear purpose.

As an alternative to using y-intercept differences to judge shape differences early in postnatal growth, a modification of the standard regression model (Equation 6) is utilized. Instead of testing for differences in regression line elevation when $\ln(x) = 0$, an "onset correction" is used to test for elevation differences at a value for x estimated to reflect the average size of an individual at postnatal growth onset (Figure 2). This is accomplished by subtracting this estimate [denoted by $\ln(x_{Onset})$] from $\ln(x)$ in Equation 6:

$$\ln(y) = \ln(b_{Onset}) + k[\ln(x) - \ln(x_{Onset})] \qquad (7)$$

The variable $\ln(b_{Onset})$ represents the modified "y-intercept", which will be used in further tests of species and sex differences in shape early in postnatal development. Note that when $\ln(x_{Onset})$ is subtracted from $\ln(x)$ for each specimen, the estimate of the slope is unaffected; only the intercept is changed.

The value for $\ln(b_{Onset})$ used in this study is defined as follows: species-specific means values for LENGTH were calculated for all animals with complete eruption of the deciduous dentition. Initial examination of the size distributions indicated no sexual dimorphism in either species at this developmental stage.* Sexes were therefore pooled to obtain a mean value of length for each species; because of the lack of dimorphism, unsexed specimens of the same stage were also included for estimating LENGTH, but were not included in regression calculations. Table 2 shows the descriptive statistics of LENGTH for these subsamples. The mean LENGTH for *C. apella* was slightly higher than that of *C. albifrons*, but a t-test indicated no significant difference between species. Therefore, the species means were pooled, so that the same value of LENGTH (39.63 mm or 3.679 in log space) was used for all regressions.

Analytical comparisons of RMA intercepts (regardless of the x-axis value at which they are examined) are not currently possible because their statistical distributions are largely unknown (Plotnick, 1989). However, tests of differences among species and sexes may be

---

*Galliari (1985) noted no sexual dimorphism in body weight of captive *Cebus apella* at 1 year of age, with the first permanent teeth erupting at 12–15 months.

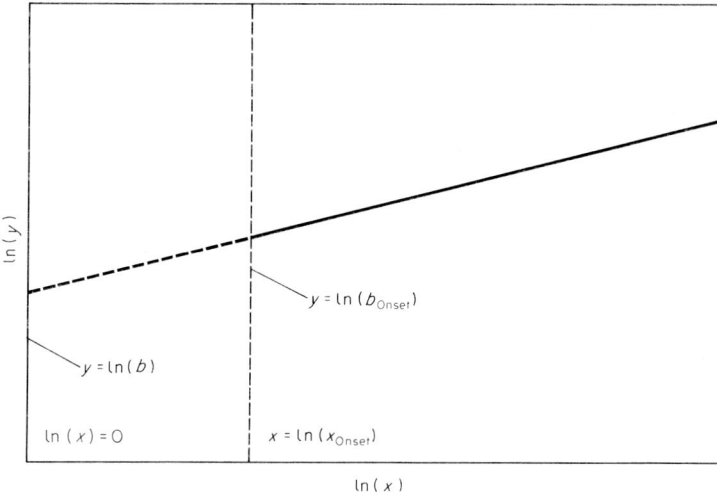

Figure 2. Schematic illustration of the modified regression model used in the analysis. The solid line represents a reduced major axis regression and spans the range of the data; the broken line is an extrapolation of the regression to an ordinate value of $\ln(x) = 0$. At $\ln(x) = 0$, the y-intercept equals $\ln(b)$. The modified regression calculates an "onset" value of the ordinate $(x = x_{\text{Onset}})$ at or near the lower size limit of the sample. The regression intercept corresponding to $x = x_{\text{Onset}}$ is $y = \ln(b_{\text{Onset}})$.

Table 2 Species- and sex-specific statistics for LENGTH (mm) at "growth onset" (see text for definition)

Species	Sex	n	Mean	SD
Cebus apella	Male	14	40·44	1·88
	Female	3	39·23	3·01
	Unknown	2	37·03	—
	Pooled	19	39·89	2·19
Cebus albifrons	Male	9	39·94	1·99
	Female	5	39·32	3·07
	Unknown	3	37·17	2·49
	Pooled	17	39·33	2·48
Grand mean			39·63	

performed on non-parametric "bootstrap" estimates of $\ln(b_{\text{Onset}})$. The bootstrap is a randomization technique that can estimate the standard errors and confidence intervals of statistics with either unknown or extremely complex analytical distributions (Efron & Tibshirani, 1986). Theoretical arguments supporting the use of the bootstrap are provided by Efron (1979, 1981), Efron & Tibshirani (1986) and Manly (1991). If a data vector contains n observations, then the bootstrap extracts a random sample of the same n from that vector. Sampling is done with replacement, so that some individuals may be sampled more than once, while others may not be sampled at all. The statistic(s) of interest [in this case, $\ln(b_{\text{Onset}})$] is then calculated from the random sample. This process is repeated for a large number of bootstrap samples, with the statistic being saved each time. When the iterations are complete, the mean of the bootstrap estimates is the estimated mean value of the statistic, while the standard deviation of the bootstrap estimates approximates the statistic's standard error.

In the study sample, bootstrap estimates of $\ln(b_{\text{Onset}})$ for species- and sex-specific samples were generated using 100 iterations, the number suggested by Efron & Tibshirani (1986) for the calculation of standard errors. Generation of bootstrap estimates were performed using the regression program mentioned previously. Examination of the resulting distributions showed the estimates for $\ln(b_{\text{Onset}})$ to be highly skewed (with a bias toward more positive values), mirroring the asymmetrical analytical distribution of the slope estimates (Plotnick, 1989). Because of these departures from normality, Kruskal–Wallis tests (Kruskal & Wallis, 1952; Sokal & Rohlf, 1981) were used to test for overall heterogeneity of mean values of $\ln(b_{\text{Onset}})$. Dunn's non-parametric multiple comparisons test (Dunn, 1964; Zar, 1984) were then used to assess pair-wise differences in mean values.

Assessing size at cessation of growth
The timing of growth cessation cannot be measured directly in the study sample. However, the sizes of adult individuals can be used as indicators of whether species or sexes differ in shape as the result of hypermorphosis. Recall that hypermorphosis involves the extrapolation of an ancestral growth trajectory into larger size ranges. If shape is size-related (as is usually the case), then shape changes may result in the "overgrown" descendants, providing the size differences are great enough. Size differences may also contribute to adult shape differences when the growth trajectories are not coincident. For example, the growth trajectory of a descendant species could be pre-displaced with respect to its ancestor, as well as extending into a greater size range. Such a combination could be described as "pre-displacement with size increase".

In determining whether size differences occur among adults, species- and sex-specific means of the log-transformed linear measurements will be compared using analysis of variance. As with the adult shape comparisons, Tukey's test will be used to identify significant differences among means. If a significant difference in size is present when comparing "ancestors" and "descendants", then hypermorphosis or (in the case of dissociated growth patterns) "size increase" may be considered as a possible explanation for adult shape differences. Conversely, lack of a size difference could discount hypermorphosis or other increases in size as explanations.

Results

Table 3 presents descriptive statistics for the raw measurements of the adult sample. Also shown are the results of Tukey's tests of species- and sex-specific means (calculated from log-transformed data). These results indicate that *C. apella* males are significantly larger in all dimensions than *C. albifrons* males, except IFS. This is not an unexpected result, as Jungers & Fleagle (1980) report that the body mass of *C. apella* males (up to 4 kg) is one-third greater than that of *C. albifrons* males (3 kg). *Cebus apella* females are larger than *C. albifrons* females in all measurements except IFS, LEVER and LENGTH. When comparing sexes within species, males are significantly larger than females in all dimensions.

Table 4 presents the results of the analyses of variance between adult samples for the long-transformed shape ratios. For the shape variables related to the production of external forces, there is a pattern of significant sexual dimorphism, but without sex-specific differences between species. For RIFS, males have relatively larger foramina (and, by extension,

Table 3 Species- and sex-specific descriptive statistics for adult measurements (see text for abbreviations)

Measurement[1]		n	mean	SD	Tukey's test[2]
LENGTH	ALM	31	61·81	4·08	APM > ALM >
	ALF	37	53·88	1·95	APF > ALF
	APM	49	64·91	3·15	
	APF	39	55·43	2·44	
IFS	ALM	31	19·99	1·79	APM > ALM >
	ALF	37	16·34	1·18	APF > ALF
	APM	49	20·80	1·70	
	APF	39	16·50	1·19	
LEVER	ALM	31	28·78	2·90	APM > ALM >
	ALF	37	22·77	1·51	APF > ALF
	APM	49	30·38	2·44	
	APF	39	23·45	1·93	
CD	ALM	31	15·02	1·54	APM > ALM >
	ALF	37	12·82	1·00	APF > ALF
	APM	49	16·81	1·12	
	APF	39	14·34	1·05	
CT	ALM	31	6·29	0·65	APM > ALM >
	ALF	37	5·54	0·51	APF > ALF
	APM	49	7·44	0·63	
	APF	39	6·19	0·49	
CS	ALM	31	8·61	0·84	APM > ALM >
	ALF	37	7·47	0·57	APF > ALF
	APM	49	9·90	0·69	
	APF	39	8·34	0·52	
SD	ALM	31	21·53	1·60	APM > ALM >
	ALF	37	18·76	1·30	APF > ALF
	APM	49	23·20	1·28	
	APF	39	19·87	1·20	
ST	ALM	31	8·71	0·81	APM > ALM >
	ALF	37	7·41	0·56	APF > ALF
	APM	49	10·09	0·61	
	APF	39	8·31	0·53	
SS	ALM	31	12·13	0·93	APM > ALM >
	ALF	37	10·44	0·65	APF > ALF
	APM	49	13·55	0·69	
	APF	39	11·38	0·64	

[1] All measurements are in millimeters.
[2] Results of Tukey's multiple comparisons test (performed on log-transformed data) for significant differences among means; groups spanned by horizontal lines are not significantly different.
ALM = *Cebus albifrons* males, ALF = *C. albifrons* females, APM = *C. apella* males, APF = *C. apella* females.

relatively larger jaw adductors), but there are no significant differences between species for either sex. For MAI, males have higher values, indicating greater mechanical adavantage when loading the anterior dentition.

When the shape variables related to the dissipation of internal forces are considered, there is a change in pattern, so that there are significant species differences, irrespective of sex. The

Table 4 **Species- and sex-specific descriptive statistics for adult shape variables**

Shape variable		n	mean	SD	Tukey's test
IFS/LENGTH (RIFS)	ALM	31	0·268	0·019	ALM > APM >
	ALF	37	0·240	0·019	ALF > APF
	APM	49	0·262	0·022	
	APF	39	0·233	0·016	
LEVER/LENGTH (MAI)	ALM	31	0·465	0·020	APM > ALM >
	ALF	37	0·423	0·020	APF > ALF
	APM	49	0·468	0·022	
	APF	39	0·423	0·021	
CD/LENGTH (RCD)	ALM	31	0·242	0·019	APM > APF >
	ALF	37	0·238	0·015	ALM > ALF
	APM	49	0·259	0·015	
	APF	39	0·259	0·015	
CT/LENGTH (RCT)	ALM	31	0·102	0·010	APM > APF >
	ALF	37	0·103	0·009	ALF > ALM
	APM	49	0·114	0·009	
	APF	39	0·112	0·009	
CS/LENGTH (RCS)	ALM	31	0·139	0·011	APM > APF >
	ALF	37	0·139	0·009	ALM > ALF
	APM	49	0·152	0·009	
	APF	39	0·151	0·009	
SD/LENGTH (RSD)	ALM	31	0·348	0·016	APF > APM >
	ALF	37	0·347	0·020	ALM > ALF
	APM	49	0·356	0·018	
	APF	39	0·359	0·018	
ST/LENGTH (RST)	ALM	31	0·141	0·009	APM > APF >
	ALF	37	0·138	0·009	ALM > ALF
	APM	49	0·155	0·010	
	APF	39	0·150	0·008	
SS/LENGTH (RSS)	ALM	31	0·196	0·009	APM > APF >
	ALF	37	0·194	0·010	ALM > ALF
	APM	49	0·208	0·011	
	APF	39	0·206	0·009	

Data were transformed to natural logarithms and Tukey's multiple comparisons test for significant differences among means was performed; groups spanned by horizontal lines are not significantly different.

relative corpus depths (RCD), thickness (RCT), and cross-sectional sizes (RCS) for the *C. apella* samples are greater than for *C. albifrons*, although there is no sexual shape dimorphism for either species.

Relative symphysis depth is the only shape variable that does not exhibit heterogeneity of sample means according to the Tukey test, although the overall heterogeneity was significant ($F = 3·9$; df = 3, 152; $P = 0·01$). Finally, relative symphysis thickness (RST) and the relative cross-sectional size of the symphysis (RSS) show a pattern similar to the corporal shape variables: the symphyses of *C. apella* are relatively thicker and larger than *C. albifrons* for both sexes and there is no sexual dimorphism in either species.

Significant changes in shape ratios can result from perturbations in either or both variables; these changes are not obvious when the shape variables are studied by themselves.

Table 5 Percentage differences in measurement means for raw measurements of the adult samples

Variable	Species % differences (by sex)		Sex % differences (by species)	
	APM vs. ALM	APF vs. ALF	ALM vs. ALF	APM vs. APF
LENGTH	**5·0**	3·0	**7·9**	**17·0**
IFS	4·1	1·0	**22·3**	**26·1**
LEVER	**5·7**	3·0	**26·2**	**29·4**
CD	**12·4**	**11·9**	**16·5**	**8·1**
CT	**18·0**	**11·7**	**12·9**	**19·3**
CS	**15·0**	**11·6**	**15·3**	**18·7**
SD	**7·6**	6·3	**14·7**	**16·2**
ST	**15·7**	**12·1**	**17·4**	**21·2**
SS	**11·7**	**9·0**	**16·2**	**19·1**

Instances in which means are significantly different (see Table 3) are in boldface.

Therefore, it is instructive to "dissect" the ratios, so that differing trends in the numerators and denominators can be examined. Table 5 expresses differences in the mean values of the raw size variables (from Table 3) in terms of percentages (Borgognini, Tarli & Repetto, 1986). In each sex-specific species comparison, the *C. albifrons* mean was subtracted from the *C. apella* mean, the difference divided by the *C. albifrons* mean and the resulting ratio multiplied by 100. For the species-specific sex comparisons, the differences in sex means were divided by the corresponding female means. In any case, a value of zero would indicate no difference in mean size.

In the comparisons of males, where all of the *C. apella* means were significantly greater, the percentage difference was least for LENGTH, which is the denominator for each shape variable. Therefore, although there is a significant difference in LENGTH between males, the significant shape differences described in Table 4 appear to be more the result of mean values for the other measurements (except for IFS) that are disproportionately larger in *C. apella* males. For the females, it is readily apparent that the significant shape differences between *C. apella* and *C. albifrons* stem from disproportionate increases in the numerator variables because there is no significant difference in LENGTH.

For the *C. albifrons* samples, the percentage differences between male and female means are least for LENGTH, although LENGTH is significantly greater in the males of both species. Therefore, size differences are potential contributors to sexual shape dimorphism in both species. With the exception of CD, the *C. apella* sample tends to be more size dimorphic, although the significance of this tendency is not tested.

Results of the RMA regression analyses are presented in Table 6. The sex- and species-specific regressions of IFS versus LENGTH (Figure 3) indicate positive allometry in each case, such that the jaw adducting muscles (as inferred from IFS) grow disproportionately faster than the mandible. The T-statistics indicate that there are no significant differences among slopes, while the results of Dunn's test on the bootstrap estimates of $\ln(b_{\text{Onset}})$ indicate that all of the pair-wise comparisons are significant. At growth onset, *C. albifrons* males exhibit the greatest absolute and relative muscle size, followed by *C. apella* females, *C. albifrons* females and *C. apella* males. The regression of LEVER versus LENGTH (Figure 4) are all positively allometric, indicating that the mechanical advantage of the jaw adductors (when loading the anterior dentition) increases with growth in each case. There are no significant differences

Table 6 Reduced major axis (RMA) regressions for sex- and species-specific, non-adult samples

		n	Slope (95% CI)	Int	r	$\ln(b_{Onset})$	Slope comparisons (Clarke's T)	$\ln(b_{Onset})$ comparisons (Dunn's test)
IFS vs. LENGTH	ALM	69	1·823 (1·708–1·945)	−4·764	0·96	1·942	ALF > APM > APF > ALM	ALM > APF > ALF > APM
	ALF	41	2·160 (1·853–2·518)	−6·061	0·88	1·886		
	APM	90	1·934 (1·849–2·023)	−5·269	0·98	1·845		
	APF	48	1·906 (1·742–2·085)	−5·112	0·95	1·898		
LEVER vs. LENGTH	ALM	69	1·201 (1·133–1·272)	−1·632	0·97	2·785	ALF > APM > ALM > APF	ALM > APF > ALF > APM
	ALF	41	1·278 (1·146–1·425)	−1·954	0·94	2·749		
	APM	90	1·261 (1·209–1·316)	−1·902	0·98	2·738		
	APF	48	1·200 (1·119–1·287)	−1·643	0·97	2·772		
CD vs. LENGTH	ALM	69	1·210 (1·127–1·300)	−2·281	0·96	2·171	APM > ALM > ALF > APF	APF > APM > ALF > ALM
	ALF	41	1·206 (1·024–1·421)	−2·261	0·86	2·177		
	APM	90	1·251 (1·184–1·324)	−2·390	0·96	2·215		
	APF	48	1·033 (0·935–1·142)	−1·524	0·94	2·276		
CT vs. LENGTH	ALM	69	0·635 (0·514–0·784)	−0·739	0·49	1·597	ALF > ALM > APM > APF	APM > APF > ALM > ALF
	ALF	41	0·672 (0·495–0·913)	−0·886	0·30	1·587		
	APM	90	0·536 (0·462–0·620)	−0·241	0·72	1·729		
	APF	48	0·531 (0·405–0·697)	−0·227	0·37	1·727		
CS vs. LENGTH	ALM	69	0·835 (0·747–0·937)	−1·283	0·88	1·789	APM > ALM > ALF > APF	APF > APM > ALM > ALF
	ALF	41	0·813 (0·660–1·001)	−1·206	0·77	1·784		
	APM	90	0·848 (0·788–0·913)	−1·256	0·94	1·864		
	APF	48	0·665 (0·578–0·765)	−0·541	0·88	1·904		
SD vs. LENGTH	ALM	69	0·932 (0·874–0·994)	−0·757	0·96	2·673	APM > APF > ALF > ALM	ALM > ALF > APF > APM
	ALF	41	0·965 (0·805–1·156)	−0·893	0·83	2·656		
	APM	90	1·102 (1·039–1·169)	−1·423	0·96	2·632		
	APF	48	1·008 (0·896–1·134)	−1·059	0·92	2·649		
ST vs. LENGTH	ALM	69	0·886 (0·801–0·980)	−1·469	0·91	1·790	APM > ALM > ALF > APF	APF > APM > ALM > ALF
	ALF	41	0·867 (0·705–1·065)	−1·419	0·77	1·768		
	APM	90	1·019 (0·957–1·085)	−1·926	0·96	1·824		
	APF	48	0·843 (0·729–0·974)	−1·255	0·87	1·845		
SS vs. LENGTH	ALM	69	0·891 (0·830–0·957)	−1·164	0·96	2·116	APM > APF > ALM > ALF	APF > ALM > APM > ALF
	ALF	41	0·844 (0·720–0·990)	−1·000	0·87	2·105		
	APM	90	1·044 (0·994–1·096)	−1·730	0·97	2·111		
	APF	48	0·901 (0·803–1·010)	−1·182	0·92	2·131		

All data have been transformed to natural logarithms. Also shown are multiple-comparisons test for differences in slopes and $\ln(b_{Onset})$. Underlines span groups that are not significantly different.

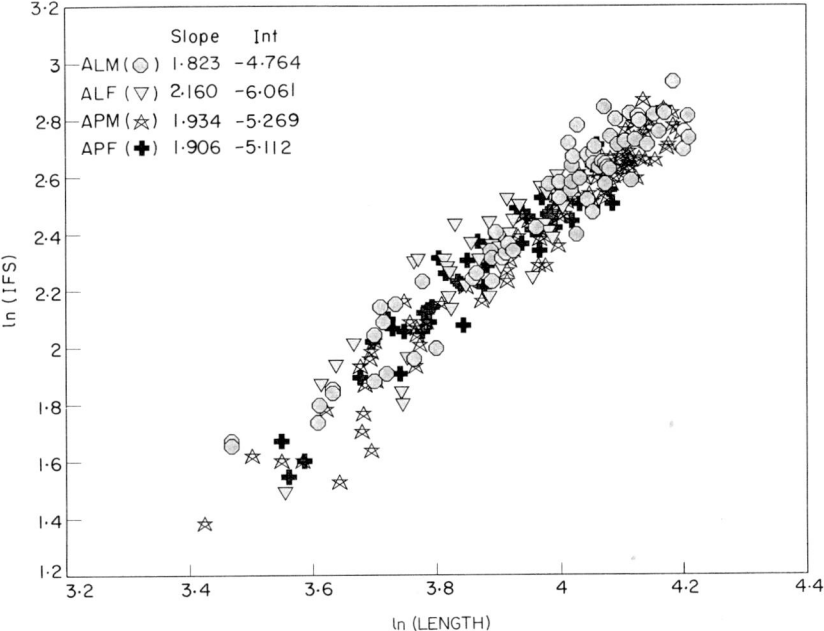

Figure 3. Reduced major axis regressions of ln (IFS) versus ln (LENGTH). ALM = *C. albifrons* males, ALF = *C. albifrons* females, APM = *C. apella* males and APF = *C. apella* females.

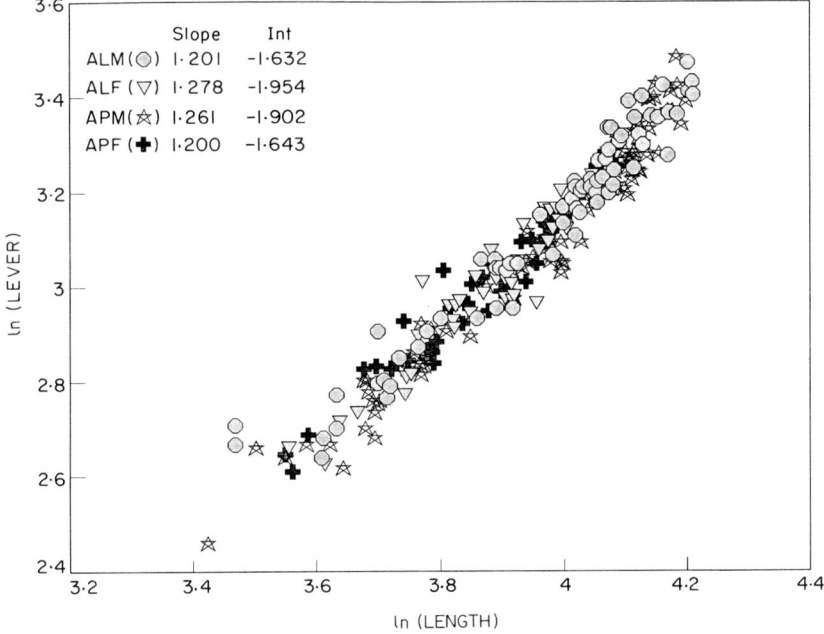

Figure 4. Reduced major axis regressions of ln (LEVER) versus ln (LENGTH).

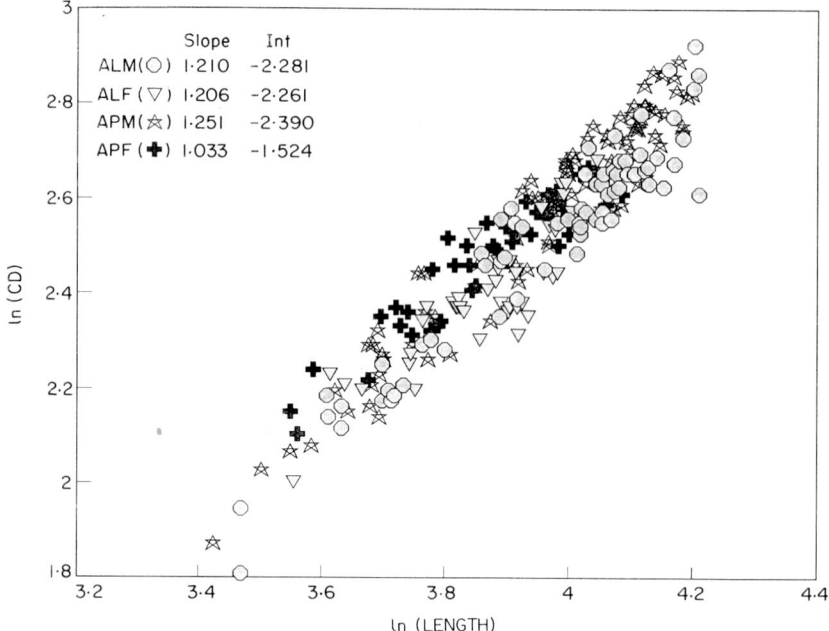

Figure 5. Reduced major axis regressions of ln (CD) versus ln (LENGTH).

among slopes. The Dunn's tests on $\ln(b_{Onset})$ are all significant. The *C. albifrons* male sample has the greatest value at growth onset, followed by *C. apella* females, *C. albifrons* females and *C. apella* males. Despite the higher elevation of the *C. albifrons* male regression line at growth onset, there is no difference in MAI between *C. albifrons* adult males and *C. apella* adult males. This lack of an adult shape difference is perhaps attributable to the slightly (but not significantly) greater slope in *C. apella* males, coupled with their greater size.

When CD is regressed against LENGTH (Figure 5), each sample scales with positive allometry, with the exception of *C. apella* females, which scale isometrically. The Clarke's *T*-statistics indicate that there is a significant pair-wise difference between the most extreme slope estimates (1·251 in *C. apella* males versus 1·033 in *C. apella* females); there are no other significant differences in slopes. The Dunn's tests on $\ln(b_{Onset})$ indicate that relative corpus depth is largest in *C. apella* females, followed by a significantly lower value of *C. apella* males and, finally, by significantly lower values for both *C. albifrons* samples (which do not differ from each other).

In each sex- and species-specific case, CT scales negatively with respect to LENGTH (Figure 6), indicating that the relative thickness of the corpus decreases with growth in each sample. Clarke's *T*-statistics indicate no differences in slope estimates. The Dunn's tests indicate a pattern that is broadly similar to what was seen for CD versus LENGTH; however, in this case, the *C. apella* males possess the greatest estimated value of RCT at growth onset.

In three of four cases, the relative cross-sectional size of the corpus (CS) scales negatively with respect to LENGTH (Figure 7). The exception is the *C. albifrons* female sample, which scales isometrically. There is a significant difference between the extremes of the slope estimates (0·848 in *C. apella* males versus 0·665 in *C. apella* females). Results of the Dunn's tests indicate that $\ln(b_{Onset})$ is greatest in *C. apella* females, followed by a significantly lower value

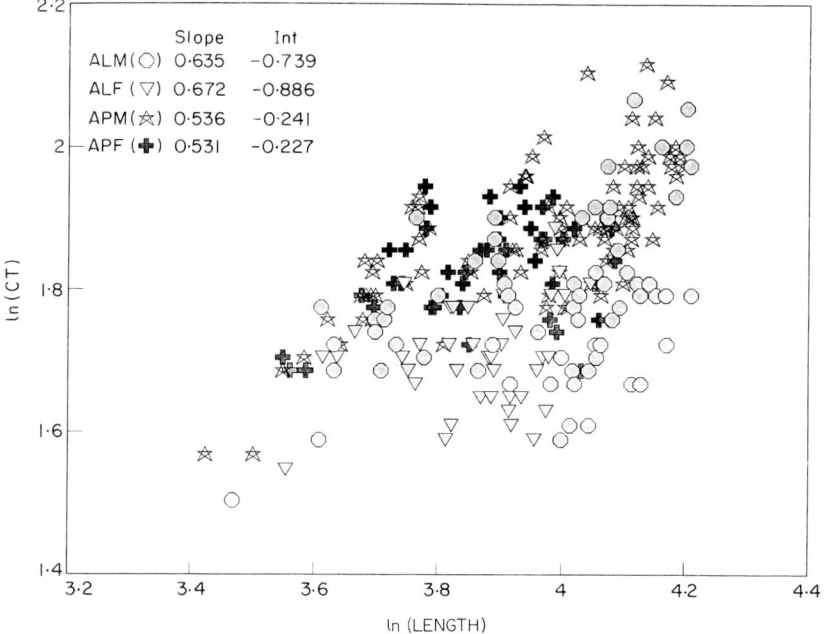

Figure 6. Reduced major axis regressions of ln (CT) versus ln (LENGTH).

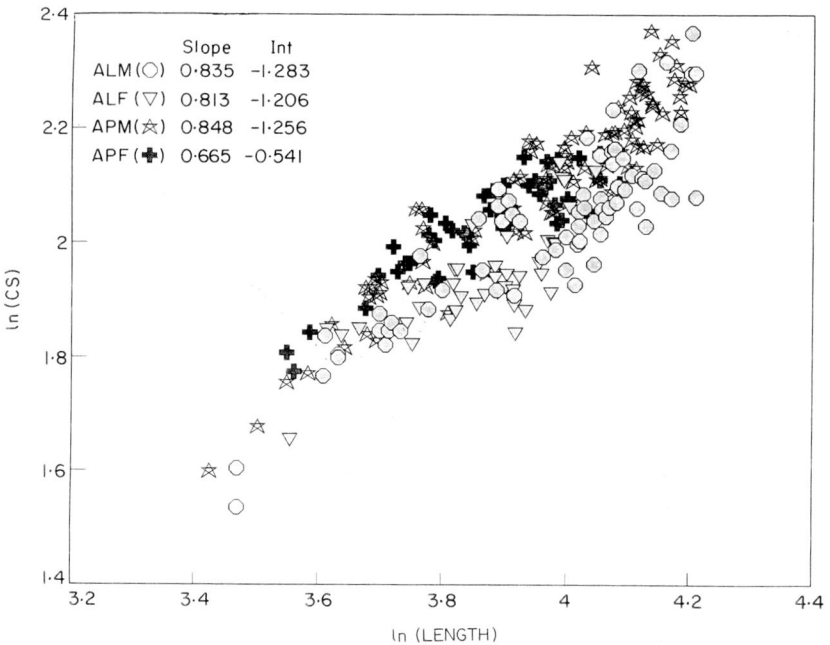

Figure 7. Reduced major axis regressions of ln (CS) versus ln (LENGTH).

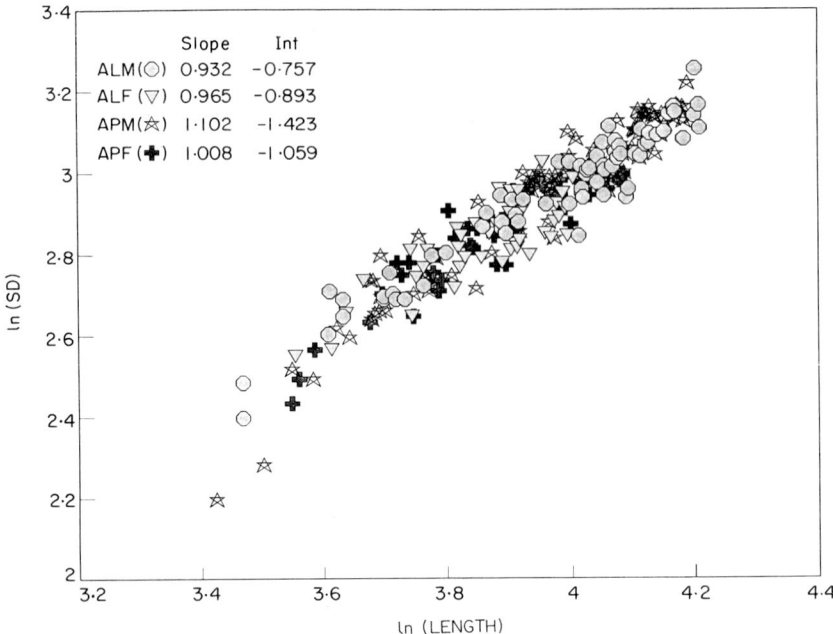

Figure 8. Reduced major axis regressions of ln (SD) versus ln (LENGTH).

for the *C. apella* male sample, followed by significantly lowest values for the *C. albifrons* samples (with sexes not significantly different from each other).

When SD is regressed against LENGTH (Figure 8), both female samples exhibit isometry, such that the relative depth of the symphysis does not tend to change during growth. The *C. albifrons* male sample is barely negatively allometric, so that the relative depth of the symphysis decreases slightly with growth. In contrast, there is a positively allometric trend in *C. apella* males, where relative and absolute symphyseal depth increases with growth. Dunn's test shows that $\ln(b_{Onset})$ is greatest in *C. albifrons* males at growth onset. The female estimates of both species are significantly lower, but are not different from each other. Finally, the elevation of the *C. apella* male sample is significantly lower than each of the other samples.

The regressions of ST versus LENGTH (Figure 9) show no significant differences in scaling patterns. In the male *C. albifrons* and female *C. apella* samples, there is negative allometry, so that the relative thickness of the symphysis decreases with growth. In contrast, the female *C. albifrons* and male *C. apella* slopes scale with isometry, indicating no significant growth-related changes in shape. Dunn's test shows that the elevations of the samples are all significantly different from each other. At growth onset, the elevation of the *C. apella* female sample is greatest, followed by *C. apella* males, *C. albifrons* males and *C. albifrons* females.

The regressions of SS versus LENGTH (Figure 10) indicate that there is slight negative allometry for the *C. albifrons* samples and isometry for the *C. apella* samples. The only heterogeneous pair of slope estimates is for the male samples, where the *C. apella* slope is significantly greater. Dunn's test shows that all of the regression elevations at growth onset are significantly different from each other. At growth onset, the relative size of the symphysis is greatest in *C. apella* females, followed by *C. albifrons* males, *C. apella* males and *C. albifrons* females.

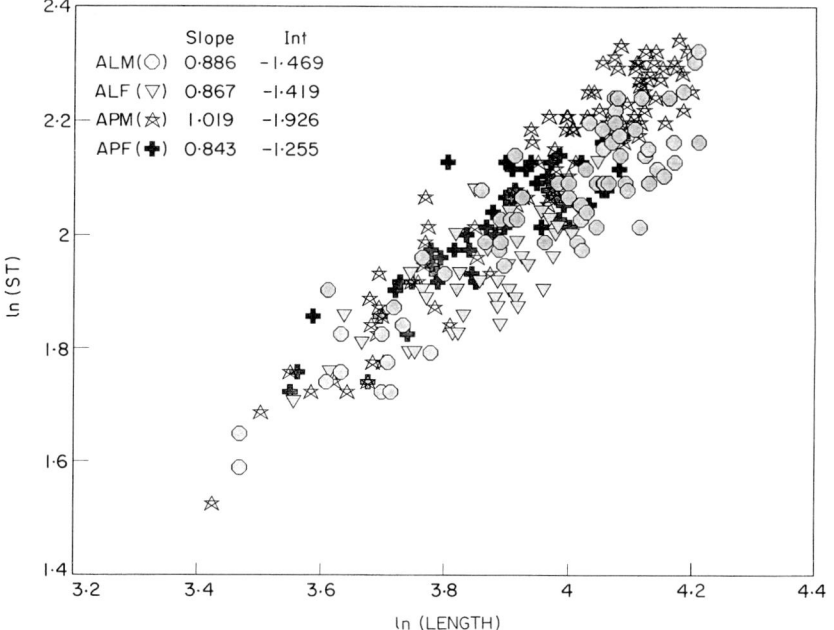

Figure 9. Reduced major axis regressions of ln (ST) versus ln (LENGTH).

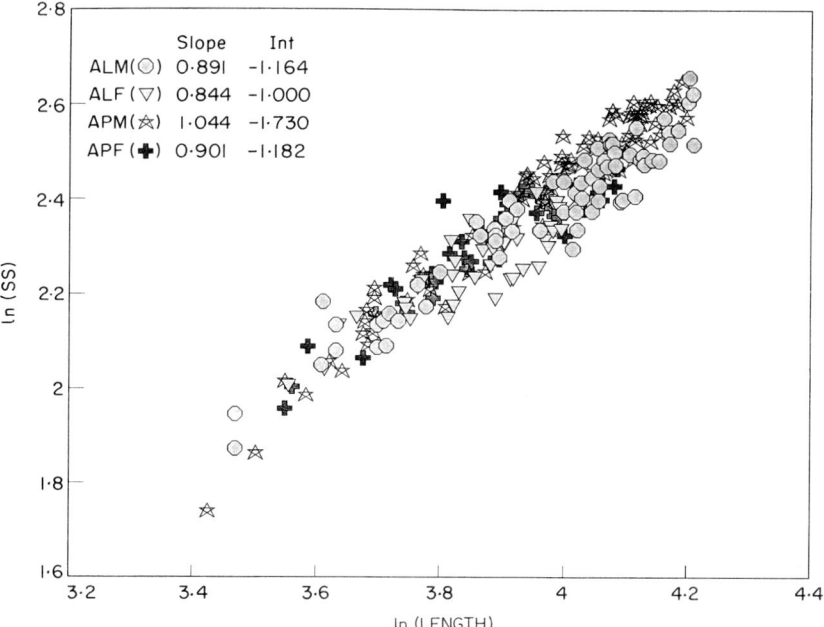

Figure 10. Reduced major axis regressions of ln (SS) versus ln (LENGTH).

Discussion

Biomechanical differences among adults

The biomechanical implications of the results shown in Tables 3 and 4 may be summarized as follows: in terms of the generation of external forces, *C. apella* males have roughly the same as *C. albifrons* males. In addition, there is no difference in the MAI. Therefore, the potential for producing high bite forces appears similar (as least as far as can be determined from skeletal measurements). For the females, there are also no indications of significant differences in force production, at least as estimated here from IFS and MAI. With respect to sexual dimorphism, males of both species appear capable of generating higher bite forces as the result of both absolutely larger muscle size and greater mechanical advantage.

For internal forces affecting the corpus, biomechanical differences are apparent between species, with no sexual shape dimorphism in either case. The corporal shapes of the *C. apella* samples suggest greater resistance to both parasagittal bending and twisting (as predicted previously). In terms of relative shear resistance, *C. apella* males and females appear more resistant than *C. albifrons* males and females, respectively. Within both species, the greater values of CS suggest greater absolute shear resistance in males.

The importance of resistance to corporal twisting in hard-object feeding becomes more apparent on further consideration of *C. apella* cranial anatomy and feeding behavior. According to Hylander (1988:56), twisting of the mandibular corpus "is apparently the most important [factor] for causing the occurrence of large stress regimes". In a member that is twisted, the greatest stresses are concentrated at the portion of the external surface that is closest to the longitudinal axis (Popov, 1976; Hylander, 1988). Therefore, when twisting loads are increased, the most effective means of increasing stress resistance is to increase the minimum diameter of the member cross-section (i.e., corpus thickness).

It is interesting to note that Hylander (1988) believes that resistance to twisting was an especially important consideration in the mandibular design of the fossil hominid *Paranthropus*, where corpus thickness is greatly increased, as compared to the form of the jaw in the more "gracile" australopithecines. Two of the points used by Hylander to argue that greater twisting stresses would be expected in *Paranthropus* could have analogues in the design of the *C. apella* mandible. The first reason for expecting greater twisting moments in *Paranthropus* is a lateral positioning of the masseter origin on widely-flaring zygomatic arches. As noted previously, an analogous bizygomatic expansion is observed in *C. apella*, relative to other *Cebus* species (Hershkovitz, 1949; Kinzey, 1974). Hylander's second reason for expecting higher twisting moments in *Paranthropus* is heavy masticatory loading of the premolars (inferred from large premolar size and heavy wear). The twisting moments resulting from canine and premolar loading are greater than for the molars because of the greater distance from the bite point to the longitudinal axis of the corpus (Demes *et al.*, 1984: Figure 16.10; Hylander, 1988). In a study of jaw form in living and fossil prosimians, Ravosa (1991, pers. comm.) found a similar degree of increased relative corporal thickness and heavy premolar wear in the subfossil archaeoleumrine *Hadropithecus*. Ravosa also interpreted this pattern as an adaptation to the resistance of heavy twisting loads.

Cebus apella has been observed using its canines in opening hard fruits and nuts (Izawa, 1975; Izawa & Mizuno, 1977). Terborgh (1983) has noted (largely unsuccessful) attempts by *C. albifrons* to open nuts with the premolars; he notes that *C. apella* is more successful at the same activity, presumably using the premolars as well. By analogy to *Paranthropus* and *Hadropithecus*, it appears that the increased corpus thickness seen in *C. apella* may be related not only to greater magnitudes of jaw adducting forces, but to greater twisting moment arms as well.

If the forces that tend to twist the corpus are in fact higher in *C. apella*, then the lack of sample heterogeneity in RSD runs counter to *a priori* expectations. Apart from greater dorsoventral shear stress, greater twisting of the corpus is expected to produce greater coronal bending at the symphysis, where the superior aspect of the section experiences compression, while the inferior aspect is placed in tension. One possible explanation lies in a simplifying assumption made in this analysis. Namely, when discussing the relative depth and thickness of the symphysis, it is assumed that these maximum and minimum dimensions correspond with superoinferior and anteroposterior *anatomical* dimensions, respectively. Because of the inclination of the symphyseal section, it is possible that the increased RST (see below) might also contribute to resisting coronal bending. However, as documented by Daegling (1990, 1991) in a comparative study of *C. apella* and *C. capucinus*, using computed tomography, the resulting biomechanical difference may only be a minor one: "In terms of biomechanical shape, then, the only difference between *C. apella* and *C. capucinus* is that the symphysis in the former is relatively broader than it is deep. . . . Thus, for resistance of masticatory stress, differences . . . at the symphysis appear to be inconsequential" (Daegling, 1990:318).

The RST was expected to be greater in *C. apella* because of increased magnitudes of forces that would tend to cause wishboning. These expectations were largely met; *C. apella* males have the relatively deepest symphyses, indicating the greatest relative resistance to wishboning, followed by the *C. apella* female mean, and by both *C. albifrons* means (not significantly different from each other). The comparisons of the RSS follows a similar pattern: *C. apella* males have the most robust symphyses, followed by *C. apella* females, then by both *C. albifrons* sexes. While RSS is meant to be a general indicator of the relative size of the symphysis, the absolute size of the symphysis (SS) is a more biomechanically relevant measure of resistance to dorsoventral shearing. For sex-specific species comparisons, the *C. apella* means are larger, indicating the ability to withstand greater shearing loads. Within species, males have absolutely larger mean values, as would be expected from the overall degree of sexual dimorphism in *Cebus* species.

In summary, the results of the mandibular size and shape comparisons are largely in keeping with the predictions based on Hylander's (1979, 1981, 1984, 1986) macaque experiments. Recall that the predictions made in this study were that the internal forces in the mandibles of both *Cebus* species differ in magnitude, but that the basic patterns of stress are the same. As mentioned previously, there are no *skeletal* indicators of greater force production between sex-specific samples. However, the more robust morphologies of the *C. apella* mandibles suggest that higher forces are being generated in some other manner, perhaps through alterations in muscle architecture, changes in the proportions of muscle fiber types, or differences in muscle-force recruitment patterns (Dechow & Carlson, 1990; Ravosa & Hylander, n.d.). Finally, the question of whether or not the mandibular shape differences among samples are determined by some criterion of "biomechanical similarity", such as the maintenance of comparable levels of internal stresses, remains open; specific hypothesis testing of this nature would require an *in vivo* experimental approach.

Growth-related biomechanical trends within samples
The following biomechanical trends are common to all of the samples. In terms of external force production, each sample exhibits disproportionately faster increases in the "muscular" component of the size of the masticatory apparatus, compared to the "skeletal" component

(as inferred by the strongly allometric regressions of IFS against LENGTH). The positively allometric scaling of LEVER to LENGTH suggests that growth-related increases in mechanical advantage are common to all samples. In general, the combination of positively allometric muscle growth and increased mechanical advantage provides greater force input than would be expected if older animals were geometrically similar versions of infants.

When the internal forces that affect the corpus are considered, scaling patterns are, with a few exceptions, broadly similar across samples. In each of the samples except *C. apella* females, CD scales to LENGTH with positive allometry. This suggests that the relative strength of the corpus in parasagittal bending increases with growth (which would be expected, based on patterns of muscle growth). *Cebus apella* females do not depart significantly from isometry, so that corporal strength in parasagittal bending does not change significantly during growth. For every sample, CT scales to LENGTH with negative allometry, so that the estimated relative resistance to twisting decreases with age. This does not necessarily imply that twisting stresses become relatively reduced with age; it is perhaps more probable that the relatively greater corporal thickness in younger animals is determined, to a great extent, by the developing permanent molars.* In terms of the scaling of the absolute CS to LENGTH, all of the samples except *C. albifrons* females demonstrate negative allometry, where the corpus becomes less robust with age. Again, the greater corporal robusticity of younger animals is most likely due to spatial demands of the developing permanent dentition, not biomechanical demands.

For SD versus LENGTH, the regression line elevation at growth onset for *C. apella* males was significantly lowest, while *C. albifrons* males were highest. However, these early shape differences are apparently "cancelled out" by differing patterns of allometric growth (negative in *C. albifrons* males, positive in *C. apella* males and isometric for the female samples), with the final result being no significant difference in RSD among adult samples. When the relative thickness of the symphysis is examined, there are no slope differences in the regression of ST versus LENGTH; however, *C. apella* symphyses are relatively thicker at growth onset, with the interspecific pattern of RST being maintained until adulthood. Finally, the lack of adult differences in RSD suggests that differences in the RSS are largely due to increased thickness (see also Daegling, 1990, 1991).

Biomechanics, heterochrony and species differentiation
The results of this study indicate consistent differences in masticatory form between the two *Cebus* species. These differences are largely in keeping with the predictions made in the Introduction; however, the results may be more meaningful when placed into an ecological context. The ability to routinely exploit hard food items imparts an ecological separation between *C. apella* and other (often sympatric) capuchin species (Terborgh, 1983). Specialization for feeding on hard fruits and nuts by *C. apella* seems to be a particularly effective way of lessening competition between *Cebus* species during seasonal periods of low food abundance (Terborgh, 1983). Adaptations for hard-object feeding with a presumably similar purpose have been documented in sympatric mangabeys (Gautier-Hion & Gautier, 1979) and peccaries (Kiltie, 1982). It seems clear that the biomechanical differences between the *Cebus* species have fairly straightforward ecological implications. However, whether this

---

*The effects of molar crown growth on the size of the corpus have been confirmed by a computed tomography of this same sample (Cole, in prep.). Another characteristic of the mandibles of young animals is that the cortical bone is both absolutely and relatively thinner.

ability actually served as the "key adaptation" (*sensu* Simpson, 1953; see also Futuyma, 1986) for the speciation of *C. apella* from a more generalized stock is unknown.

Given that the divergence of growth patterns in the *C. apella* masticatory apparatus can be related to dietary adaptation, we may attempt to discuss the observed growth patterns in a (more explicitly) evolutionary context. The predominant development sources of adult shape differences (having biomechanical implications) in the study sample are extensions or transpositions of ancestral growth trajectories (Table 7). In comparison, alterations in patterns of growth-related shape change (such as acceleration) are relatively rare. While making explicit inferences about underlying genetic covariance from phenotypic scaling patterns is probably ill advised (Cheverud, 1982; Atchley, 1983), we can make the empirical observation that there is apparently more variation in the elevations of postnatal growth trajectories than in their orientations.

There is ample empirical evidence (Table 7) to demonstrate that *C. albifrons* and *C. apella* are not simply different sized versions of the "same animal" (*sensu* Pilbeam & Gould, 1974; see also Shea, 1988). Otherwise, hypermorphosis would play a far greater role in contributing to shape differences among adults. In fact, upward extrapolation of the *C. albifrons* growth trajectory would not converge on the *C. apella* form (at any larger size) in several cases. For example, the mean RST for adult *C. apella* males is 0·155, which is significantly higher than the mean value of 0·141 for *C. albifrons* males. However, the *C. albifrons* male regression coefficient for ST versus LENGTH is negatively allometric (0·886), indicating that the predicted value of RST would *decrease* with extrapolation to *C. apella* size. As indicated in Tables 6 and 7, the source of differences in RST between male samples is a transposition (pre-displacement) that is present early in postnatal life, rather than a shape difference that arises by simply becoming larger. Other instances in which upward extrapolation of the male *C. albifrons* trajectory would fail to converge on a mean male *C. apella* shape include CT versus LENGTH and SD versus LENGTH.

A comparison of the female samples demonstrates that most of the shape differences that distinguish *C. apella* from *C. albifrons* are present without the benefit of any significant difference in LENGTH (at either postnatal growth onset or in adults). Because the adult female samples are roughly the same size (as indicated by length), we could conclude that alterations of prenatal growth processes must be relatively more important in producing the biomechanical differences observed among female adults.

In general, it appears that selection for functional differences in the target adult phenotype of *C. apella* has produced modifications in patterns of prenatal growth, rather than modifications of postnatal growth trajectories. The pattern of postnatal growth that appears to make the most significant contributions to adult male form differences in pre-displacement, along with size increase in *C. apella*; in other words, some of the observed differences in form are correlated with differences in adult male size (see Table 7). However, it is interesting to note that size differences play no role in producing the differences in form observed between female samples. Similar patterns of congruence between growth patterns of closely-related primate species have been noted by Jungers & Cole (1992) and Falsetti & Cole (1992), although these studies were concerned with growth of the postcranial skeleton. The relative rarity of differences in rates of shape change (e.g., acceleration or neoteny) suggests that patterns of ontogenetic covariance may be fairly conservative, compared to alterations in form that occur earlier in development (cf. Shea, 1985*b*). Whether prenatal growth-patterns can be altered more easily than postnatal growth (in terms of genetic alterations) remains an open question worthy of further study.

Table 7 Summary of allometric relationships between species and sexes. In each comparison, the growth patterns of the "descendant" groups (D) are described relative to corresponding "ancestral" groups (A)

Regression	Species contrasts			Sexual dimorphism	
	APM (D) vs. ALM (A)	APF (D) vs. ALF (A)	ALM (D) vs. ALF (A)	APM (D) vs. APF (A)	
IFS vs. LENGTH	size increase post-displacement	acceleration pre-displacement	SIZE INCREASE PRE-DISPLACEMENT	SIZE INCREASE POST-DISPLACEMENT	
LEVER vs. LENGTH	size increase post-displacement	pre-displacement	SIZE INCREASE PRE-DISPLACEMENT	SIZE INCREASE POST-DISPLACEMENT	
CD vs. LENGTH	SIZE INCREASE PRE-DISPLACEMENT	PRE-DISPLACEMENT	hypermorphosis	size increase post-displacement	
CT vs. LENGTH	SIZE INCREASE PRE-DISPLACEMENT	PRE-DISPLACEMENT	hypermorphosis	size increase pre-displacement	
CS vs. LENGTH	SIZE INCREASE PRE-DISPLACEMENT	PRE-DISPLACEMENT	hypermorphosis	size increase acceleration post-displacement	
SD vs. LENGTH	size increase acceleration post-displacement	none	size increase pre-displacement	hypermorphosis post-displacement	
ST vs. LENGTH	SIZE INCREASE PRE-DISPLACEMENT	PRE-DISPLACEMENT	size increase pre-displacement	size increase post-displacement	
SS vs. LENGTH	SIZE INCREASE POST-DISPLACEMENT	PRE-DISPLACEMENT	size increase pre-displacement	size increase post-displacement	

Upper-case letters denote instances in which significant differences in adult shape result (see Table 4).

Summary

1. The adult shape differences observed between *C. apella* (which routinely exploits hard foods) and *C. albifrons* (which does not) are generally in agreement with expectations that were based on *in vivo* experimental data [Hylander's experimental studies of macaques (1979, 1981, 1984, 1985, 1986)]. Specifically, the masticatory apparatus of *C. apella* is designed for relatively greater resistance of the corpus to parasagittal bending and twisting, relatively greater resistance of the symphysis to wishboning and absolutely greater resistance to dorsoventral shear stresses in both the corpus and symphysis. There is no evidence from skeletal measurements of sex-specific differences in absolute muscle size.

2. Many of the differences in adult form that are interpreted as anatomical correlates of hard-object feeding behavior are present, to at least some extent, in very young animals. Postnatal growth patterns appear to be conservative, so that divergent growth patterns make only relatively minor contributions to differences in adult shape. It is concluded that selection has acted to modify prenatal patterns of growth, which are thought to be the primary sources of the differences in form that distinguish adults.

Acknowledgements

I would like to thank Matt Ravosa and Anne Gomez for inviting me to contribute to their symposium "Ontogenetic Perspectives on Primate Evolutionary Biology." Access to museum specimens was provided by Guy G. Musser and Wolfgang Fuchs (American Museum of Natural History) and by Richard W. Thorington, Jr (United States National Museum of Natural History). James Rohlf kindly provided statistical guidance, while Elizabeth McGee offered assistance in computer programming. This manuscript has been improved through critical readings by Audrone Biknevicius, Maria Cole, David Daegling, Anthony Falsetti, John Fleagle, William Jungers, David Krause, Susan Larson, Christine Wall and Matt Ravosa. The comments of the editors and anonymous reviewers were also appreciated.

This research was supported by a Research Grant from the Theodore Roosevelt Memorial Fund of the American Museum of Natural History, by National Science Foundation Grant BNS-9020562 and by Wenner-Gren Foundation Grant 5303.

References

Alberch, P. (1980). Ontogenesis and morphological diversification. *Am. Zool.* **20,** 653–667.
Alberch, P., Gould, S. J., Oster, G. F. & Wake, D. B. (1979). Size and shape in ontogeny and phylogeny. *Paleobiol.* **5,** 296–317.
Albrecht, G. H. & Gelvin, B. R. (1987). The simple allometry equation reconsidered: Assumptions, problems, and alternative solutions. *Am. J. phys. Anthrop.* **72,** 174.
Atchley, W. R. (1983). Some genetic aspects of morphometric variation. In (J. Felsenstein, Ed.) *Numerical Taxonomy.* NATO ASI Series, Vol. G1, pp. 346–363. Berlin: Springer-Verlag.
Borgognini Tarli, S. M. & Repetto, E. (1986). Methodological considerations of the study of sexual dimorphism in past human populations. In (M. Pickford & B. Chiarelli, Eds) *Sexual Dimorphism in Living and Fossil Primates*, pp. 51–66. Firenze: Il Sedicesimo.
Bouvier, M. (1986). Biomechanical scaling of mandibular dimensions in New World monkeys. *Int. J. Primatol.* **7,** 551–567.
Cachel, S. (1984). Growth and allometry in primate masticatory muscles. *Archs. oral. Biol.* **29,** 287–293.
Cheverud, J. M. (1982). Relationships among ontogenetic, static, and evolutionary allometry. *Am. J. phys. Anthrop.* **59,** 139–149.
Clarke, M. R. B. (1980). The reduced major axis of a bivariate sample. *Biometrika* **67,** 441–446.
Cock, A. G. (1966). Genetical aspects of metrical growth and form in animals. *Q. Rev. Biol.* **41,** 131–190.

Corruccini, R. S. (1980). Size and positioning of the teeth and infratemporal fossa relative to taxonomic and dietary variation in primates. *Acta anat.* **107,** 231–235.
Daegling, D. J. (1990). Geometry and biomechanics of hominoid mandibles. Ph.D. Dissertation, State University of New York at Stony Brook.
Daegling, D. J. (1991). The influence on diet on mandibular morphology: a natural experiment. *Am. J. phys. Anthrop.* Suppl. **12,** 64–65.
Dechow, P. C. & Carlson, D. S. (1990). Occlusal force and craniofacial biomechanics during growth in rhesus monkeys. *Am. J. phys. Anthrop.* **83,** 219–237.
Defler, T. R. (1979a). On the ecology and behavior of *Cebus albifrons* in eastern Colombia. I. Ecology. *Primates* **20,** 475–490.
Defler, T. R. (1979b). On the ecology and behavior of *Cebus albifrons* in eastern Colombia. II. Behaviour. *Primates* **20,** 491–502.
Demes, B. & Creel, N. (1988). Bite force, diet, and cranial morphology of fossil hominids. *J. hum. Evol.* **17,** 657–670.
Demes, B., Creel, N. & Preuschoft, H. (1986). Functional significance of allometric trends in the hominoid masticatory apparatus. In (J. G. Else & P. C. Lee, Eds) *Selected Proceedings of the Tenth Congress of the International Primatological Society. Volume 1: Primate Evolution*, pp. 297–237. Cambridge: Cambridge University Press.
Demes, B., Preuschoft, H. & Wolff, J. E. A. (1984). Strength-stress relationships in the mandibles of hominoids. In (D. J. Chivers, B. A. Wood & A. Bilsborough, Eds) *Food Acquisition and Processing in Primates*, pp. 369–390. New York: Plenum Press.
Dickinson, C., Harrison, H. & Miller, K. (1990). Phylogenetic relationships among species of the genus *Cebus*. *Am. J. phys. Anthrop.* **81,** 215–216.
Dunn, O. J. (1964). Multiple contrasts using rank sums. *Technometrics* **6,** 241–252.
Efron, B. (1979). Bootstrap methods: another look at the jackknife. *Ann. Statist.* **7,** 1–26.
Efron, B. (1981). Nonparametric estimates of standard error: the jackknife, the bootstrap, and other resampling methods. *Biometrika* **68,** 589–599.
Efron, B. & Tibshirani, R. (1986). Bootstrap methods for standard errors, confidence intervals, and other methods of statistical accuracy. *Stat. Sci.* **1,** 54–77.
Falsetti, A. B. & Cole, T. M., III (1992). Relative growth of the postcranial skeleton in callitrichids. *J. hum. Evol.* **23,** 79–92.
Fleagle, J. G., Mittermeier, R. A. & Skopec, A. L. (1981). Differential habitat use by *Cebus apella* and *Saimiri sciureus* in Central Surinam. *Primates* **22,** 361–367.
Freese, C. H. & Oppenheimer, J. R. (1981). The capuchin monkeys, genus *Cebus*. In (A. F. Coimbra-Filho & R. A. Mittermeier, Eds) *Ecology and Behavior of Neotropical Primates. Volume 1*, pp. 331–390. Rio de Janeiro: Academia Brasileira de Ciencias.
Futuyma, D. J. (1986). *Evolutionary Biology*, 2nd edn. Sunderland, Massachusetts: Sinauer Associates.
Galliari, C. A. (1985). Dental eruption in captive-born *Cebus apella*: from birth to 30 months old. *Primates* **26,** 506–510.
Gautier-Hion, A. & Gautier, J. P. (1979). Niche écologique et diversité des espèces sympatriques dans le genre *Cercopithecus*. *Terre Vie* **33,** 493–507.
Gelvin, B. R. & Albrecht, G. H. (1987). Brain weight/body weight scaling in primates: Assumptions, problems, and alternative solutions. *Am. J. phys. Anthrop.* **72,** 202.
Gould, S. J. (1977). *Ontogeny and Phylogeny*. Cambridge: Belknap Press.
Herring, S. W. (1980). Functional design of cranial muscles: comparative and physiological studies in pigs. *Am. Zool.* **20,** 283–293.
Hershkovitz, P. (1949). Mammals of northern Colombia. Preliminary report no. 4: Monkeys (Primates), with taxonomic revisions of some forms. *Proc. U.S. Nat. Mus.* **98,** 323–427.
Hofer, H. O. (1973). Crista sagittalis externa in the skull of *Pan troglodytes* and its bearing on the reconstruction of the head of the robust type of *Australopithecus*. *Folia primatol.* **19,** 469–475.
Huxley, J. S. (1924). Constant differential growth ratios and their significance. *Nature* **114,** 895–896.
Huxley, J. S. (1932). *Problems of Relative Growth*. London: Methuen.
Hylander, W. L. (1979). The functional significance of primate mandibular form. *J. Morph.* **160,** 223–240.
Hylander, W. L. (1981). Patterns of stress and strain in the macaque mandible. In (D. S. Carlson, Ed.) *Craniofacial Biology. Monograph 10, Craniofacial Growth Series, Center for Human Growth and Development*, pp. 1–37. Ann Arbor: University of Michigan.
Hylander, W. L. (1984). Stress and strain in the madibular symphysis of primates: A test of competing hypotheses. *Am. J. phys. Anthrop.* **64,** 1–46.
Hylander, W. L. (1985). Mandibular function and biomechanical stress and scaling. *Am. Zool.* **25,** 315–330.
Hylander, W. L. (1986). *In vivo* bone strain as an indicator of masticatory bite force in *Macaca fascicularis*. *Arch. oral Biol.* **31,** 149–157.
Hylander, W. L. (1988). Implications of *in vivo* experiments for interpreting the functional significance of "robust" australopithecine jaws. In (F. E. Grine, Ed.) *Evolutionary History of the "Robust" Australopithecines*, pp. 55–83. New York: Aldine.

Hylander, W. L., Johnson, K. R. & Crompton, A. W. (1987). Loading patterns and jaw movements during mastication in *Macaca fascicularis*: a bone-strain, electromyographic, and cineradiographic analysis. *Am. J. phys. Anthrop.* **72,** 287–314.

Imbrie, J. (1956). Biometrical methods in the study of invertebrate fossils. *Bull. Am. Mus. nat. Hist.* **108,** 219–252.

Izawa, K. (1975). Foods and feeding behaviour of monkeys in the upper Amazon Basin. *Primates* **16,** 295–316.

Izawa, K. (1979). Foods and feeding behavior of wild black-capped capuchin (*Cebus apella*). *Primates* **20,** 57–76.

Izawa, K. & Mizuno, A. (1977). Palm-fruit cracking behavior of wild black-capped capuchin (*Cebus apella*). *Primates* **18,** 773–792.

Jolicoeur, P. & Mosimann, J. E. (1968). Intervalles de confiance pour la pente de l'axe majeur d'une distribution normale bidimensionalle. *Biometrie-praximetrie* **9,** 121–140.

Jungers, W. L. (1990). Scaling of postcranial joint size in hominoid primates. In (F. K. Jouffroy, M. H. Stack & C. Niemitz, Eds) *Gravity, Posture and Locomotion in Primates*, pp. 87–95. Firenze: Il Sedicesimo.

Jungers, W. L. & Cole, M. S. (1992). Relative growth and shape of the locomotor skeleton in lesser apes. *J. hum. Evol.* **23,** 93–105.

Jungers, W. L. & Fleagle, J. G. (1980). Postnatal growth allometry of the extremities in *Cebus albifrons* and *Cebus apella*: a longitudinal and comparative study. *Am. J. phys. Anthrop.* **53,** 471–478.

Kay, R. F. (1981). The nut-crackers—a new theory of the adaptations of the ramapithecines. *Am. J. phys. Anthrop.* **55,** 141–151.

Kiltie, R. A. (1982). Bite force as a basis for niche differentiation between rain forest peccaries (*Tayassu tajacu* and *T. pecari*). *Biotropica* **14,** 188–195.

Kinzey, W. G. (1974). Ceboid models for the evolution of hominoid dentition. *J. hum. Evol.* **3,** 193–203.

Kruskal, W. H. & Wallis, W. A. (1952). Use of ranks in one-criterion variance analysis. *J. Am. Stat. Assoc.* **47,** 583–621.

Kuhry, B. & Marcus, L. F. (1977). Bivariate linear models in biometry. *Syst. Zool.* **26,** 201–209.

Manly, B. F. J. (1991). *Randomization and Monte Carlo Methods in Biology*. London: Chapman and Hall.

McKinney, M. L. (1984). Allometry and heterochrony in an Eocene echinoid lineage: morphological change as a byproduct of size selection. *Paleobiol.* **10,** 407–419.

McKinney, M. L. (1986). Ecological causation of heterochrony: a test and implications for evolutionary theory. *Paleobiol.* **12,** 282–289.

McKinney, M. L. (1988a). Classifying heterochrony: allometry, size, and time. In (M. L. McKinney, Ed.) *Heterochrony in Evolution*, pp. 17–34. New York: Plenum Press.

McKinney, M. L. (1988b). Heterochrony in evolution: an overview. In (M. L. McKinney, Ed.) *Heterochrony in Evolution*, pp. 327–340. New York: Plenum Press.

McNamara, K. J. (1986). A guide to the nomenclature of heterochrony. *J. Paleontol.* **60,** 4–13.

Mosimann, J. E. (1970). Size allometry: Size and shape variables with characterizations of the lognormal and generalized gamma distributions. *J. Am. Stat. Assoc.* **56,** 930–945.

Mosimann, J. E. & James, F. C. (1979). New statistical methods for allometry with application to Florida red-winged blackbirds. *Evolution* **33,** 444–459.

Moynihan, M. (1976). *The New World Primates*. Princeton: Princeton University Press.

Nomura, E. (1926). An application of $a = kb^x$ in expressing the growth relation in the freshwater bivalve *Sphaerium heterodon* Pils. *Sci. Rep. Tohoku Univ. Biol.* **2,** 57–62.

Pilbeam, D. R. & Gould, S. J. (1974). Size and scaling in human evolution. *Science* **186,** 892–901.

Plotnick, R. E. (1989). Application of bootstrap methods to reduced major axis line fitting. *Syst. Zool.* **38,** 144–153.

Popov, E. P. (1976). *Mechanics of Materials*, 2nd edn. Englewood Cliffs, New Jersey: Prentice-Hall.

Ravosa, M. J. (1991). Structural allometry of the prosimian mandibular corpus and symphysis. *J. hum. Evol.* **20,** 3–20.

Ravosa, M. J. & Hylander, W. L. (n.d.). Function and fusion of the mandibular symphysis in primates: stiffness or strength? In (J. G. Fleagle & R. F. Kay, Eds) *Anthropoid Origins*. New York: Plenum Press.

Rayner, J. M. V. (1985). Linear relations in biomechanics: the statistics of scaling functions. *J. Zool.* **206,** 415–439.

Ricker, W. E. (1984). Computation and uses of central trend lines. *Can. J. Zool.* **62,** 1897–1905.

Rosenberger, A. L. & Kinzey, W. G. (1976). Functional patterns of molar occlusion in platyrrhine monkeys. *Am. J. phys. Anthrop.* **45,** 281–298.

SAS Institute, Inc. (1985). *SAS User's Guide: Statistics*. Cary, North Carolina: SAS Institute, Inc.

Schumacher, G. H. (1961). *Funktionelle Morphologie der Kaumuskulatur*. Jena: Gustav Fischer.

Shea, B. T. (1983). Size and diet in the evolution of African ape craniodental form. *Folia primatol.* **40,** 32–68.

Shea, B. T. (1985a). Bivariate and multivariate growth allometry: statistical and biological considerations. *J. Zool. Lond. (A)* **206,** 367–390.

Shea, B. T. (1985b). Ontogenetic allometry and scaling: A discussion based on the growth and form of the skull in African apes. In (W. L. Jungers, Ed.) *Size and Scaling in Primate Biology*, pp. 175–205. New York: Plenum Press.

Shea, B. T. (1988). Heterochrony in primates. In (M. L. McKinney, Ed.) *Heterochrony in Evolution*, pp. 237–266. New York: Plenum Press.

Simpson, G. G. (1953). *The Major Features of Evolution*. New York: Columbia University Press.

Smith, R. J. (1984). Allometric scaling in comparative biology: problems of concept and method. *Am. J. Physiol.* **246,** R152–R160.
Sokal, R. R. & Rohlf, F. J. (1981). *Biometry*, 2nd edn. New York: Freeman.
Struhsaker, T. T. & Leland, L. (1977). Palm-nut smashing by *Cebus apella* in Colombia. *Biotropica* **9,** 124–126.
Teaford, M. F. (1985). Molar microwear and diet in the genus *Cebus*. *Am. J. phys. Anthrop.* **66,** 363–370.
Terborgh, J. (1983). *Five New World Primates: A Study in Comparative Ecology*. Princeton: Princeton University Press.
Thorington, R. W., Jr (1967). Feeding and activity of *Cebus* and *Saimiri* in a Colombian forest. In (D. Starck, Schneider, R. & Kuhn, H. J., Eds) *Progress in Primatology*, pp. 180–184. Stuttgart: Gustav Fischer.
Torres de Assumpcao, C. (1983). An ecological study of the primates of southeastern Brazil, with a reappraisal of *Cebus apella* races. Ph.D. Dissertation, University of Edinburgh.
Weijs, W. A. (1980). Biomechanical models and the analysis of form: a study of the mammalian masticatory apparatus. *Am. Zool.* **20,** 707–719.
Zar, J. H. (1984). *Biostatistical Analysis*, 2nd edn. Englewood Cliffs, New Jersey: Prentice-Hall.

Brian T. Shea

Departments of Cellular, Molecular and Structural Biology, and Anthropology, Northwestern University, 303 East Chicago Avenue, Chicago, IL 60611, U.S.A.

Received 4 September 1991
Revision received 14 May 1992
and accepted 21 May 1992

Keywords: Cercopithecus, skeleton, ontogeny, allometry, heterochrony.

Ontogenetic scaling of skeletal proportions in the talapoin monkey

The talapoin monkey is distinct from its close relatives in the genus *Cercopithecus* in a variety of morphological and ecological features, inclining some to place it in its own genus, *Miopithecus*. It is also the smallest of the extant catarrhines, and in this investigation the extent to which the morphological distinctions of the talapoin monkey are the result of allometric factors is analysed. A series of cranial and postcranial measurements were taken on the skeletons of young and adult talapoin monkeys (*C. talapoin*) and moustached monkeys (*C. cephus*). Ontogenetic scaling (i.e., allometric extrapolation) was utilized as a criterion of substraction in both bivariate and multivariate comparative allometric analysis. Adult cranial and postcranial proportions do differ significantly between *C. talapoin* and *C. cephus*, but in almost all cases these adult shape differences result from the sharing of common underlying patterns of ontogenetic allometry. Use of additional data from Verheyen's (1962) monograph suggests that the marked differences in skull form within the *Cercopithecus* group (from talapoin to patas monkeys) may also be primarily a product of ontogenetic scaling. The specific postcranial proportions seen in adult talapoins do not accord with expectations derived from several recent biomechanical models for primates which engage in frequent leaping and climbing behaviors, though the comparative growth allometric perspective utilized here does clarify these adult shape differences. The predominantly paedomorphic morphology of the talapoin monkey appears to be a direct and correlated allometric consequence of a decrease in overall growth rates and terminal body size, which may be related to increased propensities for leaping behavior and a more insectivorous diet.

Journal of Human Evolution (1992) **23**, 283–307

Introduction

In the field of evolutionary biology, the decades prior to 1940 witnessed intense interest in the relationships between developmental and evolutionary transformations. This focus on ontogeny encompassed both early development (e.g., DeBeer, 1930; Child, 1941; see Haraway, 1976) and later postnatal morphogenesis (e.g., Huxley, 1932). A large number of important papers and several key synthetic treatises (e.g., Thompson, 1917; Huxley, 1932; Huxley & DeBeer, 1934; DeBeer, 1930, 1940; Goldschmidt, 1938) attest to the significance and fruitfulness of that focus. For a number of years after this period, interest in these issues waned somewhat, no doubt due in part to important stimuli within other arenas of evolutionary biology, such as the study of systematics and adaptation as evidenced in the growing fossil record (e.g., Simpson, 1944, 1953), or genetics and population biology as investigated in living forms (e.g., Dobzhansky, 1970; Mayr, 1963).

The 1960s and 70s were characterized by a resurgence of interest in development and evolution. Much of the credit for this recent catalyzation within evolutionary morphology must go to S. J. Gould, considering his seminal work in both allometry (e.g., 1966, 1968, 1971, 1975) and heterochrony (e.g., 1968, 1977). The investigation of ontogenetic allometry or relative growth (Huxley, 1932) is central in these interrelated concepts and approaches. In recent years, a great many papers in primatology and other areas of evolutionary biology have utilized comparative ontogenetic allometric investigations to elucidate various aspects of morphological diversification in living and extinct groups, and the present symposium attests to the vitality of this approach. My own work has relied heavily on such investigations of ontogenetic allometry and heterochrony (see Shea, 1988, 1990, for recent reviews). In the present contribution, this perspective is applied to a study of cranial and postcranial morphology in some African monkeys.

Due to their significant taxonomic and ecological diversity, the species of the tribe Cercopithecini represent an intriguing group for the study of a variety of evolutionary questions (Gautier-Hion *et al.*, 1988). This is certainly true for investigations of allometry and heterochrony, since average adult body size ranges from over 11 kg in the male patas monkey, to only 1·2 kg in the talapoin, which is the smallest of the extant catarrhines. Several previous studies have hinted at or hypothesized an allometric basis to the morphological distinctiveness exhibited by the talapoin monkey. Verheyen's (1962) extensive studies of cranial form have demonstrated substantial differences in skull size and shape in talapoins compared to larger members of the Cercopithecini radiation (see Figure 1). While concluding that morphometrically the skull of the talapoin was "the most aberrant and most specialized" of the *Cercopithecus*, Verheyen explicitly suggested that "it is without doubt evident that allometric growth has strongly influenced these results" (1962: 109–111). Expressing this allometric hypothesis in the terminology of heterochrony, Verheyen (1962: 88) claimed that "the entire cranial structure of the talapoin suggests that it is truly a paedomorphic branch of the genus *Cercopithecus*". Napier & Napier (1967: 30) described the talapoin as a "neotenous baboon". More recently, Delson (1975; Szalay & Delson, 1979; Strasser & Delson, 1987) has also suggested that many of these morphological distinctions might be interpreted as an interrelated complex of features related to the fact that the talapoin monkey is "a neotenous, phyletically dwarfed *Cercopithecus*" (Strasser & Delson, 1987).

Scaling of brain size has also been invoked in support of the claim that talapoin monkeys are paedomorphic and the product of rapid size decrease. Bauchot & Stephan (1969) suggested that the relatively large brain size of the talapoin (the highest encephalization value for any cercopithecoid) is a product of miniaturization along the intraspecific scaling curve of approximately 0·33, in contradistinction to the typical interspecific value of approximately 0·66–0·75. Gould (1975) has expanded this argument for the talapoin and considered a variety of other such examples within the Primates.

In terms of postcranial anatomy, talapoin monkeys do exhibit some significant differences in both limb proportions (Fleagle, 1988) and patterns of quadrupedal locomotion (Rollinson & Martin, 1981) relative to other cercopithecines. Talapoins are purported to be among the most adept and frequent of arboreal leapers within the Cercopithecidae (Fleagle, 1988), and thus they provide an opportunity to test for the presence of certain skeletal and body proportions observed in a number of other anthropoid taxa characterized by increased leaping propensities as compared to their more typically quadrupedal relatives.

The preliminary study described here is part of a broader investigation designed to examine the role of allometry and heterochrony in the morphological diversification of the Cercopithecini radiation (see Shea, 1988, 1990, for additional details). I am essentially testing the hypothesis given above that adult proportion differences between talapoins and other *Cercopithecus* monkeys are allometric consequences of overall size changes. A null hypothesis of ontogenetic scaling (Gould, 1966, 1975; Shea, 1981, 1983*a*, 1985*a*) represents the appropriate test of this prediction and the proper "criterion of subtraction" with which to assess deviations from expected allometric baselines.

It should be stressed at this point that the author follows most previous workers (e.g., Verheyen, 1962; Strasser & Delson, 1987) in interpreting the small body size of the talapoin monkey as a derived feature. This is difficult to establish definitively, but the talapoin's unique

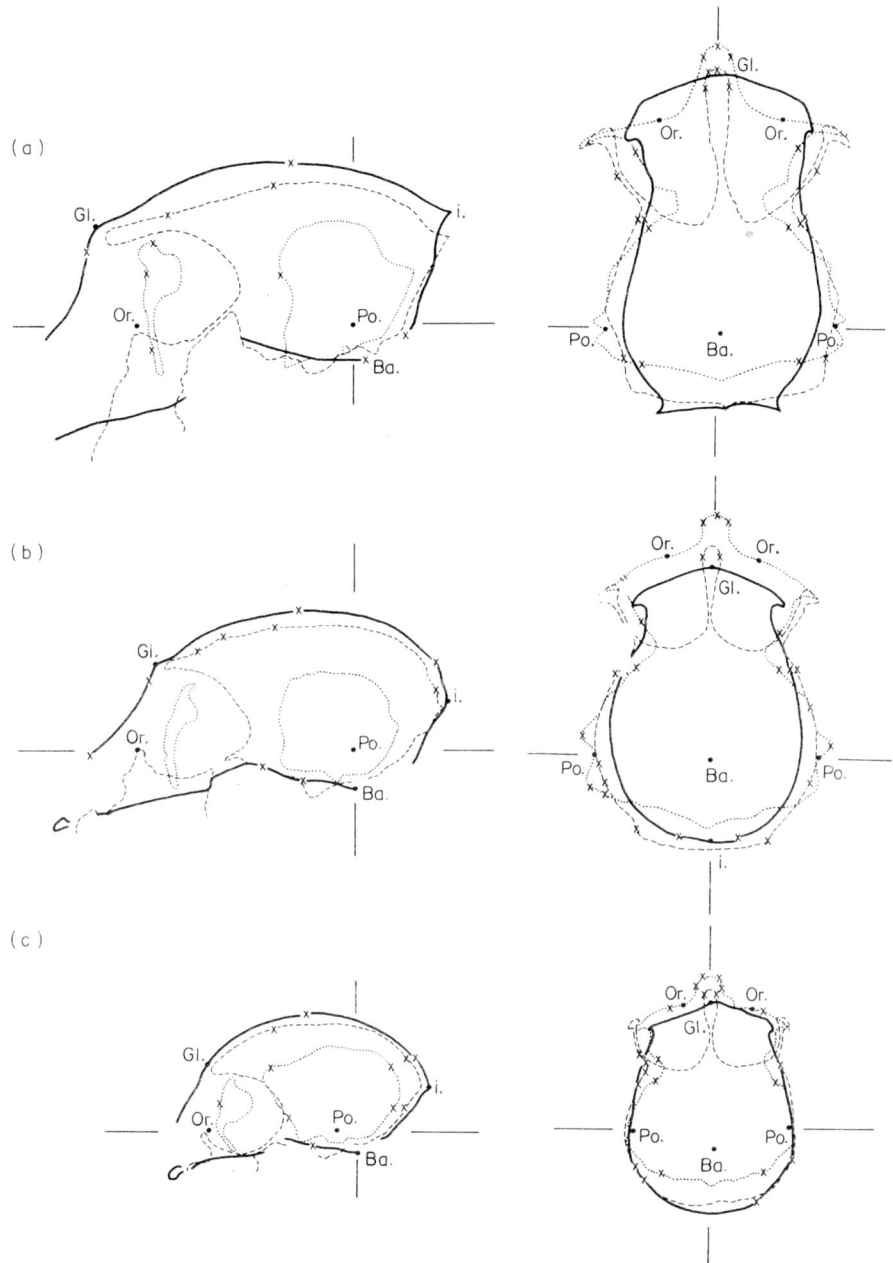

Figure 1. Lateral (left) and superior (right) views of cranial outlines for (a) *C. patas*, (b) *C. mitis* and (c) *C. talapoin*. After Verheyen, 1962.

and extreme position within the cercopithecine size range, the likely link to presumably derived ecological and dietary adaptations (see below) and the larger size of other species usually accepted as more primitive (e.g., patas and swamp monkeys), all argue for such an interpretation. Moreover, even if the directionality or polarity of size change is reversed, the nature of the morphological transformations determined through these studies of allometry and heterochrony remains unaltered (see Shea, 1983b).

Materials and methods

Samples

For this project, the moustached monkey, *Cercopithecus cephus*, was selected as the species to compare with the talapoin, *Cercopithecus (Miopithecus) talapoin*. There is no compelling reason for selecting this particular species, other than that it is a reasonably typical species of *Cercopithecus* which is considerably larger than *C. talapoin*. Verheyen (1962) did conclude that it was generally similar and perhaps closely related to talapoins, though the various authorities in Gautier-Hion *et al.* (1988) give a range of divergent opinions on this matter.

The sample contained 48 specimens of talapoin monkeys (7 young), 25 of which had partial or complete (for these purposes) postcranial skeletons along with crania. The *C. cephus* sample included 66 specimens (23 young), with 27 having associated postcranial skeletons (all of which were adult). Sample sizes in Tables and Figures will, however, depart from these levels depending on missing variables. Sexes were combined in these analyses.

All specimens for both species came from various localities in west-central Africa; due to the relatively small sample sizes and preliminary nature of this investigation, no subspecific or geographic breakdown was utilized (see Dorst & Dandelot, 1970). Nothing in the subsequent analyses suggested problematic intraspecific heterogeneity relative to the between-group comparisons.

Measurements taken on the ontogenetic series of *C. talapoin* and *C. cephus* skeletons were supplemented by literature data from Verheyen's (1962) monograph on cranial morphology in *Cercopithecus* and *Colobus* monkeys. These values were adult male and female means for the following species (Verheyen's classification): *Cercopithecus ascanius*, *C. mona*, *C. cephus*, *C. aethiops*, *C. diana*, *C. neglectus*, *C. nictitans*, *C. mitis*, *C. l'hoesti*, *C. talapoin*, *C. nigroviridis*, *C. patas*, *Colobus badius*, *Colobus kirkii*, *Colobus polykomos*, *Colobus satanus*, *Colobus abyssinicus* and *Colobus verus*. Additional data on these samples can be found in Verheyen (1962).

Measurements

In addition to descriptive data on taxon, sex and dental age, a series of 11 cranial and six postcranial linear measurements was taken. The cranial measurements were taken according to Verheyen's (1962) definitions where possible in order to permit direct comparisons. The cranial dimensions (taken to the nearest 0·1 mm) were basion–nasion, glabella–inion, basion–bregma, basion–orale, biorbital width, bizygomatic width, bizygomaxillary width, nasion–prosthion, staphylion–nasion, palate length and palate breadth. Postcranial dimensions (taken to the nearest 0·1 mm) were maximum lengths of the humerus, radius, femur and tibia, scapula length (Shea, 1986) and pelvis height (Jungers & Susman, 1984).

Analyses

Two sequential hypotheses of *ontogenetic scaling* (Gould, 1975; Shea, 1981) were tested. The first was that ontogenetic allometries of *C. talapoin* and *C. cephus* coincide, so that proportions observed in the latter result from an extension of allometric trajectories characterizing the former. The second was that the adult means of other *Cercopithecus* species measured by

Verheyen (1962) fall along an extension of this common pattern of allometric change. These hypotheses were tested from both bivariate and multivariate perspectives.

Two statistical approaches were utilized to evaluate concordance of bivariate allometric trajectories. The first of these was analysis of covariance (ANOCOVA) on log-transformed values. ANOCOVA is a robust and well-tested approach for assessing between-group similarities and differences in allometric patterns, allowing one to test for homogeneity of slopes between two groups and then, if no slope differences are found, to test for shifts in position (y-intercept values) between the samples. Analyses of covariance were supplemented by comparing reduced major axis slopes following Clarke (1980) and then testing for position differences as outlined by Tsutakawa & Hewitt (1977). Differences between groups were only considered statistically significant if *both* bivariate approaches yielded concordant results. Those cases where the results between least-squares and reduced major axis analyses are discrepant are noted in Tables 2, 4 and 6. These statistical analyses were followed by careful visual examination of all plots in order to assess linearity of relationships and determine that regression or reduced major axis results were applicable to point scatters in the observed data ranges.

Because the Verheyen (1962) data are static adult interspecific means and contain no ontogenetic samples, it is not completely appropriate to make direct statistical comparisons between these data and allometric trajectories from *C. talapoin* and *C. cephus* given here. Here, concordance of allometric patterns was assessed by both ANOCOVA and determining whether the adult Verheyen data fell within the 90% probability ellipses fit to the *Cercopithecus* scatters here. This is adequate for a first-approximation type of test, especially since the measurement techniques (if not definitions) may vary between the two studies. Obviously, a more detailed study utilizing growth series for all species concerned would permit the most robust statistical testing.

Multivariate allometric comparisons were made using principal components analysis (PCA) of the covariance matrix of log-transformed variables, following Jolicoeur (1963). Detailed discussion of my application of this technique to comparative growth allometric studies can be found in Shea (1985*b*). When dealing with ontogenetic samples from similar and closely related species such as *C. cephus* and *C. talapoin*, the first principal component can be interpreted as a vector of allometric concordance and thus an appropriate test for ontogenetic scaling, while subsequent components summarize the effects of departures from ontogenetic scaling (Shea, 1985*b*). In the PCA's combining my data with those from Verheyen (1962), interpretation of the component loadings and patterning was based on cross-checks with the bivariate comparisons, as well as "experimental" variations such as inclusion of colobines or non-*Cercopithecus* cercopithecines.

Results

Postcranial allometry
One issue of interest is whether adult talapoin monkeys differ significantly in postcranial proportions from other members of the *Cercopithecus* group, or in the present case, *C. cephus*. In terms of gross proportion contrasts among the limb elements of adult talapoin and moustached monkeys, three out of four ratios tested (humerus/femur, radius/tibia and tibia/femur) are significantly different, while the fourth (radius/humerus) is nearly so (Table 1). Relative to pelvis length, both elements of the upper limb are longer in adult talapoins than in the moustached monkeys, while the lower limb elements are not

Table 1 Significance tests for selected postcranial comparisons between adult *C. talapoin* and *C. cephus*

Ratio comparison	*C. talapoin* Mean (S.D.)/n	*C. cephus* Mean (S.D.)/n	Probability[1]
Humerus/femur	0·827 (0·019)/14	0·794 (0·026)/23	0·001
Radius/tibia	0·826 (0·022)/14	0·798 (0·027)/23	0·002
Radius/humerus	1·006 (0·020)/16	0·989 (0·036)/23	NS[2]
Tibia/femur	1·007 (0·015)/14	0·984 (0·026)/23	0·002
Humerus/pelvis	1·090 (0·052)/12	1·050 (0·042)/23	0·031
Radius/pelvis	1·098 (0·046)/12	1·038 (0·041)/23	0·001
Femur/pelvis	1·317 (0·055)/12	1·322 (0·045)/23	NS
Tibia/pelvis	1·326 (0·067)/12	1·300 (0·046)/23	NS
Scapula/pelvis	0·483 (0·023)/10	0·473 (0·031)/23	NS

[1] Based on *t*-value derived for separate variances.
[2] $P = 0.068$.
NS = not significant at $P > 0.05$.

significantly different. Talapoins have a relatively elongated radius, as indicated by a significantly higher ratio when compared to pelvic length, and a somewhat higher brachial index ($P = 0.068$). There is some evidence from the data in Table 1 for a differential elongation of the tibia in talapoins, since the ratio of tibia-to-pelvis lengths is absolutely (though not significantly) greater and the ratio of tibia-to-femur (crural index) is significantly greater in *C. talapoin* than *C. cephus* adults. Therefore, selected between- and within-limb proportions do differ in the two species, though of course in isolation these results say nothing about whether such differences are produced by simple ontogenetic scaling or more fundamental allometric dissociations of growth trajectories.

Results of bivariate regression analyses, ANOCOVA and reduced major axis comparisons for the postcranial dimensions are given in Table 2. A representative example of ontogenetic scaling of humerus versus femur lengths is seen in Figure 2a. Both species' trajectories exhibit moderate negative allometry, indicating a relative lengthening of the hindlimb compared to the forelimb during growth and in a contrast between adult means. Figure 2b illustrates that the significantly higher crural index (tibia/femur) seen in adult talapoins (Table 1) is a simple consequence of ontogenetic scaling. Slopes are slightly negatively allometric in both species, though significantly so only for the talapoin, indicating that as size increases during growth the crural index progressively decreases. Each of the individual limb elements regressed against pelvic height yields negative allometry to varying degrees, as well as somewhat lower correlation coefficients. All comparisons indicate ontogenetic scaling, as does scapula length versus pelvic height (Table 2).

Results of principal components analyses on the postcranial measurements reveal that virtually all (99·9%) of the variance is accounted for by the first component, with the individuals of the two species arrayed along a common trajectory in a plot of the two components. There is no significant separation of the taxa along the second principal component. Variable loadings on the first principal component are fairly comparable, suggesting relatively weak allometries.

Cranial allometry

Ratios of the adult cranial dimensions versus basicranial length were compared in the two species prior to allometric analysis (Table 3). This provides a gauge of shape differences in

Table 2 Results of bivariate regressions and ANOCOVA for selected postcranial comparisons (see notes for comparisons with reduced major axis results when ANOCOVA values indicate differences between the groups)

Comparison		Slope (S.E.)/intercept (S.E.)/r/n	ANOCOVA Slope	Intercept
Humerus*femur	T	0·887 (0·035)/ 0·325 (0·158)/0·986/20	NS[1]	NS
	C	0·798 (0·087)/ 0·786 (0·439)/0·881/26		
Radius*tibia	T	0·941 (0·046)/ 0·076 (0·207)/0·978/21	0·005[2]	—[3]
	C	0·683 (0·076)/ 1·359 (0·379)/0·879/26		
Radius*humerus	T	1·020 (0·030)/−0·081 (0·131)/0·990/24	0·005[2]	—[3]
	C	0·734 (0·106)/ 1·263 (0·506)/0·817/26		
Tibia*femur	T	0·938 (0·025)/ 0·288 (0·110)/0·994/21	NS	NS
	C	0·976 (0·076)/ 0·100 (0·383)/0·934/26		
Humerus*pelvis	T	0·810 (0·076)/ 0·889 (0·323)/0·940/17	NS	NS
	C	0·741 (0·109)/ 1·282 (0·515)/0·812/26		
Radius*pelvis	T	0·854 (0·080)/ 0·707 (0·338)/0·941/17	NS	NS
	C	0·673 (0·095)/ 1·589 (0·451)/0·822/26		
Femur*pelvis	T	0·893 (0·081)/ 0·722 (0·343)/0·944/17	NS	NS
	C	0·872 (0·103)/ 0·888 (0·487)/0·866/26		
Tibia*pelvis	T	0·815 (0·089)/ 1·063 (0·378)/0·921/17	NS	NS
	C	0·917 (0·105)/ 0·656 (0·500)/0·871/26		
Scapula*pelvis	T	0·851 (0·095)/−0·109 (0·403)/0·928/15	NS	NS
	C	0·962 (0·208)/−0·573 (0·988)/0·686/26		

[1]NS = not significant at $P<0·01$.
[2]RMA slope not significantly different at $P<0·01$.
[3]Position differences not significant using RMA and Tsutakawa & Hewitt's (1977) method.
T = *C. talapoin*; C = *C. cephus*; r = Pearson correlation coefficient.

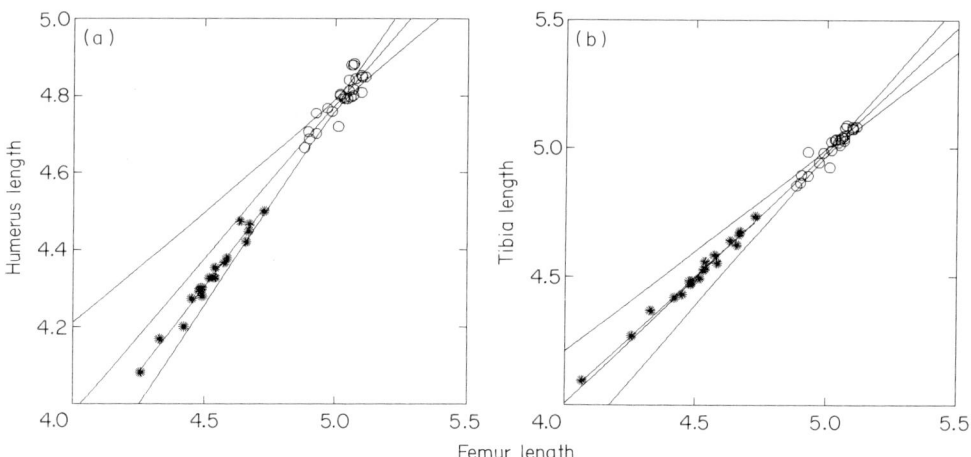

Figure 2. Ontogenetic scaling of hindlimb proportions (a) and proximal limb elements (b) in talapoin (∗) and moustached (○) monkeys. The 95% confidence interval and long regression line are for the *C. cephus* scatter; the short regression line is for the talapoins.

(adult) terminal growth phases, regardless of whether these differences are produced by ontogenetic scaling or allometric dissociations. Six of nine of the cranial ratio comparisons exhibited significant values, indicating substantial differences in skull shape as well as size between the two species.

Table 3 Significance tests for selected cranial ratio comparisons between adult *C. talapoin* and *C. cephus*. All are relative to basion–nasion (bn) length

Ratio comparison	*C. talapoin* Mean (S.D.)/*n*	*C. cephus* Mean (S.D.)/*n*	Probability[1]
Basion–bregma/bn	0·920 (0·037)/33	0·839 (0·026)/42	0·001
Basion–orale/bn	1·044 (0·036)/33	1·140 (0·037)/42	0·001
Biorbital width/bn	1·026 (0·033)/33	0·959 (0·040)/42	0·001
Bizygomatic width/bn	1·162 (0·039)/31	1·161 (0·042)/42	NS
Bizygomaxillary width/bn	0·804 (0·036)/33	0·795 (0·043)/42	NS
Nasion–prosthion/bn	0·542 (0·038)/33	0·661 (0·046)/42	0·001
Staphylion/nasion/bn	0·518 (0·022)/33	0·560 (0·034)/42	0·001
Palate length/bn	0·469 (0·024)/32	0·541 (0·031)/42	0·001
Palate width/bn	0·592 (0·028)/33	0·602 (0·029)/42	NS

[1]NS = not significant at $P < 0.01$.

Table 4 Results of bivariate regressions and ANOCOVA for selected cranial comparisons. All *y* variables are regressed on basicranial length. See notes for comparisons with reduced major axis results when ANOCOVA values indicate differences between the groups

y Variable		Slope (S.E.)/intercept (S.E.)/*r*/*n*	ANOCOVA Slope	ANOCOVA Intercept
Basion–bregma	T	0·429 (0·072)/ 2·029 (0·266)/0·694/40	NS[1]	0·001[2]
	C	0·428 (0·038)/ 2·113 (0·150)/0·817/65		
Basion–orale	T	1·372 (0·064)/−1·335 (0·236)/0·961/40	NS	0·001[3]
	C	1·446 (0·041)/−1·660 (0·162)/0·976/65		
Biorbital width	T	1·005 (0·064)/ 0·007 (0·233)/0·932/40	NS	0·001[3]
	C	0·912 (0·044)/ 0·302 (0·175)/0·933/65		
Bizygomatic width	T	1·166 (0·069)/−0·465 (0·255)/0·942/38	NS	NS
	C	1·041 (0·045)/−0·021 (0·177)/0·946/65		
Bizygomaxillary width	T	0·982 (0·093)/−0·154 (0·341)/0·864/40	NS	NS
	C	0·942 (0·066)/−0·007 (0·261)/0·874/65		
Nasion–prosthion	T	0·917 (0·154)/−0·323 (0·565)/0·695/40	NS	0·001[2]
	C	1·376 (0·089)/−1·932 (0·352)/0·890/65		
Staphylion–nasion	T	0·801 (0·094)/ 0·075 (0·346)/0·814/39	NS	0·001[2]
	C	0·886 (0·061)/−0·133 (0·241)/0·877/65		
Palate length	T	1·289 (0·116)/−1·833 (0·426)/0·880/38	NS	NS
	C	1·332 (0·086)/−1·953 (0·340)/0·890/65		
Palate width	T	0·974 (0·115)/−0·432 (0·423)/0·812/39	NS	0·001[2]
	C	1·024 (0·060)/−0·610 (0·235)/0·908/65		

[1]NS = not significant at $P < 0.01$.
[2]Position differences not significant using RMA and Tsutakawa & Hewitt's (1977) method.
[3]Position differences also significantly different using RMA and Tsutakawa & Hewitt's (1977) method.
T = *C. talapoin*; C = *C. cephus*; *r* = Pearson correlation coefficient.

Table 4 summarizes results of bivariate regressions, ANOCOVA and reduced major axis comparisons of log-transformed allometric trajectories of cranial measurements in *C. talapoin* and *C. cephus*. No significant differences in least-squares slope values are observed in the ANOCOVA. Significant differences in *y*-intercept values are indicated for basion–bregma (vault height), basion–orale (skull length), biorbital width, nasion–prosthion (face length), staphlion–nasion (face height) and palate width. However, only basion–orale and biorbital

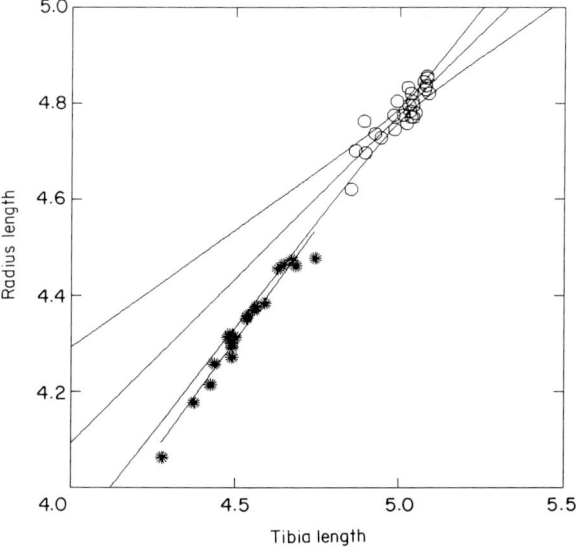

Figure 3. A plot of radius *vs.* tibia lengths in the two species. Symbols and regression fits as in Figure 2.

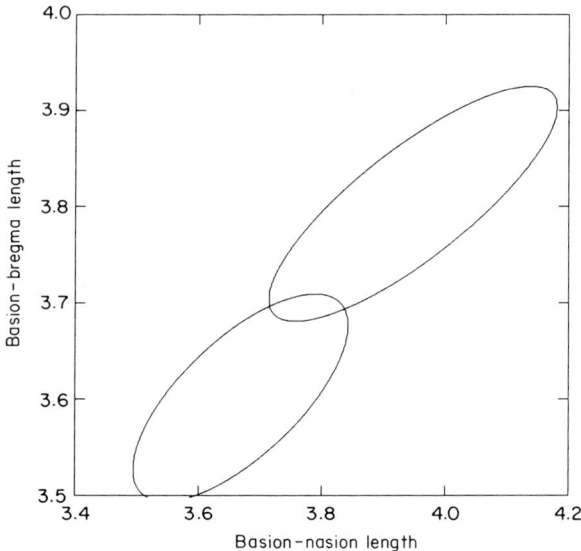

Figure 4. Ninety per cent confidence ellipsoids for a plot of vault height against skull size in talapoin and moustached monkeys. Note the upward transposition of the *C. cephus* scatter.

width reach statistical significance for position differences in reduced major axis comparisons.

The remainder of the cranial measurements describe predominantly the facial region. Here a strong pattern of ontogenetic scaling emerges from the statistical results (Table 4) and comparisons of point scatters. Figure 5 illustrates two such representative examples, for bizygomaxillary breadth and palate length. For at least two of the facial comparisons,

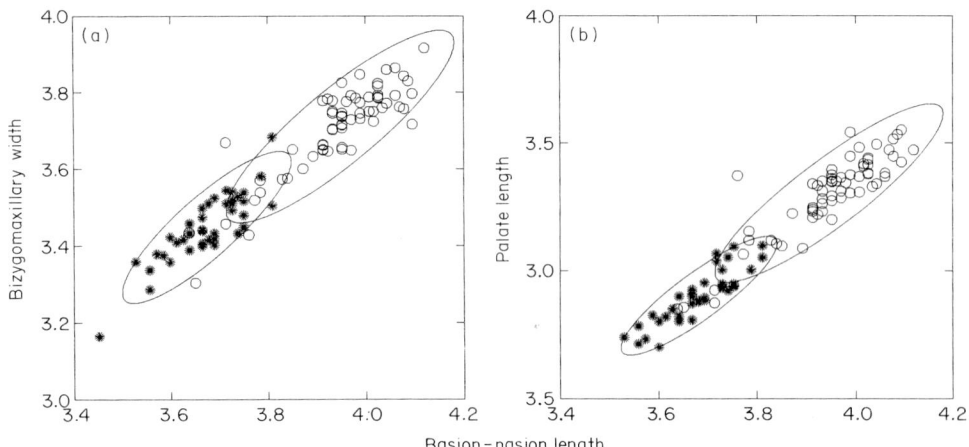

Figure 5. Ontogenetic scaling of craniofacial proportions in the two species, with 90% confidence ellipsoids illustrated. Symbols as in Figure 2.

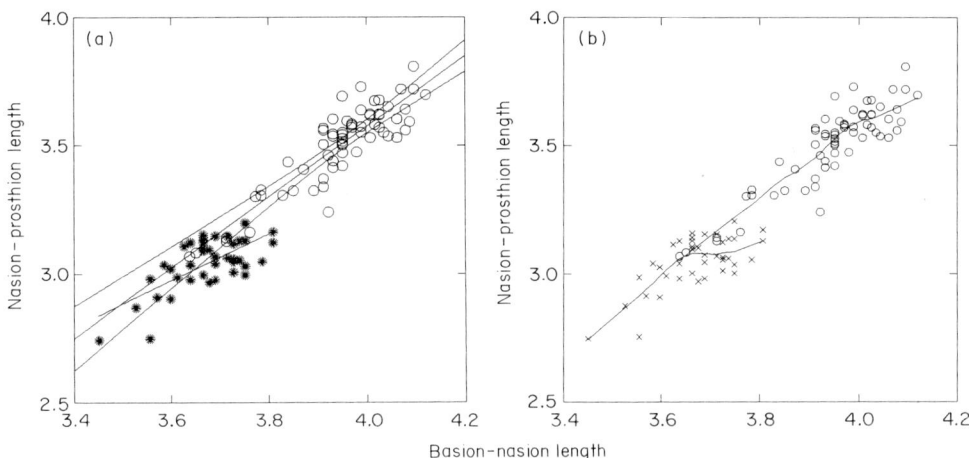

Figure 6. Plots of nasion–prosthion length vs. basion–nasion length. For (a), linear regression fits are illustrated; in (b), locally-weighted least squares fits are provided. See text for discussion.

significant curvilinearity or "complex allometry" make direct tests using linear regression problematic. Figure 6a, nasion–prosthion length versus basion–nasion length, provides a good example. Although visual comparison of linear regressions and the ANOCOVA results indicate slope or position differences in the allometric patterns, reduced major axis results suggest no significant differences (Table 4). Moreover, use of a locally-weighted least squares algorithm yields marked curvilinearity. Both species' patterns are quite similar in being characterized by an early period of high slope followed by a later period of much lower slope, when basion–nasion length is continuing to grow while nasion–prosthion has slowed or ceased its growth. Figure 6b illustrates that this overall similarity is even more fundamental, however, since the early periods of high slope fall along a common trajectory. The slopes

Table 5 **Principal component loadings (I–III) from an analysis of cranial variables in the combined ontogenetic samples of *C. talapoin* and *C. cephus***

Variable	I (95·1%)	II (1·5%)	III (0·9%)
Basion–nasion	0·157	0·008	0·010
Basion–bregma	0·100	−0·001	0·025
Basion–orale	0·208	0·019	−0·010
Biorbital width	0·129	0·023	−0·001
Bizygomatic width	0·161	0·019	0·004
Bizygomaxillary width	0·151	0·021	0·014
Nasion–prosthion	0·247	−0·048	−0·015
Staphylion–nasion	0·181	−0·026	0·032
Palate length	0·259	−0·004	−0·029
Palate breadth	0·164	0·007	0·000

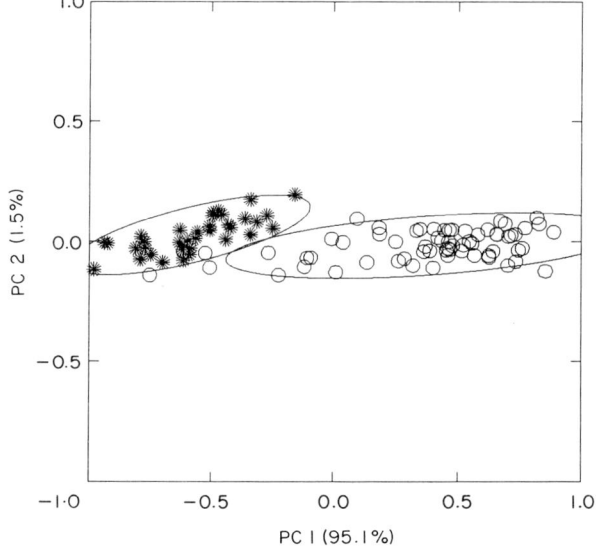

Figure 7. A plot of individual scores on principal components I and II. Symbols as in Figure 2. See text for discussion.

after the "break" appear to differ somewhat in the two species, though with these sample sizes its difficult to determine whether this is particularly meaningful. A plot of facial height (staphylion–nasion) shows a somewhat similar pattern of complex curvature in both species and again the reduced major axis results indicate no differences in position between the species trajectories.

Table 5 and Figure 7 provide results of the principal components analysis on the covariance matrix of log-transformed cranial dimensions in the ontogenetic series of *C. talapoin* and *C. cephus*. Variable loadings (eigenvectors) on the first principal component, which accounts for over 95% of the total variance, are all positive (Table 5). Since these loadings exhibit considerable variance, principal component one is summarizing both overall size and allometrically related shape and may be interpreted in this two-species case as a vector of ontogenetic scaling (Shea, 1985b). Variable loadings on the second principal component include a mix of both positive and negative values, as expected, and this component accounts for approximately 1·5% of the total variance. Though small, this is statistically significant, and in fact ANOVA and ANOCOVA results reveal that the two species can be distinguished on this component (Figure 7). The variables loading most strongly on the second component include biorbital width (demonstrating a significant shift difference in bivariate comparisons against basion–nasion length) and nasion–prosthion and staphylion–nasion, both of which exhibit considerable curvature and significant differences in the regression comparisons.

The quantitative results summarized above provide a rigorous approach to the problem of assessing patterns of covariation and the correspondence of dissociation of allometric trajectories. However, this quantitative comparison can be supplemented by qualitative assessment of key morphological differences. In this case, the following comparison is between a subadult *C. cephus* (with first permanent molars in) and an adult *C. talapoin*, of roughly equivalent overall skull size. The primary goal here was simply to pick out apparent differences in morphology between the skulls, so this should not be seen as a complete analysis designed to rigorously identify differences and similarities.

In lateral view, the talapoin has a somewhat lower cranial vault, especially in the area of the frontal. Muscle markings such as the superior temporal and nuchal are more rugose on the talapoin adult as compared to the *C. cephus* subadult. Basal view indicates that the talapoin is characterized by a relatively larger infratemporal fossa, and the palate appears to be relatively narrower and more U-shaped (due to the presence of adult incisors and canines). The occipital region in this view is especially inflated in the *C. cephus* subadult. Proportionately more of the orbital cones appear "tucked under" the anterior neurocranium in the young *C. cephus* than in the adult talapoin. Postorbital constriction is less marked in the *C. cephus* skull and of course the first permanent molar is much larger than in *C. talapoin*. From a frontal perspective, the orbits appear to have a higher height/width ratio in the young *C. cephus* and the frontal is more domed. The orbital rims are more rugose and marked in the talapoin adult. The nasal bones are somewhat more prominent and more horizontally inclined in the adult talapoin than the subadult *C. cephus*. Apart from these points, the faces appear quite similar in overall size and shape, with the exception that the talapoin has more strongly developed canine jugae (not surprisingly, in an adult with permanent canines erupted).

Verheyen adult data
Table 6 summarizes the results of regression analyses and ANOCOVA run on the pooled cranial dimensions from Verheyen's adult means and my individual specimens of *C. talapoin* and *C. cephus*. It is clear that these two groups exhibit very similar scaling patterns. Neither of the two differences emerging from the regression analysis are significantly different according to the method of Tsutakawa & Hewett (1977). One of these (nasion–prosthion length) exhibits the greatest curvilinearity, though careful visual comparison reveals a very close correspondence of the Verheyen data with the adult talapoins and moustached monkeys.

Figure 8 is a representative example of the similarity of the bivariate scaling patterns in my two groups as compared to Verheyen's adult means. The overall similarity in scaling patterns is all the more surprising considering that this is a comparison of a combined ontogenetic sequence with a static adult cluster (Verheyen's data). In this case, the marked size range plus the central role of (ontogenetic) allometric factors apparently accounts for this concordance.

Table 6 Results of bivariate regressions and ANOCOVA for selected cranial comparisons between the combined ontogenetic data for *C. talapoin* and *C. cephus* (O) and Verheyen's (1962) adult means for *Cercopithecus* (V). All *y* variables are regressed on basicranial length

y Variable		Slope (S.E.)/intercept (S.E.)/*r*/*n*	ANOCOVA Slope	Intercept
Basion–bregma	O	0·622 (0·024)/ 1·335 (0·092)/0·932/105	NS[1]	NS
	V	0·652 (0·037)/ 1·212 (0·148)/0·967/ 24		
Biorbital width	O	0·812 (0·022)/ 0·702 (0·086)/0·963/105	NS	NS
	V	0·784 (0·039)/ 0·804 (0·157)/0·974/ 24		
Bizygomatic width	O	1·015 (0·022)/ 0·083 (0·086)/0·977/103	NS	NS
	V	0·971 (0·054)/ 0·265 (0·218)/0·968/ 24		
Bizygomaxillary width	O	0·946 (0·031)/−0·022 (0·118)/0·950/105	NS	NS
	V	0·852 (0·045)/ 0·335 (0·181)/0·971/ 24		
Nasion–prosthion	O	1·522 (0·049)/−2·521 (0·189)/0·950/105	NS	0·002[2]
	V	1·662 (0·089)/−3·020 (0·358)/0·970/ 24		
Staphylion–nasion	O	1·139 (0·035)/−1·145 (0·134)/0·956/104	NS	NS
	V	1·152 (0·042)/−1·213 (0·168)/0·986/ 24		
Palate length	O	1·415 (0·040)/−2·288 (0·156)/0·961/103	NS	0·009[2]
	V	1·379 (0·071)/−2·102 (0·288)/0·972/ 24		

[1] NS = not significant at $P < 0.01$.
[2] Position differences not significant using RMA and Tsutakawa & Hewitt's (1977) method.
r = Pearson correlation coefficient.

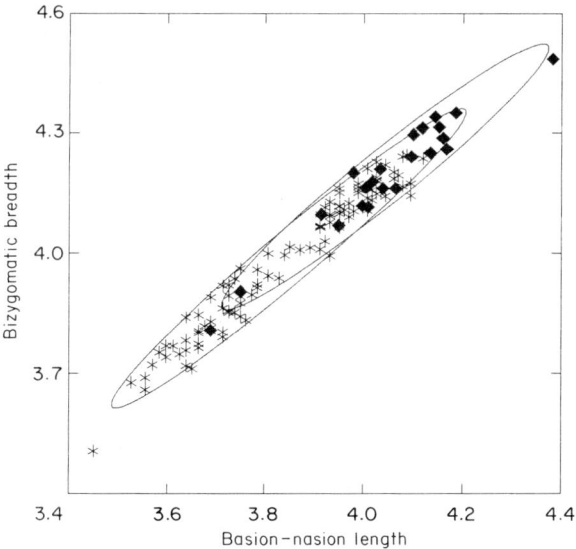

Figure 8. A representative example of ontogenetic scaling of facial proportions in a sample combining my ontogenetic sequences of talapoin plus moustached monkeys (∗) and Verheyen's (1962) means for adult *Cercopithecus* monkeys (◆). Ninety per cent confidence ellipsoids are included.

Results of principal components analysis of the pooled database support this similarity in scaling patterns. Figure 9 illustrates the results of an analysis combining my ontogenetic sequences with the Verheyen adult means. The first and second principal components account for 95·9 and 1·36% of the variance, respectively. Component loadings (Table 7)

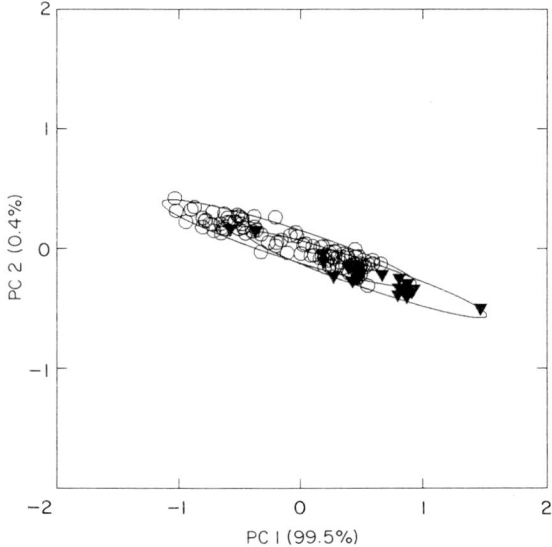

Figure 9. A plot of individual scores on principal components I and II for an analysis based on pooled samples of my ontogenetic sequences of talapoin plus moustached monkeys (○) and Verheyen's (1962) mean values for adult *Cercopithecus* monkeys (▼). Ninety per cent confidence ellipsoids are included.

Table 7   **Principal component loadings (I–III) from an analysis combining ontogenetic data for *C. talapoin* and *C. cephus* with Verheyen's (1962) adult means for *Cercopithecus***

Variable	I (95.9%)	II (1.4%)	III (0.9%)
Basion–nasion	0.172	0.013	0.005
Basion–bregma	0.109	0.013	0.017
Biorbital width	0.139	0.023	−0.008
Bizygomatic width	0.176	0.020	−0.005
Bizygomaxillary width	0.159	0.028	0.005
Nasion–prosthion	0.287	−0.048	0.002
Staphylion–nasion	0.197	−0.003	0.035
Palate length	0.257	−0.002	−0.035

reflect expected patterns of differential growth, i.e., highest for palatal and facial lengths. The large eigenvalue of the first component, plus the strong coincidence of the multivariate trajectories (Figure 9) suggests a major component of ontogenetic scaling underlying the shape differentiation within adult cercopithecines. However, it should also be stressed that an ANOCOVA reveals a small but significant ($P < 0.001$) shift or transposition on the second principal component, separating Verheyen's adult cercopithecine means from my pooled ontogenetic sequence of *C. talapoin* and *C. cephus*. The variable loading most strongly on the second component (and exhibiting the greatest curvature in bivariate plots) is nasion–prosthion length.

An additional PCA was run on the data set described above, but with the addition of the colobine means from Verheyen's (1962) monograph. Figure 10 illustrates scores on the first

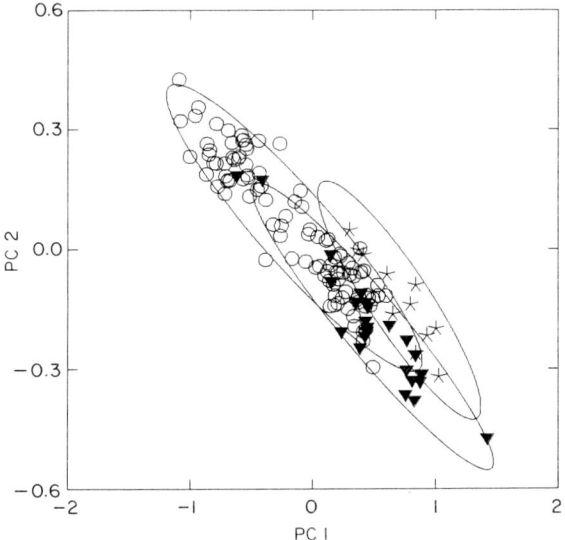

Figure 10. A plot of individual scores on principal components I and II for an analysis including Verheyen's (1962) mean values for various adult colobine monkeys (∗) in addition to the sample depicted in Figure 9. Note: axes are not standardized in this example. See text for discussion.

two components for my growth data on *C. talapoin* and *C. cephus* and both the cercopithecine and colobine means from Verheyen. The colobine scatter is quite distinct from both the Verheyen adult cercopithecine scatter, as well as the *C. talapoin* and *C. cephus* ontogenetic scatters. Subfamily differences in skull form within the Cercopithecoidea are well-established (e.g., Verheyen, 1962; Delson, 1975), and therefore the appreciable shift of the colobine scatter away from the cercopithecines indicates that the measurements and analyses are accurately summarizing key differences in craniofacial morphology.

Discussion

Postcranial proportions
Results of the present allometric study raise some interesting questions about traditional functional interpretations of limb proportions. Building on their contrast of *Pithecia pithecia*, a frequent leaper, and *Chiropotes satanas*, a more generalized arboreal quadruped, Fleagle & Meldrum (1988) argue that primates which include a marked component of leaping behavior in their locomotor repertoire should be characterized by a particular set of body proportions. These include relatively small body size, relatively lower intermembral and humeral/femoral indices, elongation of the hindlimb relative to external reference dimensions, and a relatively higher brachial index (Fleagle & Meldrum, 1988). These limb and body proportions are presumed to be related to either the generation of forces by the musculoskeletal system in producing adequate leaping behaviors, or to behaviors often associated with leaping, such as clinging and climbing (Fleagle & Meldrum, 1988).

This pattern of skeletal proportions should presumably also characterize a contrast between *C. talapoin* and *C. cephus*, since the talapoin is considered to be among the most frequent and adept of the arboreal leapers within the cercopithecines (Fleagle, 1988). However, while the talapoin's small body size certainly fits the predicted profile, limb proportions reveal a somewhat different pattern, and the present growth allometric investigation helps explicate the basis of this apparent discrepancy.

The first departure from the expected pattern involves the adult proportions (Table 1). Both the intermembral and humeral/femoral indices increase, rather than decrease, as we move from the more quadrupedal moustached monkey to the talapoin, which is characterized by greater frequencies of leaping. Furthermore, comparison to the external dimension of pelvic height indicates that adult talapoins have relatively longer forelimbs than adult *C. cephus*, while relative hindlimb lengths do not differ significantly. Fleagle & Meldrum (1988) suggest that leaping monkeys should have proportional or relatively shorter total size of the forelimbs, since these play a less prominent role in propulsion and in fact represent extra weight which must be displaced. Also contrary to predictions is the proportional hindlimb size of the talapoin compared to *C. cephus*. The talapoin does have a higher brachial index than the moustached monkey, though this difference does not reach statistical significance. Additionally, it should be pointed out here that this between-adult proportion contrast not only departs from the expectations of models based on leaping adaptations, but also from the general size-related pattern of increasing intermembral index observed among many primate groups (e.g., Jungers, 1985, 1986; Aiello, 1981; Shea, 1981) and attributed by Jungers (1978, 1985) to adaptations for maintaining functional competence in climbing behaviors with increasing body size.

Why do the talapoins not accord with the expectations of the functional models? One possible explanation which deserves further investigation is given by Rollinson & Martin (1981), who suggest on the basis of kinematic patterns (e.g., digitigrady, greater use of three-legged support patterns in walking and a relatively low incidence of trotting gaits) that talapoins approximate the mangabeys in certain ways the authors take to be indicative of terrestrial adaptations. Thus, they argue that "the talapoin became secondarily re-adapted for arboreal life later than the other guenons" (Rollinson & Martin, 1981: 420). This hypothesis is difficult or impossible to test.

The problematic adult proportion differences are clarified to a great degree by the comparative growth allometric approach taken here. Ontogenetic scaling is found to be a pervasive underpinning of the adult proportion differences observed between *C. talapoin* and *C. cephus*. For example, the adult difference in proximal limb proportions (humerus-to-femur ratio) results from differential travel along a common allometric trajectory. The significantly lower crural index of the moustached monkey also results from ontogenetic scaling. Adult differences in limb lengths relative to pelvic height also result from ontogenetic scaling.

The simple fact is that what we seem to be seeing in the postcranial skeletons of these monkeys is a continuous and shared pattern of regularly changing proportions, with basically unaltered patterns of covariation amongst the different features. This pattern is related to and/or controlled by overall size change during growth and in the interspecific contrast. In an allometric sense, proportion differences between the two points represented by mean adult *C. talapoin* and *C. cephus* are not necessarily any more significant or informative than those between any other two points, for example, young talapoins and subadult moustached monkeys (which by the way would provide a proportion contrast comparable in direction, though not precisely identical in numeric expression, to that between the adults which are traditionally the focus of locomotor comparisons). Elsewhere is stressed (e.g., Shea, 1985*a*, 1986, 1990) the point that a finding of ontogenetic scaling by no means implies a simple rejection of functional differences claimed to be associated with changing proportions. In fact, there are many reasons to believe that natural selection acts on all ontogenetic stages and indeed, perhaps in a coordinated way on entire complex trajectories of growth (Shea, 1990). However, in the present case, the finding of ontogenetic scaling explicates adult proportion differences which run counter to predictions based on functional models and, therefore, the *C. talapoin/C. cephus* comparison does not simply entail extending our functional hypothesis from inter-adult contrasts between smaller and larger species to those characterizing adult versus young conspecifics. Therefore, it would seem that the particular body proportions of the talapoin monkey owe more to its evolutionary size decrease via allometric truncation, than they do to the results of concerted selective efforts to alter individual skeletal elements in relation to specific locomotor kinematics and frequencies. Any such conclusion must ultimately be placed in the context of a detailed examination of function and changing proportions during ontogeny in many related *Cercopithecus* species (Shea, 1990). Meldrum (1991) has provided a promising beginning for various adult *Cercopithecus* monkeys, but we need to extend such studies to all later developmental stages. Certainly I would argue that the small body size of adult talapoins is functionally related to their increased leaping behavior (see Fleagle, 1988, for general discussion of this relationship within many, though certainly not all, primate groups), regardless of whether specific limb and skeletal proportions follow predicted patterns. The small size and increased leaping may both be related to the talapoin's propensity to utilize insects and small vertebrates as a major food source (see below).

I do not conclude from the foregoing that previous assessments of the general relationship between skeletal proportions and locomotor behaviors such as frequency of arboreal leaping are necessarily incorrect. Rather, to stress that various other factors, including size change via ontogenetic scaling, clearly play a role in affecting such proportions and these must be incorporated into a complete investigation of adult form and function (Shea, 1985a, 1990). It seems likely that the greatest potential dissociation between the finer details of gross limb proportions and the specific locomotor adaptations purportedly related to them will occur in closely related groups characterized by size divergence. For example, human pygmies differ from their larger conspecifics in having relatively shorter hindlimbs and thus differing in the direction, if not the degree, of related non-bipeds. But certainly we would not assign functional significance to this pattern, which in fact results from simple size decrease via truncation of ontogenetic trajectories displaying positive allometry of hindlimb growth (Shea, 1988, 1990). One cannot make the point strongly enough that studies attempting to relate form to function by utilizing the comparative approach among closely-related adults of differing body size would be on much firmer ground if ontogenetic allometric data were incorporated in order to examine scaling patterns and utilize the appropriate criterion of subtraction for such comparisons. This position is shared by many of the papers from this symposium.

An appreciation of this possibility and the potential allometric basis for a dissociation between the normal (and perhaps generally quite justified) correlation of morphology and function is seen in the scaling of brain size. This is included here in the discussion of postcranial morphology only to make the following point. It is well known that allometric studies of brain and body size have revealed a great deal about functional specializations and behavioral adaptations in living and fossil mammals. Significantly positive and negative residuals from the general interspecific trend comprise basic data reflecting meaningful adaptive differences in relative brain size between taxonomic groups (e.g., mammals versus reptiles), dietary categories (frugivores versus folivores), reproductive strategies (precocial versus altricial) and marked behavioral autapomorphy (as in the only species which writes papers about the relative brain size of other species). These and many other such examples are well established (for discussion see Jerison, 1973; Martin, 1990). Nevertheless, many authorities recognize that this pattern of morphological and behavioral concordance does not always hold, and often does not when rapid size change and associated allometric relationships are involved. Thus, in discussing the very high positive residual of brain size in the talapoin monkey, Bauchot & Stephan (1969: 267–268, quote translated in Gould, 1975: 279) caution that "These miniaturized forms appear, in the value of their index of cephalization, as favored as juveniles of a species compared with adults. This last comparison shows that we cannot use indices of cephalization thus attained to infer that species have reached a superior level of brain evolution." Jerison (1973) dealt extensively with this issue for a variety of taxa. The following passage on relative brain size could be translated to both the general patterns of postcranial morphological and behavioral correlation, as well as the specific departure from this pattern of general concordance in the talapoin:

> "The effect of selection pressures toward specifically large or small bodies are special cases of evolution and provide many of the exceptions to the general trends that have been identified. It would be foolish error, however, to use the examples just developed as argument of any force against the use of relative brain size as a general measure with which to compare large assemblages of species. Selection pressures toward unusually large or small bodies can result in species that are 'aberrant' in brain:body relations. But these are exceptional, and in our analysis of broad samples of species, the pygmy and giant species merely increase the variability of relative brain size; they contribute to the 'error' variance in the analysis." (Jerison, 1973: 345).

The challenge, of course, lies in accurately recognizing such exceptions, but it is also important to bear in mind the existence of this "error" relative to our functional linkages. Comparative analysis of ontogenetic allometric patterns in an informed phylogenetic context is certainly a major step in both identifying the specific departures as well as understanding why this "error" exists.

Craniofacial proportions
The results here confirm and extend the finding and conjectures of Verheyen and others that (1) adult talapoin monkeys have skulls shaped quite differently than their close *Cercopithecus* relatives and (2) these differences are predominantly allometric in nature. Both the bivariate and multivariate analyses demonstrate that the predominant component of shape differentiation between adult talapoins and moustached monkeys is a result of ontogenetic scaling or the sharing of common patterns of allometric growth.

The finding of strong ontogenetic scaling in the cranial (and postcranial) comparisons demonstrates that much of the morphological distinctiveness of the talapoin monkey is correlated with, and determined by, its small size. Two significant findings in the present study remind us that proper consideration must be given to factors of growth-in-time as well as growth-in-size. First, the results of the qualitative comparison clearly reveal a contrast between a young, subadult *C. cephus* with a fully adult talapoin, for the skull of the latter differs in a number of morphological features overlaying the essentially similar overall size and gross proportions. These morphological differences of course reflect the full eruption of the dentition and the associated masticatory and nuchal musculature. Workers must keep these factors in mind when making allometric comparisons of forms which differ in both size and age, though of course the presence of such differences in animals of divergent ages is expected and in no way changes our interpretation of the concordance or discordance of underlying trajectories of relative growth.

The second significant implication of growth timing relates to the presence of significant curvilinearity in several of the comparisons. Such curvilinearity is usually termed "complex allometry" (Gould, 1966; Laird, 1965), because the coefficient of relative growth is changing. The presence and meaning of such complex allometry has been appreciated since the original formulation of the allometric approach (Huxley, 1932) and in no way requires any theoretical modifications or alternative approaches. Put quite simply, it indicates that the growth of one structure has changed its relative rate of growth (either abruptly or gradually) in comparison to a second structure or overall size. While such complex allometry presents no challenge to the theoretical basis of allometric growth (Laird, 1965), it does present some statistical difficulties. This involves not so much the modeling of a particular curvilinear fit, since there are many possible approaches here, including some recent ones specifically within the allometric literature (e.g., Chappell, 1989; Jolicoeur, 1989). Rather, the difficulty emerges in trying to compare two curvilinear patterns in an attempt to test the hypothesis that one is an extrapolated or truncated version of the other. Some aspects of this problem are dealt with elsewhere (Shea *et al.*, 1990) and my own opinion is that many supposed differences in allometric patterns among closely related species may actually result from the fact that extrapolation of a curvilinear pattern yields a trajectory which at first analysis (and with most linear statistical tests) appears quite different. This issue may be especially problematic if the curvilinear phase of allometric growth occurs early in ontogeny, in a stage not measured by the investigator. Patterns of cranial vault allometry reflect brain growth and frequently fit such a pattern.

If the results of comparisons of *C. talapoin* and *C. cephus* suggest that the primary shape differences result from size variation operating on similar underlying allometric patterns, then the inclusion of Verheyen's adult data for other *Cercopithecus* only strengthen this possibility. The strong correlation with size and the location of these adult crania along an extension of the joint *C. talapoin*/*C. cephus* allometries indicates that patterns of relative growth have been strongly conserved. Yet clearly the morphological differences resulting from this scaling phenomenon are quite marked, considering the variation in proportions exhibited by a talapoin-to-patas comparison. In fact, it is likely that the genus *Cercopithecus* (including talapoins and patas monkeys) is one of the best examples among Primates of the central role that simple allometric scaling can play. Additional studies involving more detailed measurement methodologies and large subadult samples for the other *Cercopithecus* species are needed to further corroborate these apparent trends. How such morphometric investigations might be combined with broader studies focusing on physiological growth control, functional anatomy and ecological contexts is discussed elsewhere (Shea, 1988, 1990).

Phylogenetic and functional implications of size-corrected adult differences
Martin & MacLarnon (1988) have completed an interesting allometric investigation of craniodental morphology in *Cercopithecus*, combining Verheyen's (1962) cranial data and Kay's (1978) dental data. Their analyses and results are considered in some detail here, for the following two reasons. First, as in the present paper, a portion of their database is derived from Verheyen's (1962) monograph on cercopithecoid skull form and second, this will permit some general consideration of issues involved in the utilization of size-corrected adult residuals in comparative studies of adaptation and systematics. I will suggest that the somewhat problematic results which they obtain might be greatly clarified by the comparative growth allometric approaches utilized here in analysing *C. talapoin* and *C. cephus*.

In their investigation, Martin & MacLarnon (1988) computed residuals of eighteen cranial dimensions (relative to skull length) and seven dental dimensions (relative to body weight) and then utilized various statistical approaches to assess correspondence with known dietary categories and phylogenetic groupings determined from other studies. They found little association between computed residuals and dietary categories, though they claimed some correspondence with phylogenetic groupings. While it is true that their clustering based on craniodental allometric residuals exhibits a fair similarity to that derived from chromosomal evidence (Dutrillaux et al., 1988), it departs quite markedly from other such phylogenies (e.g., Ruvolo, 1988; Gautier, 1988). Moreover, their clustering based on cranial residuals is quite different than that derived from dental variables.

The point to stress is that the ontogenetic allometric investigation completed here yields a somewhat different perspective on Martin & MacLarnon's (1988) approach and conclusions, since, here, are lumped both Verheyen's (1962) adult data and the growth trajectories of *C. cephus* plus *C. talapoin*. Given that ontogenetic scaling plays such a predominant role in determining differences in interspecific adult morphology, it is worthwhile to pause and contemplate just what the residuals computed by Martin & MacLarnon (1988) mean in a biological sense. It needs to be demonstrated, rather than assumed, that such mean adult residuals necessarily carry important information about adaptation or phylogenetic proximity. Consider an extreme and hypothetical example where all the morphological differences among adult species means are the result of differential ontogenetic scaling. In this case, adult residuals would reflect essentially random scatter about the line of best fit or the allometric regression, just as would residuals which might be computed for young and subadult specimens. Most workers would probably shrink from such an assumption if we extended this reasoning to age group means or individual subadult values, which will of course also scatter about best-fit lines and have some distribution of both positive and negative residuals, rather than lying precisely on the line. Certainly we would not automatically assume much functional, let alone phylogenetic, significance for a determination that age group 1 in a given species has a positive residual while group 2 in the same species has a negative one.

The finding here that ontogenetic scaling is pervasive in the comparison of *C. talapoin* and *C. cephus*, and that it likely also plays a predominant role in producing adult variation in the rest of the guenons, suggests a reconsideration of the problematic results of Martin & MacLarnon (1988). A general pattern of ontogenetic allometric concordance may be the reason that they get clusters which are discordant in a functional and phylogenetic sense and why their cranial and dental patterns differ so dramatically. An additional contributing factor may be that the cranial and dental residuals were determined against different y variables.

Therefore, what we need to determine before making such biological inferences is whether the adult means lie at the terminal end of trajectories which are clearly statistically distinct (i.e., not ontogenetically scaled) or whether they reflect essentially random scatter about a common pattern of ontogenetic and adult allometric concordance. Only in the first case would simple comparative studies yield reasonably firm evidence that the guenon data support the conclusion that "there is some genetic basis for 'size-free' morphological features, ... [indicating] some fundamental distinction at the genetic level between the scaling effects of body size and the specific morphological adaptations of individual species" (Martin & MacLarnon, 1988: 183). Furthermore, only detailed investigation of ontogenetic allometries can yield this information, providing additional support for the claim that the most meaningful "criterion of subtraction" in biological terms is ontogenetic scaling (Gould, 1966, 1968, 1975; Shea, 1985a). Until such extensive comparative ontogenetic studies are undertaken, the conclusions of Martin & MacLarnon (1988) on guenon craniodental morphology must remain conjectural.

Allometry, heterochrony and ecological context
It is interesting that Verheyen (1962) as well as later authors referred to the morphological profile of the talapoin as the result of paedomorphosis or neoteny, since an approach based on heterochrony has only fairly recently been common in studies of primate evolution. Overall cranial and postcranial morphology in *C. talapoin* is indeed paedomorphic, or juvenilized, though it is now well established that this morphological result can be produced by quite different growth processes (Gould, 1977; Shea, 1983b). Neoteny involves global or regional dissociation of allometric patterns, which clearly does not fit the present case of strong ontogenetic scaling or allometric concordance. Those who have characterized the paedomorphic morphology of the talapoin as a result of neoteny are incorrect in identifying the process underlying the morphological juvenilization. There is also no firm evidence that talapoins are "progenetic" (Gould, 1977), i.e., they reach sexual maturity and cease growth at a significantly earlier time than their close relatives (Shea, 1990). This has recently been confirmed by Leigh (1992) in a broad analysis of anthropoid growth and life history features. Therefore, it is most appropriate to conclude that the paedomorphic morphology of the talapoin is produced by what is termed here as *rate hypomophosis* (Shea, 1983b). Regardless of terminological taxonomy, what has occurred in the talapoin (assuming a derived status for these features, of course) is a marked decrease in overall rates of body weight growth per unit time, with no major changes in underlying patterns of allometric growth. The simplest evolutionary explanation of this pattern, though one that is not directly testable through traditional comparative analyses, is that selection acted to change overall body size and the allometric shifts in body proportions were merely correlated consequences.

The physiological basis of the growth retardation in talapoins is unknown, of course. It is worth pointing out in this regard, however, that results of a number of studies published and in press (e.g., Shea, 1988, 1990, in press; Shea *et al.*, 1987, 1990) suggest that shifts in levels of growth hormone (GH) and/or insulin-like growth factor 1 (IGF-1) might be expected to produce the observed pattern of decreased growth-rate-in-time, smaller terminal size and simple allometric truncation of ancestral trajectories of relative growth. Research currently underway will hopefully further clarify this possibility for the *Cercopithecus* radiation.

As noted previously, the very large relative brain size of the talapoin monkey fits the expectations of a form characterized by rapid size decrease produced via shifts in GH and/or IGF-1. These growth hormones have little effect on growth of the brain and prenatal growth in general. Thus, we basically end up with a monkey dwarfed predominantly or exclusively via postnatal growth mechanisms, which therefore results in a creature with a large relative brain size. As a variety of workers (e.g., Jerison, 1973; Gould, 1975; Bauchot & Stephan, 1969; Shea, 1983a) have argued, this perspective cautions against placing any great weight on such cases of high (or conversely, low) encephalization in terms of adaptive specializations or behavioral functions. Aspects of relative brain size and other features in terms of pre- and postnatal growth concentration in a variety of primates, including the talapoin, are discussed in Shea (1983a). These issues are discussed more explicitly in terms of hormonal bases and variable scaling coefficients at different taxonomic levels in Atchley *et al.* (1984) and Riska & Atchley (1985).

The ecological basis of the growth retardation and small size in talapoins is uncertain and clearly a key to any detailed understanding of their evolutionary history. However, the comparative work completed by Gautier, Gautier-Hion and their colleagues in Gabon on sympatric cercopithecines provides some intriguing information (see references in Gautier-Hion *et al.*, 1988). Gautier-Hion (1978) suggested that the small size of the talapoin was related to its marked insectivory (approximately 36% animal matter in stomach contents, by far the greatest of the species compared) and she related various other factors such as habitat preference and social organization to this dietary pattern. The benefits of smaller body size to a primate feeding on a large proportion of insects and other animal prey are well known (Kay, 1975, 1984; Fleagle, 1988). It is notable that in a comparative analysis such as the present one, we can emphasize the reasonable functional linkage of small size and insectivory, but we cannot specifically reconstruct chronological selective events or distinguish between scenarios where small size is specifically targeted in response to selection for increased insectivory and those where the dietary shifts results from an adaptive niche being opened up to a species which originally underwent selection for size change for some other reason. It is interesting that Gautier-Hion (1978) notes that in addition to the association of small size and increased insectivory in the talapoin, two other fairly insectivorous *Cercopithecus* monkeys, *C. pogonias* and *C. cephus*, are also relatively small.

The implications of the results presented here for the classification and systematics of guenons depend on what weighting is assigned to skeletal features as compared to other traits. Certainly the ontogenetic allometric approach reveals that size differences and allometric factors are the primary determinants of the marked morphological differentiation stressed by Verheyen (1962) as characterizing both the small talapoin and large patas monkeys. This is a significant conclusion and one that could only be reached by detailed comparisons of growth allometries. The "cohesiveness" revealed by this allometric study argues for placing both talapoins and patas within *Cercopithecus*, a conclusion shared by even Verheyen (1962). In contrast, Strasser & Delson (1987) chose to place talapoins in a distinct genus (*Miopithecus*), based on the fact that it is the only cercopithecinan with sexual skin cyclicity, though they grant that the talapoin's skeletal distinctions might simply reflect coordinated allometric changes. They also place patas monkeys in their own genus (*Erythrocebus*), citing cranial and postcranial distinctions. The present investigation shows that a more complete reanalysis of this question utilizing growth sequences of patas monkeys might yield new insights. Similarly, many of the cranial and postcranial features purported to link *Mandrillus* and *Papio* may be influenced by allometric factors, generating a phenetic similarity which is not indicative of phylogenetic proximity (Disotell *et al.*, 1992). The primary point to stress here is that as in other such cases (see Shea, 1985*a*, for a discussion of the African apes), growth allometric approaches provide important new data of systematic significance, even though different authors may ultimately arrive at divergent conclusions regarding taxonomic classifications.

Conclusions

This preliminary investigation of patterns of skeletal allometry in talapoin monkeys and other members of the genus *Cercopithecus* has demonstrated a central role for ontogenetic scaling in the production of morphological diversity. In both the skull and postcranium, the primary shape differences between adult *C. talapoin* and *C. cephus* result from differential extension of common patterns of growth allometry. Inclusion of Verheyen's (1962) data set strongly suggests that ontogenetic scaling may account for most of the shape differences among other *Cercopithecus* taxa as well, including the patas monkey. Previous comments by Verheyen and others have suggested the possibility that the divergent morphology of talapoin monkeys might be a result of allometric and heterochronic transformation and these results provide the first test and confirmation of this hypothesis. Assuming their small body size is derived, the paedomorphic morphology of talapoins as compared to their larger relatives results from the heterochronic process of rate hypomorphosis (Shea, 1983b), characterized by decreased rates of weight growth-in-time and simple allometric truncation with no dissociation of size/shape relationships. This case provides a good example of a cascade of integrated and coordinated morphological changes resulting from a simple shift in underlying growth controls. Additional studies of patterns of allometric growth, life-history features and the genetic and epigenetic bases of growth patterns in *Cercopithecus* monkeys would greatly clarify the morphological evolution of this important radiation of anthropoids (Shea, 1990).

Acknowledgements

I wish to thank Matt Ravosa and Anne Gomez for inviting me to participate in this symposium and for their help and patience in the preparation of this paper. The assistance of the curators and other staff at the British Museum of Natural History (London), Powell-Cotton Museum (Kent, U.K.), Museum of Central Africa (Brussels), and the National Museum of Natural History (Paris), is gratefully acknowledged. Financial support was provided by the Northwestern University Research Grants Committee.

References

Aiello, L. C. (1981). The allometry of primate body proportions. *Symp. zool. Soc. Lond.* **48,** 331–358.
Atchley, W. R., Riska, B., Kohn, L. A. P., Plummer, A. A. & Rutledge, J. J. (1984). A quantitative genetic analysis of brain and body size associations, their origin and ontogeny: data from mice. *Evolution* **38,** 1165–1179.
Bauchot, R. & Stephan, H. (1969). Encéphalisation et niveau évolutif chez les simiens. *Mammalia* **33,** 225–275.
Chappell, R. (1989). Fitting bent lines to data, with applications to allometry. *J. theor. Biol.* **138,** 235–256.
Child, C. M. (1941). *Patterns and Problems of Development*. Chicago: University of Chicago Press.
Clarke, M. R. B. (1980). The reduced major axis of a bivariate sample. *Biometrics* **67,** 441–446.
Cock, A. G. (1966). Genetical aspects of metrical growth and form in animals. *Q. Rev. Biol.* **41,** 131–190.
DeBeer, G. R. (1930). *Embryology and Evolution*. Oxford: Clarendon Press.
DeBeer, G. R. (1940). *Embryos and Ancestors*. Oxford: Clarendon Press.
Delson, E. (1975). Evolutionary history of the Cercopithecidae. *Contr. Primat.* **5,** 167–217.
Disotell, T. R., Honeycutt, R. L. & Ruvolo, M. (1992). Mitochondrial DNA phylogeny of the Old World monkey tribe Papionini. *Mol. Biol. Evol.* **9**.
Dobzhansky, T. (1970). *Genetics of the Evolutionary Process*. New York: Columbia University Press.
Doran, D. M. (1992). The ontogeny of chimpanzee and pygmy chimpanzee locomotor behavior: a case study of morphological paedomorphosis and its behavioral correlates. *J. hum. Evol.* **23,** 139–157.
Dorst, J. & Dandelot, P. (1970). *A Field Guide to the Larger Mammals of Africa*. London: Collins.
Dutrillaux, B., Muleris, M. & Couturier, J. (1988). Chromosomal evolution of Cercopithecinae. In (A. Gautier-Hion, F. Bourliere, J.-P. Gautier & J. Kingdon, Eds) *A Primate Radiation: Evolutionary Biology of the African Guenons*, pp. 150–159. Cambridge: Cambridge University Press.
Fleagle, J. G. (1988). *Primate Adaptation and Evolution*. New York: Academic Press.
Fleagle, J. G. & Meldrum, D. J. (1988). Locomotor behavior and skeletal morphology of two sympatric Pithecine monkeys, *Pithecia pithecia* and *Chiropotes satanas*. *Am. J. Primat.* **16,** 227–249.
Gautier, J.-P. (1988). Interspecific affinities among guenons as deduced from vocalizations. In (A. Gautier-Hion, F. Bourliere, J.-P. Gautier & J. Kingdon, Eds) *A Primate Radiation: Evolutionary Biology of the African Guenons*, pp. 194–226. Cambridge: Cambridge University Press.
Gautier-Hion, A. (1978). Food niches and coexistence in sympatric primates in Gabon. In (D. J. Chivers & J. Herbert, Eds) *Recent Advances in Primatology, vol. II*, pp. 270–286. New York: Academic Press.

Gautier-Hion, A., Bourliere, F., Gautier, J. P. & Kingdon, J. (1988). *A Primate Radiation: Evolutionary Biology of the African Guenons*. Cambridge: Cambridge University Press.

Goldschmidt, R. (1938). *Physiological Genetics*. New York: McGraw-Hill.

Gould, S. J. (1966). Allometry and size in ontogeny and phylogeny. *Biol Rev.* **41,** 587–640.

Gould, S. J. (1968). Ontogeny and the explanation of form: an allometric analysis. In (D. B. Macurda, Ed.) *Paleobiological Aspects of Growth and Development, A Symposium. Paleontol. Soc. Mem.* **2,** 81–98.

Gould, S. J. (1971). Geometric scaling in allometric growth: a contribution to the problem of scaling in the evolution of size. *Am. Nat.* **105,** 113–136.

Gould, S. J. (1975). Allometry in primates, with emphasis on scaling and evolution of the brain. *Contr. Primat.* **5,** 244–292.

Gould, S. J. (1977). *Ontogeny and Phylogeny*. Cambridge: Harvard University Press.

Haraway, D. J. (1976). *Crystals, Fabrics, and Fields*. New Haven: Yale University Press.

Huxley, J. S. (1932). *Problems of Relative Growth*. London: MacVeagh.

Huxley, J. S. & DeBeer, G. R. (1934). *The Elements of Experimental Embryology*. Cambridge: Cambridge University Press.

Jerison, H. J. (1973). *Evolution of the Brain and Intelligence*. New York: Academic Press.

Jolicoeur, P. (1963). The multivariate generalization of the allometry equation. *Biometrics* **19,** 497–499.

Jolicoeur, P. (1989). A simplified model for bivariate complex allometry. *J. theor. Biol.* **140,** 41–49.

Jungers, W. L. (1978). The functional significance of skeletal allometry in *Megaladapis* in comparison to living prosimians. *Am. J. phys. Anthrop.* **19,** 303–314.

Jungers, W. L. (1985). Body size and scaling of limb proportions in primates. In (W. L. Jungers, Ed.) *Size and Scaling in Primate Biology*, pp. 345–382. New York: Plenum Press.

Jungers, W. L. (1986). Aspects of size and scaling in primate biology with special reference to the locomotor skeleton. *Yearb. phys. Anthrop.* **27,** 73–97.

Jungers, W. L. & Susman, R. L. (1984). Body size and skeletal allometry in African apes. In (R. L. Susman, Ed.) *The Pygmy Chimpanzee: Evolutionary Biology and Behavior*, pp. 131–178. New York: Plenum Press.

Kay, R. F. (1975). The functional adaptations of primate molar teeth. *Am. J. phys. Anthrop.* **43,** 195–216.

Kay, R. F. (1978). Molar structure and diet in extant Cercopithecidae. In (P. M. Butler & K. A. Joysey, Eds) *Development, Function and Evolution of Teeth*, pp. 309–339. London: Academic Press.

Kay, R. F. (1984). On the use of anatomical features to infer foraging behavior in extinct primates. In (P. S. Rodman & J. G. H. Cant, Eds) *Adaptations for Foraging in Nonhuman Primates: Contributions to an Organismal Biology of Prosimians, Monkeys and Apes*, pp. 21–53. New York: Columbia University Press.

Laird, A. K. (1965). Dynamics of relative growth. *Growth* **29,** 249–263.

Leigh, S. R. (1992). Ontogeny and body size dimorphism in anthropoid primates. Ph.D. Dissertation, Northwestern University.

Martin, R. D. (1990). *Primate Origins and Evolution*. Princeton, NJ: Princeton University Press.

Martin, R. D. & MacLarnon, A. M. (1988). Quantitative comparisons of the skull and teeth in guenons. In (A. Gautier-Hion, F., Bourliere, J.-P. Gautier & J. Kingdon, Eds) *A Primate Radiation: Evolutionary Biology of the African Guenons*, pp. 160–183. Cambridge: Cambridge University Press.

Mayr, E. (1963). *Animal Species and Evolution*. Cambridge: Harvard University Press.

Meldrum, D. J. (1991). Kinematics of the cercopithecine foot on arboreal and terrestrial substrates with implications for the interpretation of hominid terrestrial adaptations. *Am. J. phys. Anthrop.* **84,** 273–290.

Napier, J. R. & Napier, P. H. (1967). *A Handbook of Living Primates*. New York: Academic Press.

Riska, B. & Atchley, W. R. (1985). Genetics of growth predict patterns of brain-size evolution. *Science* **229,** 668–671.

Rollinson, J. & Martin, R. D. (1981). Comparative aspects of primate locomotion, with special reference to arboreal cercopithecines. *Symp. zool. Soc. Lond.* **48,** 377–427.

Ruvolo, M. (1988). Genetic evolution of the African guenons. In (A. Gautier-Hion, F. Bourliere, J.-P. Gautier & J. Kingdon, Eds) *A Primate Radiation: Evolutionary Biology of the African Guenons*, pp. 127–139. Cambridge: Cambridge University Press.

Shea, B. T. (1981). Relative growth of the limbs and trunk in the African apes. *Am. J. phys. Anthrop.* **56,** 179–202.

Shea, B. T. (1983a). Phyletic size change and brain/body scaling: a consideration based on the African pongids and other primates. *Int. J. Primat.* **4,** 33–62.

Shea, B. T. (1983b). Allometry and heterochrony in the African apes. *Am. J. phys. Anthrop.* **62,** 275–289.

Shea, B. T. (1985a). Ontogenetic allometry and scaling: a discussion based on the growth and form of the skull in African apes. In (W. L. Jungers, Ed.) *Size and Scaling in Primate Biology*, pp. 175–206. New York: Plenum Press.MB

Shea, B. T. (1985b). Bivariate and multivariate growth allometry: statistical and biological considerations. *J. Zool. Lond.* **206,** 367–390.

Shea, B. T. (1986). Scapula form and locomotion in chimpanzee evolution. *Am. J. phys. Anthrop.* **70,** 475–488.

Shea, B. T. (1988). Heterochrony in primates. In (M. L. McKinney, Ed.) *Heterochrony in Evolution: A Multidisciplinary Approach*, pp. 237–266. New York: Plenum Press.

Shea, B. T. (1990). Dynamic morphology: growth, life history, and ecology in primate evolution. In (C. J. DeRousseau, Ed.) *Primate Life History and Evolution*, pp. 325–352. New York: Wiley-Liss, Inc.

Shea, B. T. (in press). A developmental perspective on size change and allometry in evolution. *Evolutionary Anthropology*.
Shea, B. T., Hammer, R. E. & Brinster, R. L. (1987). Growth allometry of the organs in giant transgenic mice. *Endocrin.* **121,** 1924–1930.
Shea, B. T., Hammer, R. E., Brinster, R. L. & Ravosa, M. J. (1990). Relative growth of the skull and postcranium in giant transgenic mice. *Genet. Res. Camb.* **56,** 21–34.
Simpson, G. G. (1944). *Tempo and Mode in Evolution.* New York: Columbia University Press.
Simpson, G. G. (1953). *The Major Features of Evolution.* New York: Columbia University Press.
Strasser, E. & Delson, E. (1987). Cladistic analysis of cercopithecid relationships. *J. hum. Evol.* **16,** 81–99.
Szalay, F. S. & Delson, E. (1979). *Evolutionary History of the Primates.* New York: Academic Press.
Thompson, D. W. (1917). *On Growth and Form.* Cambridge: Cambridge University Press.
Tsutakawa, R. K. & Hewett, J. E. (1977). Quick test for comparing two populations with bivariate data. *Biometrics* **33,** 215–219.
Verheyen, W. N. (1962). Contribution à la craniologie comparèe des primates. *Mus. r. Afr. Centr. (Tervuren), Ann., Sci. Zool.* **105,** 1–247.